A FIELD GUIDE TO THE
STARS AND PLANETS

THE PETERSON FIELD GUIDE SERIES®

A FIELD GUIDE TO THE

STARS AND PLANETS

FOURTH EDITION

JAY M. PASACHOFF

With Monthly Star Maps and Atlas Charts by
WIL TIRION

SPONSORED BY THE NATIONAL AUDUBON SOCIETY,
THE NATIONAL WILDLIFE FEDERATION, AND
THE ROGER TORY PETERSON INSTITUTE

HOUGHTON MIFFLIN COMPANY
BOSTON   NEW YORK

Dedicated to

DONALD AND FLORENCE MENZEL

who started me on astronomy
and who became my friends.

For information about permission to reproduce selections from
this book, write to Permissions, Houghton Mifflin Company,
215 Park Avenue, New York, New York 10003

PETERSON FIELD GUIDES and **PETERSON FIELD GUIDE SERIES**
are registered trademarks of Houghton Mifflin Company.

Library of Congress Cataloging in Publication Data

Pasachoff, Jay M.
A field guide to the stars and planets. —4th ed. / Jay M. Pasachoff ;
with monthly star maps and atlas charts by Wil Tirion.
p. cm. — (The Peterson field guide series)
Includes bibliographical references and index.
ISBN 0-395-93432-X (cloth). — ISBN 0-395-93431-1 (pbk.)
1. Astronomy—Observer's manuals. I. Title. II. Series.
QB64.P37   2000
523—dc21   99-27354   CIP

Book design by Anne Chalmers
Typeface: Linotype-Hell Fairfield; Futura Condensed (Adobe)

Photograph on pages ii–iii © Pekka Parviainen
Photograph on page 511 © Anglo-Australian Observatory/
Royal Observatory, Edinburgh

Printed in the United States of America

RMT 10 9 8 7 6 5 4

EDITOR'S NOTE

Among all the inhabitants of the planet Earth, humans alone have systematically considered the heavenly bodies. We have given names to the constellations and charted their relative positions and movements. Although we formerly believed that we alone were able to navigate by celestial means, now we know that nocturnal bird migrants sometimes take their direction by means of an innate ability to read the night sky and from magnetic fields.

In recent years, more people than ever before have become aware of space and want to know what is out there. The tiny dots of light in the night sky may be obscured by city fog, but on clear nights they cannot fail to stir the inquiring mind.

The first step in astronomy, as in zoology, is to put names to things, to identify them. This Field Guide will facilitate the process and is equally usable for the observer depending on the naked eye, the binocular, or a small astronomical telescope. Unlike most of the books in the Field Guide series, which tend to be regional in scope or at least confined to a single continent, it may be used at any point on the earth's surface and on any day of the year. In line with the general policy of the other Field Guides, emphasis has been put on new and simplified techniques of recognition.

The book should present no problems to beginners interested in finding their way around in the heavens, but at the same time the completeness of its charts and tables should make it a useful tool for serious amateurs and even for professionals.

It is a joy to thumb through the book while relaxing in an armchair, but inasmuch as it is basically a Field Guide, put it to practical use. Use it on clear nights to interpret the free show put on by the heavens.

ROGER TORY PETERSON

ACKNOWLEDGMENTS

It is a pleasure in this fourth edition to enhance the usefulness and beauty of *A Field Guide to the Stars and Planets*. In the book, I present tours of the stars, the planets, the sun, the moon, and other objects in the heavens. I provide introductory descriptions and figures, graphs, and tables of information in a form useful to novices, while providing this and other information in a style and quantity useful for those already quite knowledgeable in astronomy. I am very pleased to continue my lengthy collaboration with the celestial cartographer Wil Tirion. He not only redrew in color a complete set of Atlas Charts covering the whole sky but also redrew the Monthly Sky Maps to make them easier to use. He also provided supplemental charts for variable stars and other special objects.

I appreciate the assistance of Cathryn Baskin for her work on the preparation of descriptive material to accompany the Atlas Charts and on other phases of the previous editions of the book. Robert Murphy of Scientia, Inc., has not only prepared his standard Graphic Timetables showing the positions of the planets but also specially designed Graphic Timetables for this Field Guide to show which of the brightest stars, clusters, variable stars, nebulae, and galaxies are suitable for viewing in various seasons. The fruits of Ewen Whitaker's lengthy study of the moon are evident in the material he has prepared to accompany the moon maps. Robert Argyle of the Institute of Astronomy, University of Cambridge, and Robert Tanguay have provided updates and new information on double stars. I thank them all.

I appreciate the continued cooperation of Leif Robinson, Roger W. Sinnott, Alan Hirshfeld, Dennis di Cicco, Rick Fienberg, and others at *Sky & Telescope* and thank them for their permission to reprint some of the tables that they have carefully prepared as part of their *Sky Catalogue* 2000.0.

Charles Case and the National Geographic Society were kind enough to allow me to use the special moon maps beautifully prepared in a cooperative effort of the National Geographic Society and the U.S. Geological Survey. I thank Adrienne Wassermann of the USGS for assistance with photos and planetary charts and Jennifer Blue for labeling the Mars chart.

I thank many people who assisted with various phases of the preparation of the book in this and earlier editions, including Naomi Pasachoff for her work on the history and mythology of the constellations, Janet Akyüz Mattei (American Association of Variable Star Observers) for providing information on variable stars and the light curves used, Ewen Whitaker (Lunar and Planetary Laboratory, University of Arizona) for providing descriptive information and photographs of the moon, Jim Mullaney for a variety of comments, and Walter Bennett for verifying many of the Atlas Charts at the telescope. I also thank Brian Marsden and Dan Green (Harvard-Smithsonian Center for Astrophysics), John Bortle for comments on minor bodies of the solar system, and Fred Espenak (NASA Goddard Space Flight Center) for information about eclipses.

A Field Guide is obviously based on the work of many people carried out over an extended period of time. Users wanting to go beyond this book to do more observing would naturally refer to such magnificent sources as *Sky & Telescope* and *Astronomy* magazines each month. Walter Scott Houston's long run of columns in back issues of *Sky & Telescope* remain a valuable source, as does Robert Burnham's *Celestial Handbook*. We have consulted all these sources in preparation of this Field Guide. These and other sources are listed in the bibliography.

Stephan Martin has worked with me on digital astronomical technology and in indexing and selecting the many new images used and has made a new set of lunar-phase photographs. Susie Kaufman has worked with me on many aspects of the pictures and text. Thanks are still due Liz Stell for her work on the third edition. At Houghton Mifflin, Harry Foster headed the Field Guide series for some time, and Barry Estabrook has taken over as publisher for this fourth edition. Lisa White is editor of this fourth edition, and has provided expertise on many aspects of publishing. Anne Chalmers is the designer.

Nancy Pasachoff Kutner expertly prepared the index.

My wife, Naomi, and our daughters, Eloise and Deborah, continue to inspire me. I am glad that our daughters make time in their careers to continue to read proof. I still keep in mind how I can keep things clear and helpful for observers like them.

It is a pleasure for me to be associated with my late professor,

Donald H. Menzel, in this Field Guide. He had written the first edition; I only regret that he did not live to participate in the subsequent ones. I learned so much from him on a series of eclipse expeditions, and so much from his example as a scientist dealing with other aspects of astronomy, that I am forever in his debt. I also thank Florence K. Menzel, who, along with her husband, has been kind and helpful to me since my student days.

I hope that you enjoy *A Field Guide to the Stars and Planets*. It would be nice if it were error-free, but no book is. I do hope that you will write me with your comments, to point out errors, or with observing suggestions that are not in this book.

This book's Web page, which contains various links to updated information and photos as well as sources of equipment, is www.williams.edu/astronomy/fieldguide. If you do not succeed in getting through, please contact me at jay.m.pasachoff@williams.edu to find out the new URL or whatever new technology is applicable.

JAY M. PASACHOFF
Hopkins Observatory
Williams College
Williamstown, Massachusetts 01267

CONTENTS

TABLES AND APPENDIXES

How to Use This Book

I hope that with this *Field Guide to the Stars and Planets* in hand, it will be pleasant for you to find your way around the sky. I have tried to make it easy for you to identify what you see; at the same time, I will try to demonstrate the excitement of our current understanding of the universe.

GENERAL ORGANIZATION. I begin by describing in chapter 1 a framework for observing the heavens. I describe how to tell stars from planets, how to identify the brightest stars, and how to find a few of the most prominent groupings of stars in the sky. Then, in chapter 2, I give you a brief tour around the heavens, season by season. This chapter can be used together with the Monthly Sky Maps that follow in chapter 3. No special knowledge or equipment is needed to use these maps. For observers in the northern hemisphere, a set of four maps appears for each month: two maps—one with constellation outlines and one without—for use when facing north, and a similar pair for use when facing south. For observers in the southern hemisphere, a set of two maps shows stars with constellation outlines.

Next, in chapter 4, I describe the constellations and the classical myths associated with them. A list of the current constellations (appendix 1) appears on p. 512–513.

In chapter 5, I describe the types of objects that are relatively constant in their places in the sky, including stars, nebulae, and galaxies. I discuss astronomers' current understanding of these objects, including the stages in the life of a star. I also provide information about the times of year when a selection of the most interesting double and variable stars, star clusters, nebulae, and galaxies are visible. In this chapter, I include spectacular photographs of some of the most beautiful objects; more photographs are mixed into chapter 7. The celestial objects that move with respect to the stars—the moon, the sun, the planets, comets, meteors, and asteroids—have their own chapters later on. Chapter 5 ends with a discussion of a very modern problem: light pollution. Light pollution is stealing views of our beautiful skies from many people. The chapter ends with a discussion of the latest set of glints in the sky: Iridium flares.

Two types of objects of special interest to those observing the sky are double and variable stars, so chapter 6 is devoted to them. This chapter includes charts and tables that will enable you to find many examples of these stars.

Chapter 7 is an atlas of the entire sky, broken down into 52 charts, drawn by Wil Tirion. All the brightest stars and constellations are shown, as on the Monthly Sky Maps; however, the Atlas Charts provide a more detailed look at each region of the sky. Although many of the objects on the Atlas Charts can be seen with the naked eye or with binoculars, you will find the charts even more interesting if you have access to a small or medium-sized telescope. Each chart shows not only stars but also nebulae, galaxies, and a wide variety of other celestial objects. Descriptions of these objects and photographs of some of them accompany the Atlas Charts. A list of nonstellar, deep-sky objects, the Messier Catalogue, precedes the charts, along with a table showing the region of sky covered by each chart and a visual key to the Atlas Charts.

To use the Atlas, first locate an object of interest, using either the Monthly Sky Maps or the celestial coordinates listed in the Messier Catalogue or in other tables in this guide. Then turn to the Atlas Chart where your object is shown. Alternatively, you may choose to survey the whole area shown on a chart with a telescope.

As part of the introduction to the Atlas Charts, I explain the symbols used on the charts and the names used for stars and other types of astronomical objects. I also briefly explain the system of celestial coordinates—right ascension and declination—used to indicate the locations of objects in the sky. The apparent position of objects in the sky changes slightly over the years because the earth wobbles as it spins; the celestial cartographer Wil Tirion has compensated for this precession by drawing the charts and calculating the tables in this book for the year 2000, and deviations from these positions will not be noticeable for many years.

Though the positions of distant objects in the universe change only slightly in the sky, the positions of the moon, most of the planets, and the sun change quite drastically in the course of the year. The path the sun follows through the sky in the course of a year is called the ecliptic; it is indicated by a dotted line on the Monthly Sky Maps and the Atlas Charts. The moon and the planets never stray far from this line.

Chapter 8 describes the moon and includes nine maps of its craters and other features of its surface. Chapter 9 explains how to locate the planets in the sky and how to predict when they will be visible. Chapter 10 describes what each planet is like and what you may see if you observe it with binoculars or with a telescope.

Chapter 11 describes comets, with special attention to the 1997 appearance of the bright comet Hale-Bopp (officially C/1995 O1). Chapter 12 discusses asteroids, the minor planets. Chapter 13 discusses meteors and lists meteor showers that you can see in the sky at different times of the year. Meteors usually flash across the sky, and asteroids move too slowly for their motion to be apparent. Lights that appear to move slowly and steadily across the sky are usually airplanes or—particularly in the couple of hours after sunset or before sunrise—artificial satellites in orbit around the earth.

In chapter 14, I turn from the nighttime sky to the daytime sky and discuss the major object that is visible all day—the sun. I discuss not only the everyday sun and how to observe it, but also how to observe a total solar eclipse and why a total eclipse is so glorious. I also describe annular eclipses and how to observe them.

In chapter 15, I discuss technical aspects of the positions of objects in the sky, ways to tell time by the sun and the stars. I also discuss calendars and their history.

Finally, in chapter 16, I give some information about how to choose a telescope of your own.

The end of the book contains a glossary, suggestions for additional reading, an extensive set of tables, and an index.

ILLUSTRATIONS. Rendering the sky in a book is a difficult problem because it requires stretching and squashing the curve of the sky onto a flat page. Our celestial cartographer, Wil Tirion, has solved this problem in an improved way. The Monthly Sky Maps in chapter 3 are presented in a special projection that makes the maps easy to use while distorting the shapes of the constellations as little as possible.

In addition to the 72 Monthly Sky Maps and the 52 Atlas Charts by Wil Tirion, nine detailed maps of the moon's surface, prepared in a collaborative effort of the National Geographic Society and the U.S. Geological Survey, enhance *A Field Guide to the Stars and Planets*. A number of Graphic Timetables have also been provided to help you determine when stars, planets, and other celestial objects will be visible above the horizon. All photographs in this guide are oriented with north at the top (unless otherwise indicated), to make it easy for you to compare them with the Atlas Charts in chapter 7.

SOME OBSERVING HINTS. Your eyes have to be dark-adapted to see the sky well. This process may take 5 to 15 minutes after you go outside from a lighted room. As you watch the sky during this time, more and more stars will become visible. To maintain your adaptation to darkness, cover the front of your flashlight with red plastic or use one of the new devices that generates red light with a light-emitting diode (LED).

TELESCOPES. The observing suggestions in this guide are designed to help you locate interesting objects in the sky, whether or not you have a telescope. When I mention a "small telescope" in this guide, I am referring to one with a lens less than 4 inches (10 cm) in diameter. (Most telescopes that small have lenses rather than mirrors.) A "medium-sized telescope" has a lens or mirror about 4–10 in. (10–25 cm) in diameter. If you are interested in purchasing a telescope, you will find a list of some telescopes that are popular with amateurs on p. 510, along with a list of telescope manufacturers.

HOW TO USE THIS BOOK. If you want to survey the stars and constellations, use figures 3 and 4 in chapter 1 and the Monthly Sky Maps in chapter 3. The seasonal tours in chapter 2 illustrate how the constellations seem to move across the sky as the earth rotates around the sun. The Graphic Timetable of the Brightest Stars (fig. 1-3) in chapter 1 will show you when the brightest stars visible from midnorthern latitudes will be above the horizon.

If you see a bright object in the sky and want to identify it, the first step is to determine whether it is a star or a planet (see p. 7). Then check the Graphic Timetable of the Brightest Stars in chapter 1 or the Graphic Timetables of the Planets in chapter 9, to see which bright stars or planets are visible on your date of observation. Or you can refer to the Monthly Sky Maps in chapter 3. You can refer to the diagrams of the positions of the planets in chapter 10 or plot the positions of the brightest planets on the Monthly Sky Maps in chapter 3 (or on the Atlas Charts in chapter 7) using Appendix 11, which shows the planets' longitudes along the ecliptic.

If you are using binoculars or a telescope and want to look at a variety of interesting objects, such as double and variable stars, star clusters, nebulae, or galaxies, use the Graphic Timetables in chapter 5 to find out which ones will be visible on your date of observation. Then turn to the Atlas Charts in chapter 7, where observing notes supplement detailed charts of each region of the sky. Each chart is oriented with north at the top. Remember that binoculars give a right-side-up image, but most telescopes give inverted images compared with the way celestial objects appear to the naked eye.

If you want to study features on the moon, refer to the maps in chapter 8.

If you want to observe the planets with a telescope, refer to the information in chapters 9 and 10.

If a bright comet is in the sky, see chapter 11.

If you want to see a meteor shower, use the table in chapter 13.

If you want to know how to observe the sun safely, or when and where the next solar eclipse will occur, refer to chapter 14.

If you want to know how to tell time by the sun and the stars, refer to chapter 15.

Some comments on photographing the sky with still cameras or with video cameras appear in chapter 16.

So, if you work your way through this book, you will have made friends with the sun in the morning and the stars at night.

Younger readers and other novice observers may enjoy my two brief introductory shirt-pocket-size books, *First Guide to Astronomy* and *First Guide to the Solar System*. Both are in the Peterson First Guide Series. Those interested in more of the science behind astronomy will be interested in my textbooks.

See www.williams.edu/astronomy/fieldguide for updates and links.

(Overleaf) *A Hubble Space Telescope view showing several stages of the life cycles of stars. A hot blue supergiant star has ejected a ring of gas. Near it in the sky, a cluster of hot stars shines and has blown a large cavity around itself. High-energy radiation has caused gas to glow at lower right. Two reddish, compact, tadpole-shaped emission nebulae are thought to be gas and dust evaporation from solar systems in formation. (Wolfgang Branduer, JPL/IPAC; Eva K. Arebel, U. Washington; You-Hua Chu, U. Illinois Urbana-Champaign; and NASA)*

A First Look at the Sky

I hope you will use this book to become familiar with the sky. Finding your way around the sky is like finding your way around a large city—it is easy if you are familiar with the streets and have navigated there before, but otherwise it takes some time to become familiar with routes and shortcuts. In this first chapter of *A Field Guide to the Stars and Planets,* I will assume that you are new to observing the heavens. I will start from scratch and show you some of the basic ways that you can orient yourself when observing. My focus here will be on some of the most prominent stars and constellations that you can observe with the naked eye or binoculars.

Before you begin to observe the nighttime sky, you should know which way north, south, east, and west are. If you don't know the compass directions for the place from which you are observing, though, you can find them with the aid of the Big Dipper and the North Star, Polaris (see p. 19).

One of the first things you will notice when you start to study the sky is that stars and other objects are of different brightnesses. Perhaps the easiest way to determine what is what in the sky is to take advantage of this fact. Except for the moon, the brightest objects in the nighttime sky are some of the planets. The planets change their positions slightly from night to night with respect to the stars in the background; in chapter 9 I show you how to locate the planets on any given night.

Three characteristics tell you quickly if an object is a planet.

1. **BRIGHTNESS.** Some of the planets simply appear too bright to be stars. Venus, the brightest planet, is an example. It can never be very far away from the sun in the sky, so whenever an extremely bright dot of light—the "evening star"—appears in the sky toward the west after sunset, or in the morning sky toward the east before sunrise—the "morning star"—it is probably Venus (fig. 1-1).

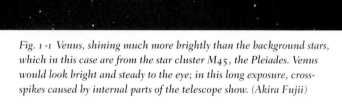

Fig. 1 -1 Venus, shining much more brightly than the background stars, which in this case are from the star cluster M45, the Pleiades. Venus would look bright and steady to the eye; in this long exposure, cross-spikes caused by internal parts of the telescope show. (Akira Fujii)

Fig. 1 -2 The Big Dipper. Note that the middle star in the handle is double; the fainter star, Alcor, is above the brighter star, Mizar. (Akira Fujii)

It is often the first bright object visible, before any stars appear in the sky. Mercury also appears in these areas of the sky around sunrise and sunset, but it never looks as bright as Venus nor gets as far from the sun as Venus does. Mercury appears only during twilight, and Venus never remains visible through the night.

Whenever a very bright yellowish white point of light appears in the sky in the middle of the night, it is probably Jupiter. Unlike Mercury and Venus, Jupiter is not always near the sun in the sky; it can appear high in the sky at midnight. Mars and Saturn can also appear far from the sun in the sky, rising well after sunset; Mars rarely outshines Jupiter, though, and the brightness of Saturn never equals that of Jupiter or Venus. Mars can often be distinguished by the fact that it has a slight but distinct reddish tinge. Saturn, on the other hand, appears to be yellowish. The other planets are too faint to be seen with the naked eye.

2. TWINKLING. Planets usually seem to shine steadily, while stars twinkle. Twinkling is an effect of turbulence in the earth's atmosphere: the atmosphere bends the starlight passing through it, and, as small regions of the atmosphere move about, the intensity of a star's light varies slightly but rapidly. Observations with a telescope would also reveal that a star appears to move around slightly. The reason why stars twinkle and planets do not is that stars are so far away that they look like points even when viewed through large telescopes; planets, though, are close enough to earth that their telescopic images are tiny disks. The light from different parts of a planet's disk averages out and makes the planet appear relatively steady in both brightness and position.

If the atmosphere is especially turbulent, or if you are looking through an especially large amount of atmosphere (when you are looking at an object low in the sky, for example, making your line of sight pass obliquely through the atmosphere), even planets can seem to twinkle. Under these conditions, the object you are observing may even seem to change in color—when Venus is low in the western sky, it is not uncommon to see it change from greenish to reddish and back again.

3. LOCATION. All the planets always appear close to an imaginary line across the sky, so objects located far from that line cannot be planets. The line is called the *ecliptic,* and it is followed (more or less) not only by the planets but also by the moon. (The ecliptic is actually the path followed by the sun across the background of stars in the course of the year.) Since the earth is but one of the planets, and since all the planets orbit the sun in approximately the same plane, from our point of view the planets and sun must follow roughly the same line across the sky. The moon orbits the earth at only a slight angle to the plane of the planets, so it too always appears close to the ecliptic. The location of the ecliptic is plotted as a dotted line on the Monthly Sky Maps in chapter 3, which show how the sky looks to the naked eye at different times.

From northern temperate latitudes, including the continental U.S., Canada, and Europe, the ecliptic crosses the southern part of the sky. This means that any bright objects at the *zenith*—the point directly over your head—or in the northern sky cannot be planets. (There are occasional exceptions to this if you are observing from the southernmost parts of the U.S.)

Now that you know how to tell whether you are looking at a star or a planet, you can look around the sky and identify some of the brightest stars. Some people find it easier to identify a few individual bright stars. Others prefer to locate a few favorite constellations or color photo asterisms—a few stars, also roughly in the same direction from us, that are parts of one or more constellations.

Many people can identify one or two specific constellations or asterisms, even though they don't know any other constellations. (This statement holds true for many professional astronomers.) The most prominent asterism in the sky is the Big Dipper, whose seven stars trace out the shape of a dipper in the northern sky (fig. 1-2). The Big Dipper is an asterism rather than a constellation because it makes up only part of the constellation Ursa Major, the Big Bear (fig. 2-2, p. 23).

The four stars in the bowl of the Big Dipper make a squarish (actually trapezoidal) shape about 10° across. (Ten degrees is about the width of your fist, if you hold it up at arm's length against the sky.) Curving away from the bowl are the three stars in the handle, which cover another 15°. All the stars in the Big Dipper except the one that connects the handle to the bowl are of about the same brightness, which makes it easy to single out the Dipper in the sky.

Sky observers—including both professional and amateur astronomers—usually express star brightness in magnitudes, the scale of which is described in detail in chapter 3. The lower the magnitude, the brighter the star. The brightest stars in the sky are magnitude zero (0), or in two cases, magnitudes −0.7 and −1.4. Figures 1-4 and 1-5 on pp. 14–17 show the brightest stars in the sky; the faintest star shown is magnitude 3.5. The naked eye can see stars about 10 times fainter than this, down to those as dim as 6th magnitude under perfect sky conditions.

One difference between the maps or charts in this guide and the real sky is that all the stars in the sky look like points, even though they have different brightnesses. The charts and maps in this guide represent these different brightnesses (magnitudes) as circles of different sizes.

It is often interesting to begin by identifying the brightest star near the zenith. Table 1 (p. 12) lists the 21 brightest stars in the sky. Following the table is a display—a Graphic Timetable (fig. 1-3)—that shows when the brightest stars visible from midnorthern latitudes are passing their highest points in the sky. On any given date, different stars will be overhead at different times of night; the whole sequence changes with the seasons as the earth orbits the sun. The positions of the stars repeat from year to year.

To use the Graphic Timetable of the Brightest Stars (fig. 1-3), run your finger down the side to find your date of observation, then move across the page to find the time of night when you are observing. You will see the names of the brightest stars that are transiting at about that time. An object *transits* when it passes your *meridian*—the imaginary line passing from the point due north on the horizon through the zenith to the point due south on the horizon.

TABLE 1. THE BRIGHTEST STARS IN THE SKY

RANK	STAR	CONSTELLATION	MAGNITUDE	R.A. (EPOCH 2000.0)	DEC.
1	Sirius	Canis Major	−1.46	6^h45^m	−16°43'
2	Canopus*	Carina	−0.72	6^h24^m	−52°42'
3	Rigil Kent*	Centaurus	−0.27 (dbl)	14^h40^m	−60°50'
4	Arcturus	Boötes	−0.04	14^h16^m	+19°11'
5	Vega	Lyra	+0.03	18^h37^m	+38°47'
6	Capella	Auriga	+0.08	5^h17^m	+46°00'
7	Rigel	Orion	+0.12 (dbl)	5^h15^m	−8°12'
8	Procyon	Canis Minor	+0.38	7^h39^m	+5°14'
9	Achernar*	Eridanus	+0.46	1^h38^m	−57°14'
10	Betelgeuse	Orion	+0.50 (var)	5^h55^m	+7°24'
11	Hadar*	Centaurus	+0.61	14^h04^m	−60°20'
12	Altair	Aquila	+0.77	19^h51^m	8°52'
13	Aldebaran	Taurus	+0.85 (var)	4^h36^m	+16°31'
14	Acrux*	Crux	+0.87 (dbl)	12^h27^m	−63°05'
15	Antares	Scorpius	+0.96 (var)	16^h29^m	−26°26'
16	Spica	Virgo	+0.98	13^h25^m	−11°10'
17	Pollux	Gemini	+1.14	7^h45^m	+28°02'
18	Fomalhaut	Piscis Austrinus	+1.16	22^h58^m	−29°37'
19	Deneb	Cygnus	+1.25	20^h41^m	+45°17'
20	Mimosa*	Crux	+1.25	12^h47^m	−59°41'
21	Regulus	Leo	+1.35	10^h08^m	+11°58'

NOTES: (dbl) = double star; combined magnitude of components given.
(var) = variable star; brightest magnitude given.
r.a. = right ascension, in hours and minutes (see p. 495)
dec. = declination, in degrees and minutes (see p. 496)
* = star not visible from midnorthern latitudes.

The Graphic Timetable (opposite) shows when the brightest northern stars can be seen above the horizon from midnorthern latitudes; the maximum altitude each star reaches above the northern or southern horizon (for observers at 40° N latitude) is also given, in parentheses.

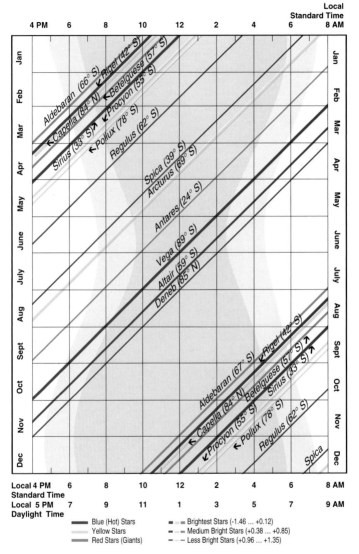

Local Standard Time

Local 4 PM 6 8 10 12 2 4 6 8 AM
Standard Time

Local 5 PM 7 9 11 1 3 5 7 9 AM
Daylight Time

- ▬▬▬ Blue (Hot) Stars
- ▬▬▬ Yellow Stars
- ▬▬▬ Red Stars (Giants)

- ▬ ▬ Brightest Stars (-1.46 ... +0.12)
- ▬ ▬ Medium Bright Stars (+0.38 ... +0.85)
- ▬ ▬ Less Bright Stars (+0.96 ... +1.35)

Fig. 1-3 Graphic Timetable of the Brightest Stars, showing when they transit. (© 2000 Scientia, Inc.)

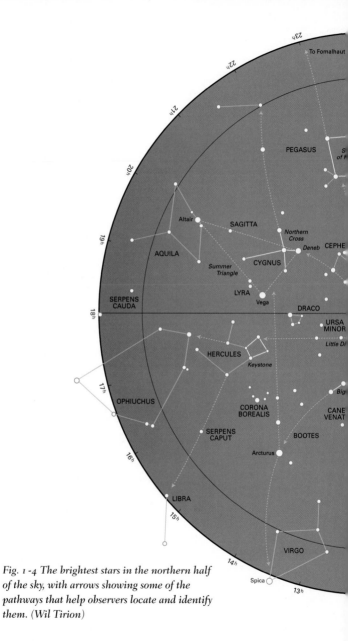

Fig. 1-4 The brightest stars in the northern half of the sky, with arrows showing some of the pathways that help observers locate and identify them. (Wil Tirion)

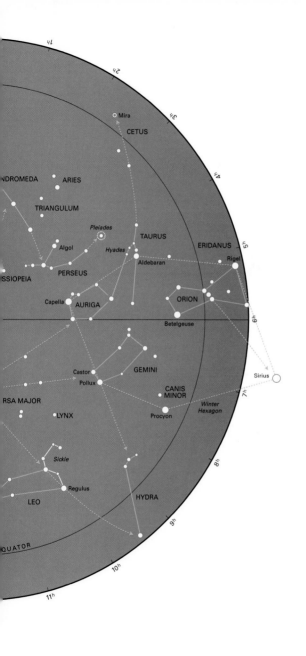

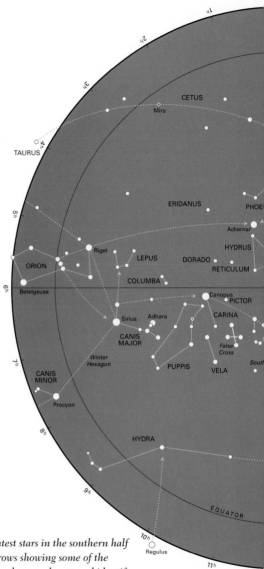

Fig. 1-5 The brightest stars in the southern half of the sky, with arrows showing some of the pathways that help observers locate and identify them. (Wil Tirion)

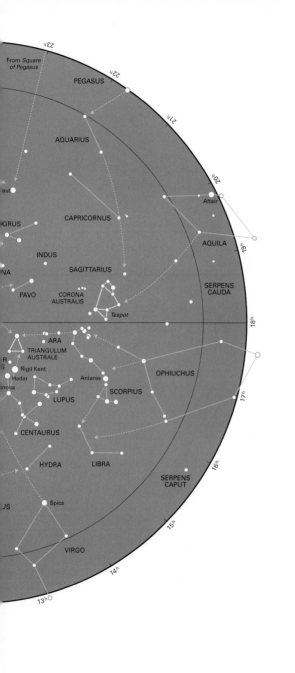

Figure 1-3 also shows how high the stars are in the sky, in degrees above the horizon, at their time of transit, for an observer at 40° N latitude. This *altitude* above the horizon is the highest point that each star reaches in the arc it traces across the sky. For example, Sirius, the brightest star in the sky, reaches a maximum of 33° above the southern horizon—slightly more than one-third the altitude of the zenith. Since your fist covers about 10° of sky (when you place your thumb flat on the outside of the fist and hold it at arm's length), you can mark off the altitude above the horizon in 10° segments. You may want to verify first that about nine of your fists indeed cover 90° from horizon to zenith.

In the region near the ecliptic, a bright object could be a star or a planet. When looking in this part of the sky, do make sure you know which planets are up (above the horizon). The Graphic Timetables in chapter 9 provide this information.

Figures 1-4 and 1-5 are pairs of sky maps centered on the north celestial pole and on the south celestial pole, respectively. The *celestial poles* are the imaginary points where the earth's axis, if extended, would meet the celestial sphere. The north and south celestial poles lie above the earth's north and south poles, respectively. As the earth rotates, the sky appears to rotate in the opposite direction around the celestial poles. The sky thus seems to rotate once around the celestial poles every 24 hours. Midway between the celestial poles is the celestial equator, which lies on the celestial sphere, above the earth's equator. The *celestial equator* separates the northern and southern halves of the sky.

Figure 1-4 shows the northern half of the celestial sphere, and is spread across two pages with some overlap between. This map is centered on the north celestial pole. Below that pole is the Big Dipper. Since the sky appears to rotate around the north or south celestial pole (depending on which hemisphere you are in), whichever pole you can see always remains at a constant height in the sky. (If you are observing from a latitude of 40° N on earth, the north celestial pole will always be tilted up 40° above due north on the horizon; if you are observing from a latitude of 30° N, the pole will always be tilted up 30°, etc.) Observers at mid-northern latitudes will see the Big Dipper appear to revolve around the north celestial pole every 24 hours. For these observers, the Big Dipper is close enough to the north celestial pole that it will never set, and is thus an example of a *circumpolar* asterism.

Figure 1-5 shows the southern half of the celestial sphere. It includes some stars (near the celestial equator) that midnorthern observers can sometimes see and some stars that never rise above the horizon at northern latitudes.

The Big Dipper is a particularly handy asterism to know because you can follow lines marked out by its stars and trace them across the sky to other interesting objects. Best known is the line marked by the two stars at the end of the bowl, which are known as the *Pointers*. These two stars point to the North Star, Polaris. To find Polaris, follow a straight line from the Pointers upward from the bowl of the Dipper and try to imagine the line curving slightly as it follows the curve of the sky for about 30°. (This is three fists' width, or about five times the distance between the Pointers, which are separated by 5½°.) Polaris is at the end of an asterism known as the Little Dipper. None of the stars in the Little Dipper is as bright as the five brightest stars of the Big Dipper; the back two stars of the bowl and the two stars between the bowl and Polaris may be hard to see with your naked eye.

Polaris is not an especially bright star, but it is bright enough to be visible ordinarily. It is the brightest star in that region of the sky, so it is not easily confused with other stars. Polaris is within 1° of the true north celestial pole and is thus of help not only to navigators at sea but also to land-based amateurs navigating around the sky. If you face Polaris, you are facing north. Thus it is best to find Polaris in order to orient yourself before you use any of the charts or maps in this guide.

If you continue along the arc from the Pointers through Polaris, you will come to the Great Square of Pegasus. This pathway and others that you can follow from one constellation to another are marked with dotted lines on Figs. 1-4 and 1-5. For example, instead of following the Pointers to Polaris, you can follow the curve of the Big Dipper's handle over about 30° (three fists' width, thumb included) of sky to the bright star Arcturus. If you can follow the same arc for another 30° without hitting the horizon, you will come to the bright star Spica. To remember this, think of "arc to Arcturus," and then "spike to Spica."

If, instead of finding Polaris, you follow the Pointers or the two stars that form the rear of the Big Dipper's bowl in the opposite direction, you will come to the constellation Leo, the Lion, about 35° away. Leo contains the bright star Regulus, which is located at the base of the "sickle" in figure 1-4. You can find other stars and constellations using the pathways marked on figures 1-4 and 1-5; the angles between some of the stars and constellations are listed on page 20.

TABLE 2. ANGLES IN THE SKY

top stars of bowl of Big Dipper	10°
Pointers of the Big Dipper	5½°
Castor and Pollux (in Gemini)	4½°
Great Square of Pegasus, width	17°
end stars of Orion's belt	3°
Orion's belt to Betelgeuse or Rigel	9°
end of Orion's belt to Sirius	21°

NOTE: One fist (thumb included) covers about 10° of the sky.

One other asterism is particularly striking and may grab your attention as you scan the sky. This prominent grouping is the straight line marked by three bright stars separated by a total of only 3°; it appears at the extreme right of figure 1-4 and at the extreme left of figure 1-5. These stars mark the belt of Orion, the Hunter, and appear in the evening sky in winter (fig. 2-8). One of the stars lies on the celestial equator, which explains why Orion is well into the southern part of the sky for an observer at midnorthern latitude.

About 10° to the north of Orion's belt is the bright reddish star Betelgeuse, and almost 10° to the south of Orion's belt is the bright bluish star Rigel. (Many people pronounce Betelgeuse as "Beetle-juice"; others say "beh-tel-jooz." Rigel is pronounced "ry-jel," with the accent on the first syllable.) We shall see in chapter 5 that the colors of stars reveal their temperatures; Betelgeuse is a cool star and Rigel is a hot one. If you follow the line made by Orion's belt to the east (or left, since you are facing south), you will soon come to the bluish white star Sirius, the brightest star in the sky.

Those who use this Field Guide extensively will quickly find themselves interested not only in stars and constellations but also in double and variable stars, star clusters, nebulae, and galaxies, all of which are described in chapter 5. (A set of Graphic Timetables, showing when a few of the most prominent examples of these objects are visible above the horizon also appears in that chapter.)

2

A TOUR OF THE SKY

The sun dominates the daytime sky. Sunlight scatters throughout the atmosphere, making the sky blue. This blue sky is brighter than the stars behind it, so we cannot usually see the stars during the daytime, as we shall discuss later in this chapter. When the sky is clear enough, the moon can often be seen even in the daytime, especially if you know where to look. Chapter 8 discusses the phases of the moon and where to find the moon in the sky.

Shortly before sunset or just after sunrise, the sky becomes dark enough for us to see the brightest planets and stars. Venus is the brightest of these objects and is sometimes bright enough to

Fig. 2-1 A conjunction of Venus, Jupiter, and the Moon.
(Jay M. Pasachoff)

cast noticeable shadows. If you see an exceedingly bright object in the west at or after sunset—the "evening star"—it is usually Venus (fig. 2-1). If you see an exceedingly bright object in the east before or at sunrise—the "morning star"—it is also usually Venus. A second object in the sky near the sun, not as bright but usually even closer to the sun, may be Mercury.

Jupiter, Mars, and Saturn can also be prominent in the sky and can be quite far from the sun, so planets visible late at night are from this trio. Mars's reddish tinge is subtle, yet not hard to see even with the naked eye. Saturn's rings are noticeable with a small telescope but are not visible to the naked eye. Jupiter's moons, belts, and Great Red Spot also require a small telescope to be seen. To tell at a glance which planets are visible in the sky on your night of observation, you can use one of the Graphic Timetables in chapter 9. The characteristics that will help you distinguish the planets from the bright stars nearby are described in chapter 1.

BEYOND THE SOLAR SYSTEM

As the sky darkens, the brightest stars become visible. In a city, only a few dozen stars may become visible because the sky remains very bright even at night. But far from the haze of pollution and the competing glimmer of city lights, about 3,000 stars may be visible to the naked eye. An equal number of stars are hidden beyond the horizon, and most can be viewed by waiting through the night or for a different time of year. The rest of the stars become visible only to observers closer to the equator or in the hemisphere opposite to that from which you are observing. The Atlas Charts and descriptions in chapter 7 show the entire sky—the stars and constellations that are visible from the southern hemisphere as well as the ones that can be seen from the northern hemisphere. However, since most users of this guide are at midnorthern latitudes, much of the following discussion is designed for them.

The region of the sky near Polaris, the North Star, is visible from midnorthern latitudes all through the year. In this region you can easily see the Big Dipper, the asterism that makes up part of the constellation Ursa Major, the Big Bear. In the autumn evening sky, the bowl of the Big Dipper appears right side up, while in the evening sky in the spring, the bowl appears upside-down. Figure 2-2 shows the Big Bear, including the Big Dipper, from the celestial atlas Johann Bayer published in 1603.

On the front side of the bowl of the Dipper, the Pointers point to Polaris, which lies at the end of the handle of the Little Dipper.

Fig. 2-2 Ursa Major (left) *and Ursa Minor* (right) *in the sky. The foreground was added with a computer program. (Tony and Daphne Hallas)*

To find Polaris, follow a line that extends north from the Pointers for about 30°. Thirty degrees is one-third of the distance between the horizon and the zenith. Extend your arm upward from horizontal one-third of the way toward the zenith to get an idea of 30°. (Also, 30° is about three widths of your hand at the end of your outstretched arm.) The altitude of Polaris above the horizon is equal to your latitude on the earth. For example, if you are at latitude 90°, at the north pole, Polaris is 90° above the horizon, that is, directly overhead.

Polaris remains at this same location in the sky all year. Because its position is stationary, Polaris is visible throughout the year and appears on all the Monthly Sky Maps for the northern hemisphere in chapter 3. You can use Polaris in all seasons as a reference point for locating other stars. The Big Bear, the Little Bear, and the other circumpolar constellations near Polaris also remain visible above the horizon all year.

Other groups of stars, as we will see below, are visible only during certain seasons of the year. Each night these stars rise above the eastern horizon, travel above and around Polaris, and then set below the western horizon. As the seasons progress, you can see successive groups of constellations making this journey across the sky.

All the stars are so far away that we measure the distance in terms of how long it takes their light to reach us. Light travels very rapidly: 186,000 miles/sec (300,000 km/sec). This is 1,080,000,000 km/hr, 10 million times faster than a car. Light can travel seven times the distance around the earth in a single second, can travel from the moon to the earth in slightly more than a second, and can travel from the sun in eight minutes. Yet light takes over four years to reach us from the nearest star, Proxima Centauri. We call the distance that light travels in a year *one light-year* and say that Proxima Centauri is thus four light-years away. Only a few dozen stars are within 20 light-years of our sun (see appendix 4).

To compare stars, it is useful and important to know how bright they would be if they were all at the same distance from us, so

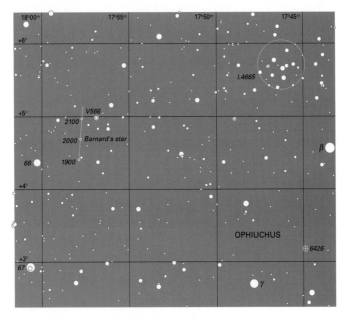

Fig. 2-3 *The path of Barnard's star across the sky. This star appears to move across the sky with respect to more distant stars, as marked with the arrow, covering a distance as great as our moon's diameter in about 180 years. (Wil Tirion)*

that only their intrinsic differences in brightness—how bright they actually are—and not merely their positions affect the result. The standard distance that astronomers use for this purpose is about 32.6 light-years.

All stars move around in space. Astronomers can measure the speed at which they are approaching us or receding from us by measuring their *Doppler shifts,* the slight shifts in color that result from such motion. (You can hear the Doppler shift for sound waves as a change in pitch as an object passes. Doppler shifts for light waves are similar.) But most stars are too far away for us to detect their motions across the sky. For the nearest stars, though, we can detect the *proper motions,* the motions across the sky. The star with the largest proper motion is Barnard's star. It is six light-years away from us. It moves across the sky sufficiently fast that it covers a distance as great as our moon's diameter in only about 180 years (fig. 2-3). Using the Doppler shift to measure a star's motion toward us or away from us together with the proper motion for the side-to-side direction gives the star actual velocity in space.

MEASURING DISTANCES AND ANGLES

Astronomers measure the distance to the nearest stars by a method of triangulation called *trigonometric parallax:* the nearest stars appear to move slightly with respect to the more distant stars. To see a similar effect, hold your thumb up with your arm outstretched, and look at it first with one eye closed and then, without moving your head, with the other eye closed. Notice how your thumb seems to move slightly across the background. Now do the same thing with your arm bent so that your thumb is closer to your eye—your thumb will appear to move more. Similarly, the more a star appears to move as the earth orbits the sun, the closer it is.

Astronomers figure out how far away from the earth and the sun they would have to be before the distance between the earth and the sun covered an angle of only 1 arc second (where 1 arc second is 1/60 of 1 arc minute, 1 arc minute is 1/60 of one degree [1°], and 1° is 1/360 of the way around the sky). Thus 1 arc second is a tiny angle, 1/3,600 of a degree of arc. The distance at which the angle between the earth and the sun is only 1 arc second is called 1 *parsec* (from the words "parallax," for the method of measuring distance, and "second," for the actual angle). The value of 1 parsec works out to be 3.262... light-years.

The standard distance astronomers use to determine the inherent brightnesses of stars is 10 parsecs. The brightness a star would have at this distance of about 32.6 light-years from us is called its *absolute magnitude*. The absolute magnitude is a measure of the intrinsic brightness of the star, how bright the star actually is. By contrast, a star's apparent brightness, how bright it appears to us as opposed to how bright it is intrinsically, is its *apparent magnitude*. A star's apparent magnitude is affected not only by how bright the star is intrinsically but also by how far the star is away from us. In the following chapter, we discuss practical aspects of the magnitude scale.

At the turn of the twentieth century, Ejnar Hertzsprung and Henry Norris Russell plotted many stars on the same graph, removing the effect of distance on brightness by either plotting a group of stars that were all at the same distance or by using the absolute magnitudes. When they plotted the temperature on one axis and the magnitude on the other, they found that the points representing most stars fell on the graph in a narrow band—the *main sequence*. The brighter stars are hotter for the most part, and the fainter stars are cooler.

A few stars are brighter than main-sequence stars of the same temperature. These exceptional stars are thus known as *giants,* and a few even more exceptional stars are brighter still and are known as *supergiants*. Giants and supergiants turn out to be especially large in addition to being especially bright. A few stars are fainter than main-sequence stars of the same temperature. These exceptionally faint stars are known as *white dwarfs*. They turn out to be extraordinarily small for stars—only the size of the planet earth.

PHOTOGRAPHING THE SKY

Many people like to photograph the sky and the objects in it. Sometimes sky photographs are taken through telescopes, but quite nice photographs of the sky and celestial phenomena can be made with ordinary cameras and lenses mounted on tripods.

Sensitive color films, known as "fast films," are now commonly available that allow photographs of the constellations to be taken in seconds with an ordinary camera and lens. The faster, or more sensitive, the film, though, the grainier the image usually appears, so each person must make his own tradeoff between speed and the size of the grain on the film. A normal, 50 mm, lens on a 35 mm single-lens reflex camera shows a field of view about 50° across, which shows several constellations. The key to a successful constellation photograph is having a sturdy tripod that does not vibrate. It is also usually best to use a cable release or self-timer that allows you to start the exposure without shaking the

Fig. 2-4 Star trails. (Charles Schotthoefer)

camera by touching it. Try a series of exposures, such as 1, 2, 4, and 8 seconds.

If the lens is opened to its widest, you may see more distortion in the picture—particularly toward its edges—than if the lens is "closed" somewhat. For example, a lens's widest opening may be labeled *f*/2. Each f-stop (1.4, 2, 2.8, 4, 5.6, 8, 11, 16, 21) represents a change in the lens's opening by a factor of 2. So *f*/2.8 lets in half the light that *f*/2 does. Further, a lens open to *f*/4 lets in half the light of a lens opened to *f*/2.8, and thus one-fourth the light of a lens opened to *f*/2. But the *f*/4 image may prove better, even though the exposure time must be longer, because it may show less distortion.

Sometimes it is fun to take very long exposures, even some hours long, to show the paths of the stars across the sky (fig. 2-4). When the north celestial pole is in the photograph, these star trails show the circles that the stars make around the pole. To take such a picture, put your camera on a sturdy tripod. You must be as far as possible from artificial lights, which would fog—or put a uniform glow on—your film and so limit the length of your exposure. To limit the sky fog, it is often best to take star-trail photographs at about *f*/11. Here again, it is for you to experiment. There is no single correct exposure or exposure time.

The moon can be photographed with an ordinary lens. Contrary to what you might expect, it takes a similar photographic lens opening and exposure time to a sunlit scene on earth. After all, it is about the same distance from the sun as we are. It takes a telephoto lens to show features on the surface of the moon.

You can also use your video camera to photograph the moon, stars, and planets. One advantage of a video camera is that you can see the results right away. Cameras with the telephoto lenses that zoom to the longest focal lengths can fill the screen with the moon. The sensitivity of the camera to faint light levels is more important for including the planets and stars. These show only as points of light; it takes a telescope to show surface features of planets.

To take photographs like most of those in this book, you must attach your camera to a telescope. In those cases, the telescope acts as a lens, so ordinarily you remove the lens from the camera itself. The telescope is set to track the stars, which requires a simple motor that turns once a day at the rate of the stars. Such telescopes are commercially available. Quite a different hobby, though also an interesting one, is making your own telescope. If you attach an ordinary camera to a tracking telescope, your photograph will show fainter stars and can show the Milky Way.

SEASONAL TOURS

As you read each tour, please follow along with the suitable Monthly Sky Map in chapter 3. Many of the invisible pathways in the sky we will follow are shown in Figures 1-4 and 1-5 in chapter 1.

THE AUTUMN SKY (MONTHLY MAP #8)

In the darkening sky on an autumn evening, the Pointers in the bowl of the Big Dipper point upward toward Polaris. From Polaris, follow the Little Dipper, which is upside-down. A Native American legend holds that the autumn colors spill out of the Little Dipper at this time of year, making the trees turn bright colors.

Continue along the arc from the Pointers to Polaris for an equal distance on the other side of Polaris. You will find a prominent W-shaped constellation, Cassiopeia (fig. 2-5). This constellation, like most others, was known by the ancient Greeks and was named after a character in Greek mythology. Cassiopeia was married to Cepheus, the King of Ethiopia, who has his own constellation, which is shaped like a house with a peaked roof, lying west of Cassiopeia.

Fig. 2-5 Cassiopeia, from the star atlas of Bayer (1603). The brightest object shown was an exploding star—a supernova—visible only a few decades before the chart was drawn. (Mendillo Collection of Astronomical Prints)

If we continue upward from Cassiopeia, we find the constellation named after Andromeda, who in Greek mythology was Cassiopeia's daughter. In Andromeda, a faint hazy patch of light is sometimes visible to the naked eye. This light actually comes from the center of the Great Galaxy in Andromeda, the nearest galaxy to our own. (A galaxy is an enormous group of billions of stars, plus dust and gas.) The Andromeda Galaxy (M31) is much farther from us than any of the individual stars we see in the sky, so it marks the farthest that we can see with the naked eye. A telescope is necessary to reveal the spiral shape of this galaxy.

Now look high in the sky, in a direction south of Andromeda. You will see four stars marking the corners of a square, known as the Great Square of Pegasus. Pegasus was the flying horse in Greek mythology. It arose from the blood of Medusa after Perseus killed her. One of the stars of the Great Square is actually in the constellation Andromeda.

If it is very dark out, you can also see the Milky Way passing high overhead, directly through Cassiopeia. The Milky Way—a grouping of stars, dust, and gas in our own galaxy—appears as a hazy band across the sky, with ragged edges, dark patches, and rifts. It is actually the gas, dust, and stars of our own galaxy, a disk that we see from a position inside it. Wide-field photographs on fast color film, with your camera mounted on a tripod, will give nice views of the Milky Way.

Moving east from Cassiopeia along the Milky Way, you will come to the constellation Perseus. In Greek mythology, Perseus was a hero who slew the Medusa, flew off on Pegasus (who is conveniently located nearby in the sky), and then from his winged mount saw Andromeda, whom he saved from the sea monster Cetus. Binoculars or a telescope will show you a pair of *open clusters,* each a group of many stars, close to each other in Perseus. This double cluster is unusual; clusters usually appear by themselves. Thus the double cluster is relatively easy to notice. In general, open clusters, also known as galactic clusters, usually contain a few hundred stars. The stars in open clusters are fairly young by astronomical standards, between 10 million and a billion years old.

Also in Perseus is perhaps the best known of the variable stars, Algol, often called the demon star from its Arabic name. Algol is actually a pair of stars that orbit each other in a way that brings one in front of the other every 2.9 days. During these eclipses, the total brightness we see drops dramatically. I discuss Algol and give a table of its minimum brightness in chapter 6.

On the opposite side of Cassiopeia, beyond Cepheus along the Milky Way, a cross of bright stars is visible directly overhead. This

Fig. 2-6 The globular cluster M13. Atlas Chart 17 (Daniel Good)

prominent grouping of stars, known as the Northern Cross, lies in the constellation Cygnus, the Swan. This swan appears to be flying south, as birds do at this time of year. Deneb is the star at the Swan's tail. Slightly to the west of the Swan, you will see the bright star Vega in the constellation Lyra, the Lyre. Vega is the third brightest star visible from midnorthern latitudes.

Farther west, beyond Vega, you will come to the constellation Hercules, named for the Greek hero who performed 12 great labors. With binoculars or a small telescope, we can barely see an object that looks like a hazy mothball. This object is a *globular cluster,* a spherical group of thousands of stars. It is labeled M13 (fig. 2-6), from its place in a catalog of nonstellar objects, the Messier Catalogue (table 12, p. 213), compiled in the eighteenth century by Charles Messier. M13 is the brightest globular cluster that we can see from the northern hemisphere. Globular clusters are interesting because all the stars in them are very old. None is much younger than 10 billion years. Further, they are all at the same distance from us, which makes it relatively easy for astronomers to compare their properties without worrying about the effects of distance.

THE WINTER SKY (MONTHLY MAP #11)

At dusk on a winter evening, the Big Dipper is low in the northern sky. The constellations that were easiest to see in the autumn now will appear closer and closer to the western horizon at the same hour on each successive night. By January 1, Cygnus, the Swan, sets in the western sky in the early evening; Cassiopeia is high overhead to our north; Perseus is even higher in the sky.

The Milky Way, the hazy band of gas and dust that gives our galaxy its name, also appears high in the early evening sky at this time of year. Since our home galaxy, the Milky Way Galaxy, is disk-shaped, we see many stars and much dust and gas when we look along the plane of the disk, but few stars and little dust and gas when we look in other directions. The next constellation to the southeast along the Milky Way is Auriga, the Charioteer, with its bright star Capella.

To the south of this part of the Milky Way, you can see a cluster of six or seven stars that are close together in the sky. Though faint, they will catch your eye as your vision sweeps across the sky. These stars are the Pleiades, the Seven Sisters of Greek mythology, the daughters of Atlas (fig. 2-7). The Pleiades are an open cluster of over a hundred stars; the larger the binoculars or wide-field telescope you use, the more you will see. Atlas Chart 10 in

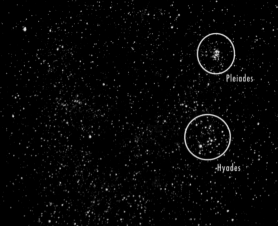

Fig. 2-7 The Pleiades, the close set of stars (above left of center), is an open cluster of stars known as the Seven Sisters. In Greek mythology, these sisters were pursued by Orion and were given refuge in the sky. Atlas Chart 10. (Richard Hill)

chapter 7 is supplemented by a special chart showing the brightest stars of the Pleiades.

Look farther south, and compare with Atlas Chart 11, facing south (page 249). (Turn this book so that its right edge is closest to you.) The most prominent group of stars in the sky in the wintertime is three bright stars in a straight line. They form the belt of Orion, the Hunter (fig. 2-8). Extending down from Orion's belt (to the right on the photograph) is his sword. On Orion's shoulder (to the left on the photograph) you will see the reddish star Betelgeuse, one of the brightest stars visible from midnorthern latitudes. Symmetrically on the other side of Orion's belt, you will find the bluish star Rigel, marking Orion's heel. In chapter 4, we will see that the colors of stars tell us their temperatures: reddish stars are relatively cool—about 3,000°C—and bluish stars are relatively hot—over 10,000°C.

In the area of Orion's sword is a hazy region, the Orion Nebula, that is visible with binoculars or small telescopes. A nebula is a hazy region of sky that contains clouds of gas or dust (see p. 164). The Orion Nebula (fig. 2-9) marks the presence of a cloud of gas and dust; in its vicinity new stars are now forming. Many nebulae are really "nurseries" for young stars that have recently been born

Fig. 2-8 The constellation Orion, with reddish Betelgeuse above Orion's belt and blue-white Rigel below it. (Akira Fujii)

Fig. 2-9 The Orion Nebula, M42 and M43 (Atlas Chart 24), in the belt. Fig. 7-27 is a key. (© Jerry Lodriguss)

Fig. 2-10 The Orion Nebula, with the Hubble Space Telescope's view of a small portion of a ground-based view. (NASA/ESA and Space Telescope Science Institute)

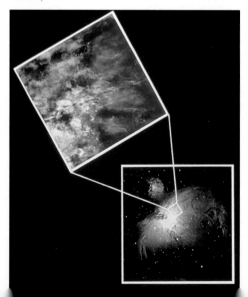

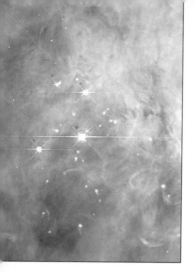

Fig. 2-11 The central part of the Orion Nebula as observed with the Hubble Space Telescope. The observations are improved in resolution by a factor of at least 5 over ground-based photographs. (Jeff Hester, IPAC, and NASA)

Fig. 2-12 The Horsehead Nebula in the constellation Orion. It descends from the star zeta Orionis (Atlas Chart 24), the leftmost star in Orion's belt. (© Anglo-Australian Observatory/Royal Observatory Edinburgh)

and are the site of ongoing stellar birth; we discuss them in more detail in chapter 4.

The Hubble Space Telescope has been turned to study the Orion Nebula (fig. 2-10). The Hubble image in figure 2-11 shows the region near the Trapezium, a set of bright stars near the center of the visible nebula that provides the energy to keep the nebula glowing. The improved resolution shows that the nebula has narrow streamers and jets. Studies of these smaller objects are telling us how stars form in regions such as this.

The Horsehead Nebula, near the leftmost star in Orion's belt, is difficult to see with a small telescope but is beautiful in photographs. Dark dust absorbs light coming from behind it to silhouette the shape of a horse's head (fig. 2-12).

Orion (fig. 2-13) seems to be warding off Taurus, the Bull, a constellation you will find by looking beyond Orion's shield. Between the top of the shield and the Pleiades lies a V-shaped group

Fig. 2-13 *Orion, from the star atlas of Bayer (1603). Orion is holding up his shield to ward off Taurus, the Bull. (Mendillo Collection of Astronomical Prints)*

Fig. 2-14 *The constellation Crux (the Southern Cross), low in the sky as viewed from the Florida Keys. The reflections of the brightest stars are visible in the ocean. The Eta Carinae Nebula appears red low in the sky at right. (R. Scott Ireland)*

of stars, the Hyades. The reddish star Aldebaran marks the end of one side of the V. The Hyades outline the face of Taurus; the Pleiades ride on the bull's shoulder. The Hyades and the Pleiades are both *open clusters* of stars (also called *galactic clusters*); these clusters are locations where perhaps 100 or more stars are close together in an irregularly shaped group.

At Orion's heel is his dog, Canis Major. Orion's belt points directly to Sirius, the brightest star in the sky and the brightest star in the photograph of the constellation of Orion. Rising soon after Orion, Sirius appears blue-white and is part of the constellation Canis Major, the Great Dog. Nearby is the yellow-white star Procyon in Canis Minor, the Little Dog. This bright star forms a nearly equilateral triangle with Sirius and Betelgeuse.

THE SPRING SKY (MONTHLY MAP #2)

As spring approaches, Orion and the V-shaped Hyades move closer to the western horizon each evening and eventually disappear from our view at sunset. Now you will see a pair of stars, the Twins (Castor and Pollux), in the western sky at dusk. Although the two stars are about the same in brightness, Pollux appears slightly reddish, while Castor does not. These stars are in the constellation Gemini, the Twins, which was named after two Roman military gods.

South of the Twins, the bright star Sirius is prominent in the western sky, in the constellation Canis Major, the Great Dog. South of Sirius is the constellation Puppis, the Ship's Stern. This constellation includes a star, zeta Puppis, that is known to be one of the intrinsically brightest and hottest stars in the sky. Zeta Puppis is so far away, though, that it does not appear prominent when we observe this part of the sky. Puppis and several other constellations in this region and farther south were once considered to form a giant constellation of a ship, Argo Navis (fig. 4-6).

Ursa Major, the Big Bear, appears high in the northern sky on spring evenings and is tipped so much that any imaginary things in the Big Dipper would spill out. If you follow the Pointers backward, you will see the constellation Leo, the Lion, just to the south of the zenith. To most people, Leo looks more like a backward question mark or sickle than a lion's head. A bright star, Regulus, is at the base of the question mark, at the lion's heart. East of Regulus, a triangle of stars marks the rest of Leo's body. Many people visualize Leo as a lion in the sky with a sickle-shaped head and triangular tail.

On the northern star map, if you follow an arc begun by the stars in the Big Dipper's handle, you will come to a bright reddish

Fig. 2-15 *The constellation Lyra, which includes the bright star Vega. Lyra is off to the side of Cygnus, which lies in the Milky Way. The brightest stars in Cygnus form the Northern Cross. Altair, in Aquila (the Eagle), is off to the side. (Akira Fujii)*

Fig. 2-16 *The Milky Way, rising almost perpendicularly to the horizon in the summer sky. (Compare, for example, Sky Map 7.) It is visible to the naked eye on dark nights if you are far from city lights. (Dennis di Cicco)*

star, Arcturus. This star lies in the constellation Boötes, the Herdsman. Farther along the arc from the Big Dipper through Arcturus, you will find another bright star, Spica. This blue-white star rises in the east-southeast in the constellation Virgo, the Virgin.

Look up again toward the northeast, to the lower left of Boötes and Hercules, which rise in the east in the evening. Here in the constellation Lyra (fig. 2-15), you will see Vega, a star that is brighter than Spica.

Fig. 2-17 M33, a spiral galaxy in Triangulum, the second closest spiral galaxy to our own and the farthest object one can see with the unaided eye. Atlas Chart 9. (© Jerry Lodriguss)

On summer evenings at sunset, Vega is the brightest star near the zenith. The constellation Cygnus, which includes the prominent stars that make up the Northern Cross, lies east of Vega along the highest part of the Milky Way. Deneb is at the head of the Swan. Deneb, Altair, and Vega make the Summer Triangle (fig. 2-15). The Milky Way rises perpendicularly at this time of year (fig. 2-16).

Arcturus, very slightly brighter than Vega, is the reddish star that is also high in the sky to the west. Hercules lies about 10° to the west of Vega.

Continuing north along the Milky Way from Cygnus, you will find the W-shaped constellation, Cassiopeia. To the southeast of Cassiopeia is Andromeda, near the horizon. Andromeda's Great Galaxy in Andromeda, M31, may be faintly visible to the naked eye as a hazy patch of sky. The spiral galaxy M33 (fig. 2-17) is in the constellation Triangulum, to the side of Andromeda. M31 and M33 are the farthest objects that you can see with your unaided eye, though it is fair to say that the glimpse of haze you can see does not reveal the beauty shown in long-exposure photographs. The Great Square of Pegasus is farther south, on the same side of the Milky Way as Andromeda.

The bright star Spica lies toward the southwest, in the constellation Virgo, the Virgin. In the south you will find the bright reddish star Antares in the constellation Scorpius, the Scorpion (fig. 2-18). (Antares means "compared with Ares," the Greek name for Mars, because Antares is also reddish.) The stars of Scorpius wind a great distance across the sky.

To the east of Scorpius, you will see the constellation Sagittarius, the Archer, in which the center of our galaxy lies. The farther south you are, the higher in the sky Sagittarius rises, and the better you can see the beautiful star clouds in the direction of the center of the Milky Way.

Moving south along the Milky Way, you will come to the star Altair in the constellation Aquila, the Eagle. Altair, Deneb (in the constellation Cygnus), and Vega make up the Summer Triangle.

Every summer around August 12, the Perseid meteor shower occurs (fig. 2-19). At the height of this shower, as many as one bright meteor per minute may be visible in pre-dawn hours. The meteors streak across the sky like shooting stars or falling stars, though they are only bits of interplanetary dust burning up in the earth's atmosphere. The dates of meteor showers are listed in a table in chapter 13 (p. 468). To photograph a meteor shower, set up an ordinary camera on a tripod, as you would to take star trails. Use a wide-angle lens to cover an appreciable region of the sky.

Fig. 2-18 The constellation Scorpius, including the bright reddish star Antares. (Akira Fujii)

Fig. 2-19 Perseid meteor trails. Though the meteors can be traced back toward Perseus, they are seen in all parts of the sky. (Akira Fujii)

Then just keep taking photographs of durations of at least a few minutes each. There is no telling how soon a meteor will cross in front of your camera. But you must have your camera lens already opened and facing the sky, since you cannot catch up with a meteor trail to take its picture.

Now that you have followed these seasonal tours, you can use the Monthly Sky Maps in the next chapter to find your way around the sky at any time of year.

THE SOUTHERN SKY (MAPS #13–24)

If you are fortunate enough to be able to see the southern sky, you will find some of the prettiest celestial objects in it. Sky Maps 13–24 show what the sky looks like from month to month from southern latitudes.

At any given location, we see stars only within 90° of the point over our head, our zenith. Stars too close to the opposite pole never rise into our view. Thus many of the southern constellations are never seen from midnorthern latitudes. A map of the southern half of the sky contains many constellations whose names are relatively unfamiliar (fig. 2-20).

Fig. 2-20 The southern (australis) half of the sky, from the Atlas Coelestis *(1661) by Andreas Cellarius. (Institute for Advanced Study, Princeton, photo by Jay M. Pasachoff)*

Among the most obvious asterisms seen from southern latitudes is the pair of stars in the constellation Centaurus, alpha Centauri and beta Centauri. Alpha Centauri is the nearest star system to us, only about 4.2 light-years away. It is a triple-star system. The nearest of the three, too faint to be easily observed, is the single nearest star to us, and so is known as Proxima Centauri.

The line from alpha to beta, the brighter to the fainter of the pair, leads to Crux, the Southern Cross (fig. 2-21). Though it is called a cross, it is actually missing a center star, so looks to me more like a kite. Three of its stars are blue and therefore hot, while the fourth is very red and therefore relatively cool. Just to the Centaurus side of the Southern Cross is the Coalsack, a dark region of absorbing dust. The Southern Cross and the Coalsack are quite visible to the unaided eye (fig. 2-22). Don't miss the globular cluster omega Centauri (fig. 2-24).

If you look up from a dark spot while you are at southern lati-

Fig. 2-21 *Alpha and beta Centauri (bottom, in the left half of the image) point toward the Southern Cross. Atlas Chart 50. (Jay M. Pasachoff)*

Fig. 2-22 *The constellations Centaurus and Crux. Alpha Centauri and beta Centauri point at the Coalsack, beyond which lies Crux, the Southern Cross. These constellations are high in the sky only from southern latitudes. (Dennis Dawson with a 40-minute guided exposure on ISO 400 film and f/1.8 with a 50 mm lens)*

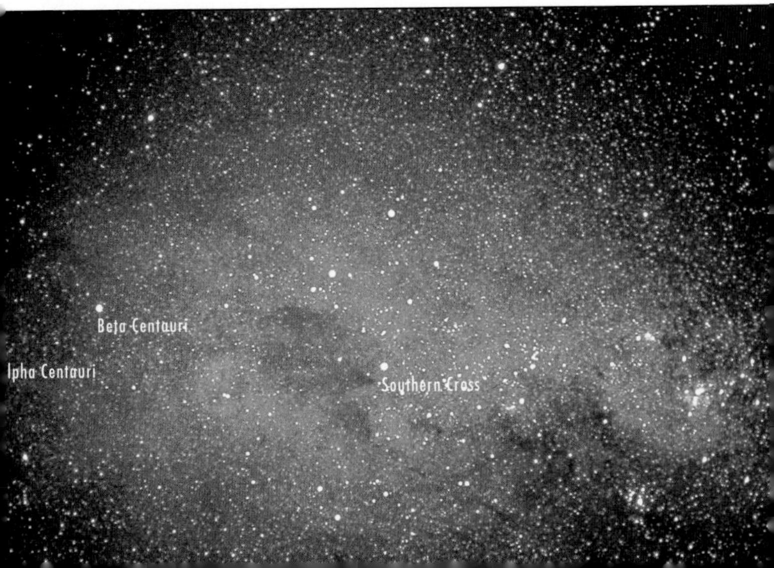

Fig. 2-23 The Large Magellanic Cloud and the Small Magellanic Cloud in the sky. (Akira Fujii)

tudes, you may seem to see a pair of faint, hazy regions of space. These objects were seen as clouds by the crew of Magellan's ship and so are known as the Magellanic Clouds. The Large Magellanic Cloud and Small Magellanic Cloud are really small galaxies that are associated with our own Milky Way Galaxy. We now know, from study of an exciting supernova that went off in it, that the Large Magellanic Cloud is 169,000 light-years away from us. The Magellanic Clouds are best seen in the months around January, which is southern-hemisphere summer (and northern-hemisphere winter).

Fig. 2-24 Omega Centauri, the most magnificent southern globular cluster, noticeable with the unaided eye. Atlas Chart 39. (Akira Fujii)

THE MONTHLY SKY MAPS

When we look at the stars at night, we have the feeling that we are underneath a giant bowl. Although the stars are really at many different distances from the earth, which spins underneath them, it is often convenient to picture the sky as the ancients did — as stars attached to a giant sphere rotating overhead. The set of constellations we see appears to rotate one-twelfth of the way around the sky each month, so at the same time of night one month later, we see an additional 30° of constellations in the east and lose sight of 30° of constellations that have set in the west.

It is, of course, very difficult to flatten out a sphere so that it can be reproduced on a flat page. The distortions in the Mercator projection of the earth, in which Greenland is made to look as large as South America merely because it is located much closer to a pole, are familiar to us all.

We have chosen a new way of drawing these star maps to minimize distortions in the regions of the sky that are most often studied. Star maps are made by projecting the positions of the stars onto a flat plane as though their shadows were being cast from a given point. We have chosen this point so that the position approximately halfway up the sky to the zenith (45° in altitude) has minimum distortion. Though the constellations near the boundaries of the maps are expanded, the amount of the expansion is not too great, and constellations near the horizon are not compressed. Also, because of the projection used, when you follow a pathway from one constellation to another using certain pairs of stars, the angles between the paths will be the same on the maps as they are in the sky.

These Monthly Sky Maps are designed to be easy to use; no knowledge of celestial coordinates is necessary. Most of the objects plotted on these maps are of magnitude 4.5 or brighter—bright enough to be seen with the naked eye. Seventy-two maps are included. The first 48 are for northern-hemisphere observers; they are specially marked for latitudes between 10° N and 50° N and can also be used somewhat outside this range.

The second set of 24 maps is for observers in the southern hemisphere; these maps are specially marked for latitudes between the equator and 40° S latitude and can also be used somewhat outside this range. The words "Sky Map" and the map number are printed in blue instead of yellow on these maps, so you can distinguish the southern-hemisphere maps from the northern-hemisphere ones at a glance.

Across the bottom of each map is a series of curved lines that represent the horizon for observers at different latitudes. In the upper middle portion of each map, the zenith (the point directly overhead) for observers at each latitude is marked with a plus sign. The directions east and west are marked at the edges.

We have provided a considerable amount of extra sky past the zenith on each map, to make it easier to orient yourself. You will soon become used to the locations of the northern and southern horizons and the zenith for your latitude. Some people will find it helpful to cut masks out of opaque paper and use them to hide the stars below the horizons for their latitude.

Each set of maps is valid for specific times on specific days, as shown in Table 6 (pp. 52–53). In particular, Sky Map 1 —which is valid in our ordinary system of timekeeping for midnight on January 1 —is drawn for *sidereal time* (time by the stars, as described in chapter 15) of 6^h40^m. Each succeeding map (Sky Map 2, Sky Map 3, and so on) is valid two hours later on the same date or one month later at the same time. Some of the times of night at which the maps are valid are shown in the lower left of the left-hand pages. Arrows near the east and west points show the direction in which the sky rotates through the night.

Each Monthly Sky Map for the northern hemisphere actually consists of two pairs of maps. The first pair has the constellation figures drawn on it; the second pair shows the same stars and omits the outlines of the constellation figures. (For the southern hemisphere, all the maps show constellation outlines.) In each pair of maps, the left-hand page shows what you will see when facing north, and the right-hand page shows what you will see when facing south. When facing north, hold the book with its left edge toward you and look down at the map. In this orientation it will correspond to the sky. When facing south, hold the book with its right edge toward you and look down at the map to have it correspond to the sky.

On the maps, stars of different brightnesses (magnitudes) are shown as dots of different sizes. I describe the magnitude scale of brightness below. The maps in this section include only stars that are bright enough to be visible to the naked eye. In addition, the Milky Way is shown, and a few other objects of special interest. These objects are also shown and described in chapter 7, where a set of Atlas Charts breaks down the sky into 52 sections. The Atlas Charts (primarily for telescope observers) are more detailed than the Monthly Sky Maps in this chapter; each Monthly Sky Map shows a view of half the sky visible at a given time.

THE MAGNITUDE SCALE

Astronomers describe the brightness of stars with a scale built on a historical base. Over two thousand years ago, the Greek astronomer Hipparchus classified stars by brightness, and by not long thereafter, stars were often divided into six classes of brightness. In about the year 140, Ptolemy, perhaps quoting Hipparchus, said that the brightest stars were "of the first magnitude," the next brightest group of stars were "of the second magnitude," and so on. The faintest stars visible to the naked eye were "of the sixth magnitude."

This scale was placed on a mathematical basis in the mid-nineteenth century. Measurements showed that a difference of 5 magnitudes corresponded to a factor of about 100 in brightness; the current magnitude scale is defined so that a factor of 100 corresponds to exactly 5 magnitudes. A few stars are even brighter than first magnitude and have been accommodated by having magnitudes of 0 and then negative numbers on the scale. The brightest star in the sky is Sirius, whose magnitude is −1.4. Canopus, the second-brightest star, is not visible north of the southern U.S. and has a magnitude of −0.7. Alpha Centauri, the third brightest star (also not visible from midnorthern latitudes), and Arcturus, the fourth brightest star, are slightly brighter than magnitude 0.0. Another dozen stars are fainter than magnitude 0.0 but brighter than magnitude 1.0. Most stars are much dimmer; the number of stars for each whole unit of magnitude increases rapidly as we go to fainter magnitudes.

The magnitude scale is different from most scales we commonly use in that increasing in brightness by one unit on the magnitude scale corresponds to multiplying by a fixed number

DIFFERENCES IN MAGNITUDES	FACTOR IN BRIGHTNESS	
1 mag	2.512	times
2 mag	6.31	times
3 mag	15.85	times
4 mag	39.81	times
5 mag	100	times
6 mag	251	times
7 mag	631	times
8 mag	1585	times
9 mag	3981	times
10 mag	10,000	times
15 mag	1,000,000	times

(about 2.51) on a scale of brightness given in units of energy. For example, if we consider first a star of 3rd magnitude and then ask how much brighter a star of 2nd magnitude is, we have subtracted 1 on the magnitude scale. The result: the star of 2nd magnitude is about 2.5 times brighter than the star of 3rd magnitude. Thus for each magnitude added or subtracted, stars are about 2.5 times fainter or brighter, respectively. By definition, for each 5 magnitudes added, stars are exactly 100 times fainter. (Thus each magnitude corresponds to the fifth root of 100, or 2.512..., which is about 2.5 times brighter or fainter.) An increase or decrease of 2 magnitudes corresponds to $(2.512)^2$, or a little over 6 times; 3 magnitudes corresponds to $(2.512)^3$, or a little over 15 times, etc., as shown in Table 3.

Although Sirius, at magnitude −1.4, is the brightest star, the moon and some of the planets get somewhat brighter in the sky. Venus can be as bright as magnitude −4.4. The full moon is magnitude −12.6, and the sun is magnitude −26.8. Note that the smaller the magnitude number (or the more negative it is), the *brighter* the object is.

Going to fainter magnitudes, 6th magnitude is the faintest that the naked eye can see under the best observing conditions. A medium-sized telescope (with a lens or mirror 6–10 inches, 15–25 cm, in diameter) will allow you to see stars of 10th or 12th magnitude. The best ground-based telescopes can observe to about 25th magnitude. The Hubble Space Telescope, because it is above the atmosphere, has a dark sky, and concentrates light exceptionally well, with long exposures can reach 30th magnitude.

You don't need a telescope to use the Monthly Sky Maps that follow; they are designed for observations with the naked eye or binoculars. Thus they show stars of the 4th magnitude and brighter. If you want to observe regions of the sky in greater detail, you can use the Atlas Charts in chapter 7, which are designed primarily for telescope observations. They show stars of the 6th magnitude, with some stars of the 7th magnitude.

TABLE 4. EXAMPLES OF MAGNITUDES

−27	Sun
−13	Moon
−5 to −2	Venus
−3 to −1	Jupiter
−2 to +2	Mars
−2 to +3	Mercury (visible range)
−1.4	*Sirius (brightest star)*
0	*Vega*
0 to +1	Saturn
+1.8 to +3.3	*stars in Big Dipper*
+2.0 to +5.0	*stars in Little Dipper*
+6	*faintest naked-eye star*
+6	Uranus
+8	Neptune
+14	Pluto
+28	*faintest ground-based telescope limit*
+30	*Hubble Space Telescope limit*

Solar-system objects are in Roman type; farther objects are in italics.

(Constellations at least 20° above the horizon from 40° North)
See constellation mythology in chapter 6 and constellation symbols in appendix 1.
Circumpolar (always visible at night):
UMi, UMa, Dra, Cep, Cas, Cam

Andromeda: 1 N, 7 N-12 N
Aquarius: 7 S-11 S
Aquila: 5 S-10 S, 5 N-10 N
Aries: 1 N, 8 N-12 N, 1 S, 8 S-12 S
Auriga: 1 N-3 N, 10 N-12 N, 1 S-3 S, 10 S-12 S
Boötes: 2 N-8 N, 2 S-8 S
Camelopardalis: on all northern maps; highest on 12 N
Cancer: 1 N-5 N, 11 N-12 N
Canes Venatici: 1 N-7 N, 1 S, 2 S, 6 S
Canis Major: 1 S-3 S, 11 S-12 S
Canis Minor: 1 S-4 S, 11 S-12 S, 1 N-3 N, 11 N-12 N
Capricornus: 7 S-9 S
Cassiopeia: 1 N, 6 N-12 N
Cepheus: 6 N-12 N
Cetus: 1 S, 9 S-12 S
Coma Berenices: 2 N-7 N, 2 S-7 S
Corona Borealis: 3 N-8 N, 3 S-8 S
Corvus: 2 S-5 S
Cygnus: 5 N-11 N
Delphinus: 5 N-11 N, 5 S-11 S
Draco: on all northern maps; highest on 6 N
Equuleus: 6 S-11 S, 6 N-11 N
Eridanus: 1 S, 10 S-12 S
Gemini: 1 N-4 N, 11 N-12 N, 1 S-4 S, 10 S-12 S
Hercules: 4 N-9 N
Hydra: 1 S-4 S, 12 S

Lacerta: 6 N-12 N
Leo: 1 N-6 N, 12 N, 1 S-6 S, 12 S
Leo Minor: 1 N-5 N, 12 N, 1 S-5 S, 12 S
Lepus: 1 S-2 S, 11 S-12 S
Libra: 4 S-5 S
Lynx: 1 N-4 N, 11 N-12 N
Lyra: 4 N-10 N, 5 S-10 S
Monoceros: 1 S-3 S, 11 S-12 S
Ophiuchus: 4 S-8 S, 4 N, 8 N
Orion: 1 S-3 S, 10 S-12 S, 2 N-3 N, 10 N-11 N
Pegasus: 7 N-12 N, 8 S-12 S
Perseus: 1 N-2 N, 9 N-12 N
Pisces: 1 N, 8 N-12 N, 1 S, 8 S-12 S
Sagitta: 5 S-10 S, 5 N-10 N
Sagittarius: 6 S-8 S
Scorpius: 5 S-6 S
Serpens Caput: 3 S-8 S
Serpens Cauda: 8 S-5 S, 3 N-9 N
Sextans: 1 S-5 S
Taurus: 1 S-2 S, 9 S-12 S, 1 N-2 N, 9 N-11 N
Triangulum: 1 N, 8 N-12 N, 1 S, 8 S-12 S
Ursa Major: on all northern maps; highest on 4 N
Ursa Minor: on all northern maps
Vulpecula: 5 N-10 N, 5 S-10 S
Virgo: 2 S-7 S, 2 N, 6 N

TABLE 6: INDEX TO MONTHLY SKY MAPS

Standard/D.S.T.	Jan.		Feb.		March		April		May		June	
	1	15	1	15	1	15	1	15	1	15	1	15
6 p.m./7 p.m. 18h/19h	10		11		12		1		2		3	
7 p.m./8 p.m. 19h/20h		11		12		1		2		3		4
8 p.m./9 p.m. 20h/21h	11		12		1		2		3		4	
9 p.m./10 p.m. 21h/22h		12		1		2		3		4		5
10 p.m./11 p.m. 22h/23h	12		1		2		3		4		5	
11 p.m./midnight 23h/24h		1		2		3		4		5		6
midnight/1 a.m. 24h/1h	1		2		3		4		5		6	
1 a.m./2 a.m. 1h/2h		2		3		4		5		6		7
2 a.m./3 a.m. 2h/3h	2		3		4		5		6		7	
3 a.m./4 a.m. 3h/4h		3		4		5		6		7		8
4 a.m./5 a.m. 4h/5h	3		4		5		6		7		8	
5 a.m./6 a.m. 5h/6h		4		5		6		7		8		9
6 a.m./7 a.m. 6h/7h	4		5		6		7		8		9	

Note: For the southern hemisphere Sky Maps, add 12 to the above Sky Map numbers.

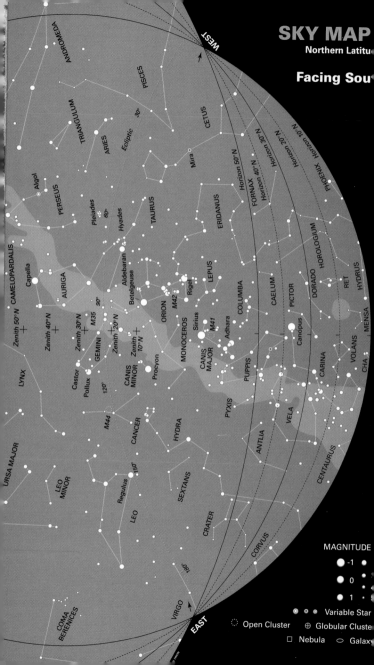

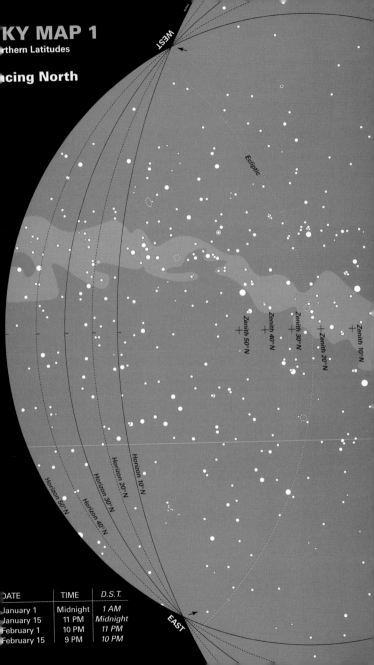

KY MAP 1
rthern Latitudes

cing North

WEST

Ecliptic

+ Zenith 50° N
+ Zenith 40° N
+ Zenith 30° N
+ Zenith 20° N
+ Zenith 10° N

Horizon 50° N
Horizon 40° N
Horizon 30° N
Horizon 20° N
Horizon 10° N

EAST

DATE	TIME	D.S.T.
January 1	Midnight	1 AM
January 15	11 PM	Midnight
February 1	10 PM	11 PM
February 15	9 PM	10 PM

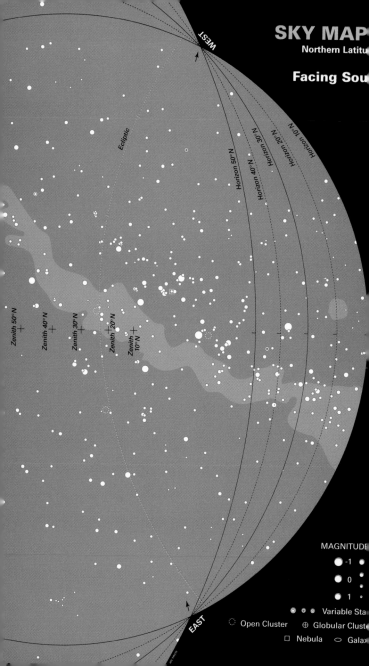

SKY MAP

Northern Latitu

Facing Sou

Ecliptic

Horizon 50° N
Horizon 40° N
Horizon 30° N
Horizon 20° N
Horizon 10° N

WEST

EAST

Zenith 50° N
Zenith 40° N
Zenith 30° N
Zenith 20° N
Zenith 10° N

MAGNITUDE

-1
0
1

Variable Sta

Open Cluster   ⊕ Globular Cluste

□ Nebula   ○ Gala

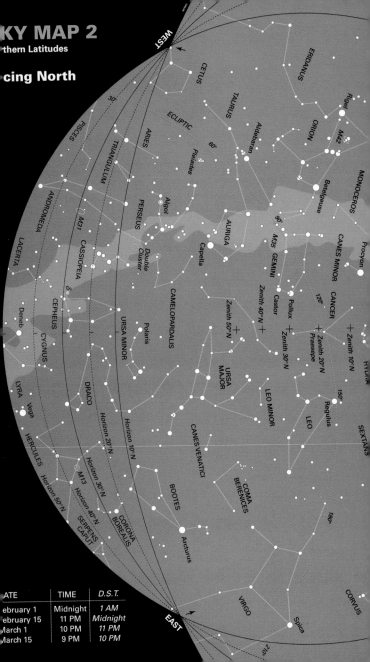

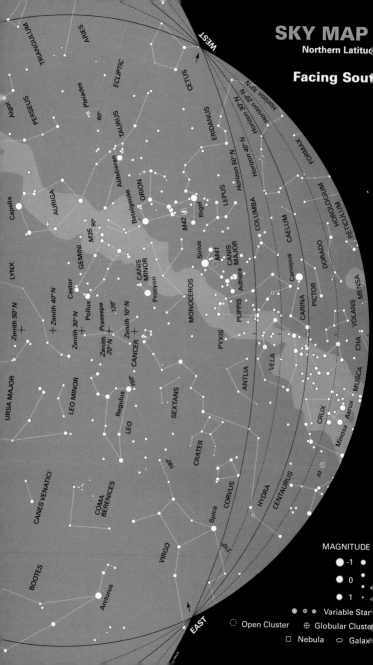

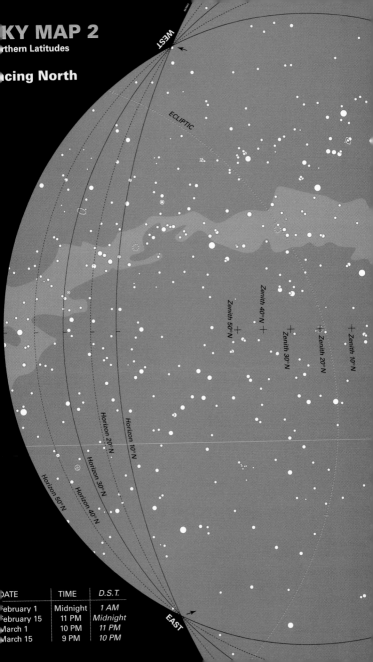

SKY MAP 2

Northern Latitudes

Facing North

ECLIPTIC

NORTH

WEST

Zenith 50°N
Zenith 40°N
Zenith 30°N
Zenith 20°N
Zenith 10°N

Horizon 50°N
Horizon 40°N
Horizon 30°N
Horizon 20°N
Horizon 10°N

EAST

DATE	TIME	D.S.T.
February 1	Midnight	1 AM
February 15	11 PM	Midnight
March 1	10 PM	11 PM
March 15	9 PM	10 PM

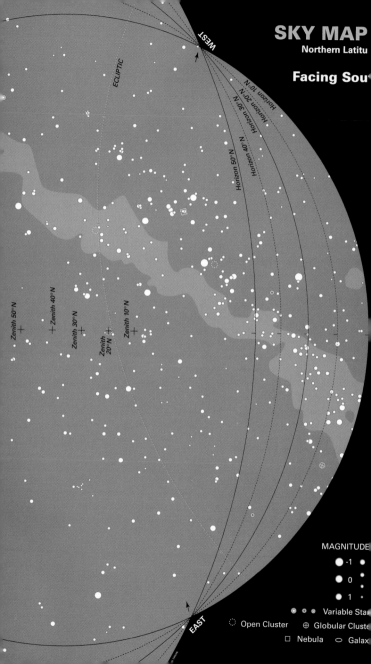

SKY MAP

Northern Latitu

Facing Sou

ECLIPTIC

WEST

Horizon 10° N
Horizon 20° N
Horizon 30° N
Horizon 40° N
Horizon 50° N

Zenith 50° N
Zenith 40° N
Zenith 30° N
Zenith 20° N
Zenith 10° N

EAST

MAGNITUDE

● -1
● 0
● 1

● ● ● Variable Sta

○ Open Cluster   ⊕ Globular Cluste

□ Nebula   ○ Galax

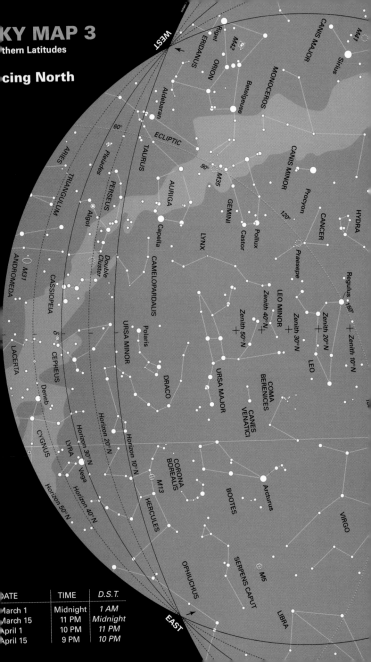

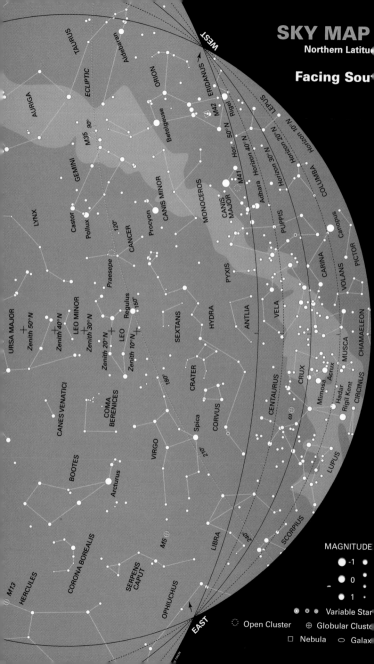

WEST

TAURUS
Aldebaran
ECLIPTIC
ORION
ERIDANUS
M42
Rigel
LEPUS
Betelgeuse
Hor. 50° N
Horizon 40° N
Horizon 30° N
Horizon 20° N
Horizon 10° N
AURIGA
M35   90°
Adhara
CANIS MAJOR
M41
COLUMBA
GEMINI
Castor
Pollux
CANIS MINOR
Procyon
MONOCEROS
PUPPIS
Canopus
CARINA
PICTOR
LYNX
120°
CANCER
Praesepe
PYXIS
VELA
VOLANS
LEO MINOR
Regulus
150°
LEO
Zenith 20° N
Zenith 10° N
HYDRA
ANTLIA
MUSCA
CHAMAELEON
URSA MAJOR
Zenith 50° N
Zenith 40° N
Zenith 30° N
SEXTANS
Acrux
Mimosa
CRUX
Hadar
CIRCINUS
CANES VENATICI
180°
CRATER
CENTAURUS
Rigil Kent
ω
COMA BERENICES
VIRGO
CORVUS
Spica
210°
LUPUS
BOOTES
Arcturus
M5
LIBRA
SCORPIUS
240°
CORONA BOREALIS
SERPENS CAPUT
M13
HERCULES
OPHIUCHUS

EAST

MAGNITUDE

● -1   ●
● 0   ●
– 1   ●

● ● ●   Variable Star
○ Open Cluster   ⊕ Globular Cluster
□ Nebula   ○ Galaxy

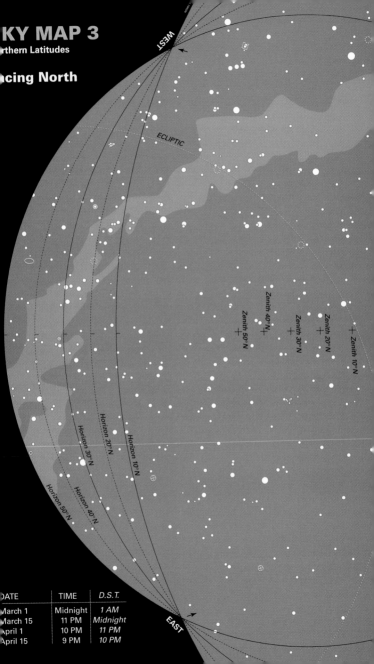

KY MAP 3
rthern Latitudes

cing North

WEST

ECLIPTIC

Zenith 50°N

Zenith 40°N

Zenith 30°N

Zenith 20°N

Zenith 10°N

Horizon 10°N

Horizon 20°N

Horizon 30°N

Horizon 40°N

Horizon 50°N

EAST

DATE	TIME	D.S.T.
March 1	Midnight	1 AM
March 15	11 PM	Midnight
April 1	10 PM	11 PM
April 15	9 PM	10 PM

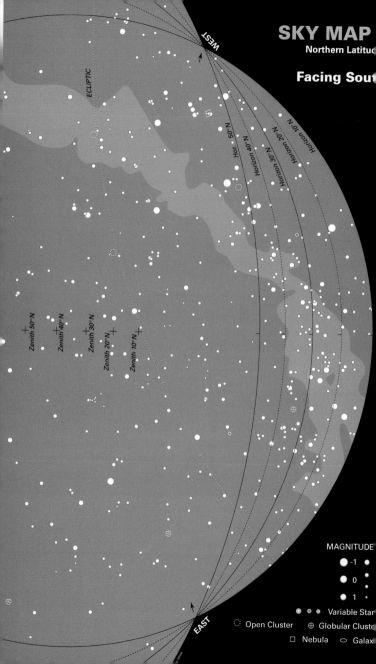

SKY MAP

Northern Latitud

Facing Sout

ECLIPTIC

WEST

Horizon 50° N
Horizon 40° N
Horizon 30° N
Horizon 20° N
Horizon 10° N

Zenith 50° N
Zenith 40° N
Zenith 30° N
Zenith 20° N
Zenith 10° N

EAST

MAGNITUDE

● -1
● 0
● 1

Variable Star

◌ Open Cluster   ⊕ Globular Cluster

□ Nebula   ○ Galax

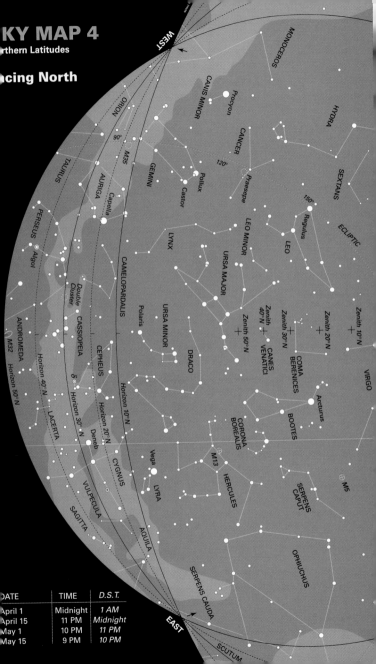

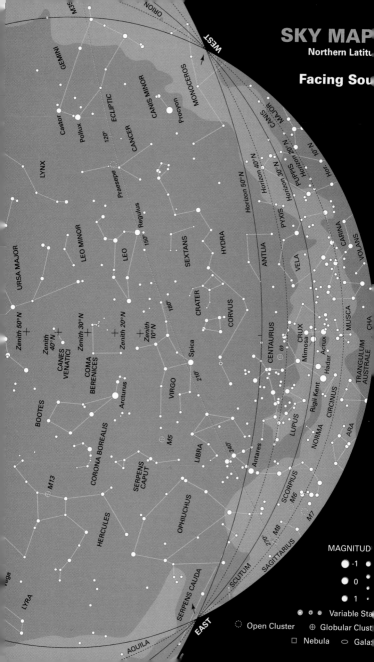

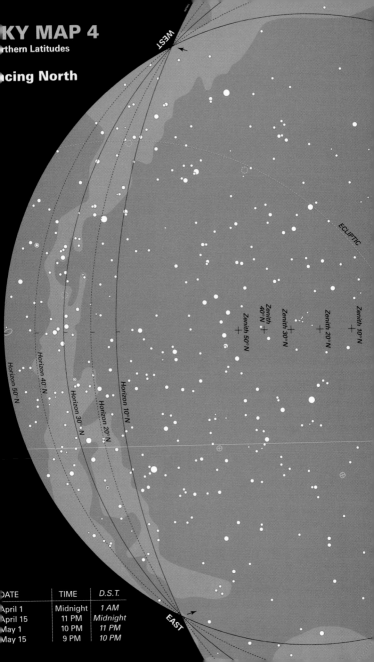

WEST

ECLIPTIC

Zenith 10° N

Zenith 20° N

Zenith 30° N

Zenith 40° N

Zenith 50° N

Horizon 50° N

Horizon 40° N

Horizon 30° N

Horizon 20° N

Horizon 10° N

EAST

DATE	TIME	*D.S.T.*
April 1	Midnight	*1 AM*
April 15	11 PM	*Midnight*
May 1	10 PM	*11 PM*
May 15	9 PM	*10 PM*

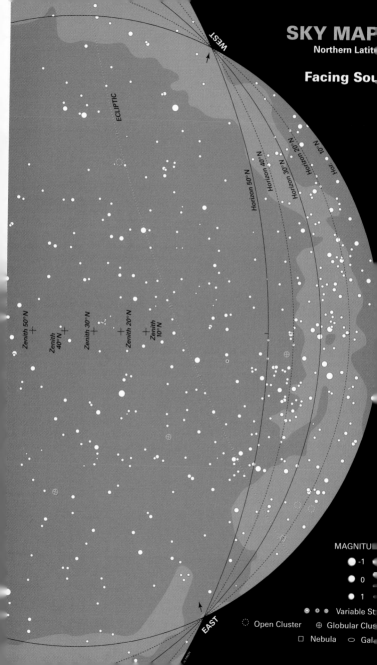

SKY MAP

Northern Latitu...

Facing Sou...

ECLIPTIC

WEST

EAST

Horizon 50° N
Horizon 40° N
Horizon 30° N
Horizon 20° N
Horizon 10°

Zenith 50° N
Zenith 40° N
Zenith 30° N
Zenith 20° N
Zenith 10° N

MAGNITU...

● -1
● 0
● 1

◉ ◉ ◉ Variable St...
◌ Open Cluster   ⊕ Globular Clus...
☐ Nebula   ○ Gal...

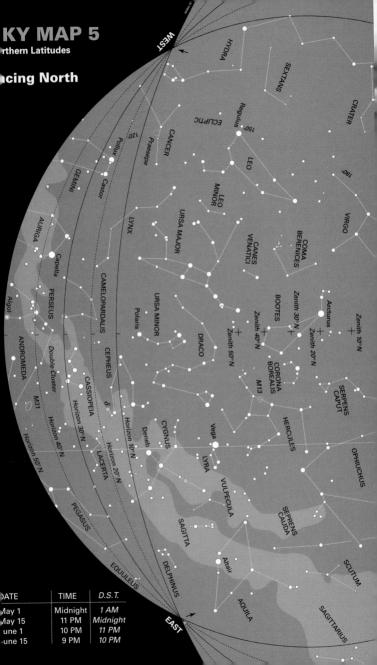

SKY MAP 5

rthern Latitudes

cing North

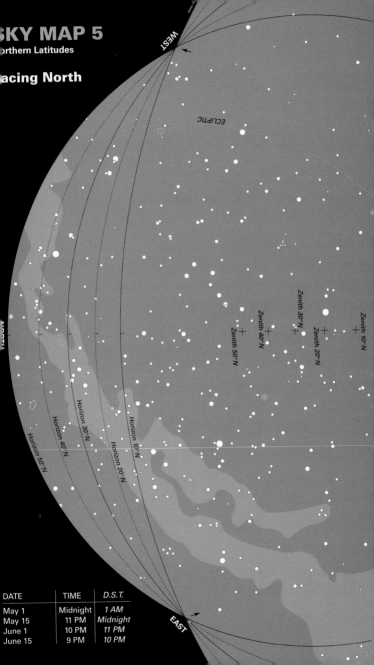

WEST

EAST

ECLIPTIC

Zenith 10° N
Zenith 20° N
Zenith 30° N
Zenith 40° N
Zenith 50° N

Horizon 50° N
Horizon 40° N
Horizon 30° N
Horizon 20° N
Horizon 10° N

DATE	TIME	D.S.T.
May 1	Midnight	1 AM
May 15	11 PM	Midnight
June 1	10 PM	11 PM
June 15	9 PM	10 PM

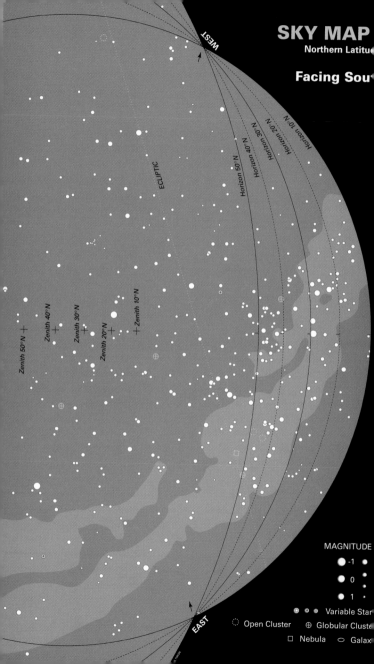

SKY MAP

Northern Latitude

Facing Sout

ECLIPTIC

WEST

EAST

Horizon 50° N
Horizon 40° N
Horizon 30° N
Horizon 20° N
Horizon 10° N

Zenith 50° N
Zenith 40° N
Zenith 30° N
Zenith 20° N
Zenith 10° N

MAGNITUDE

● -1
● 0
● 1

● ● ● Variable Star
◌ Open Cluster   ⊕ Globular Cluster
□ Nebula   ○ Galaxy

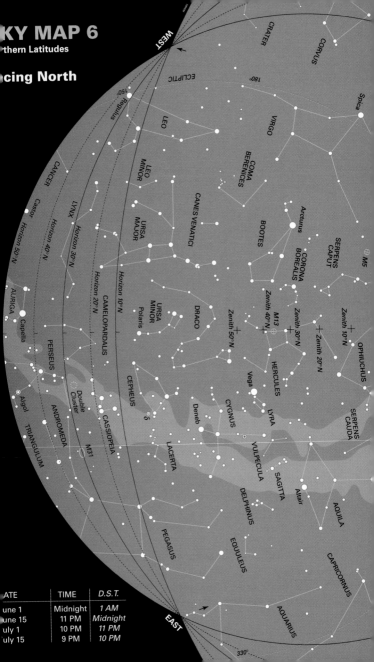

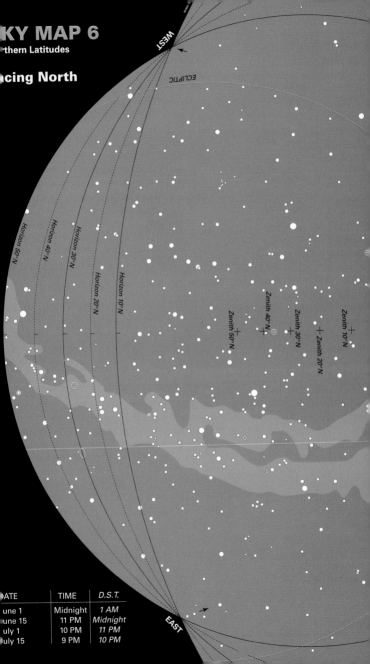

SKY MAP 6
thern Latitudes

cing North

WEST

ECLIPTIC

Horizon 50°N

Horizon 40°N

Horizon 30°N

Horizon 20°N

Horizon 10°N

Zenith 50°N

Zenith 40°N

Zenith 30°N

Zenith 20°N

Zenith 10°N

EAST

ATE	TIME	D.S.T.
une 1	Midnight	1 AM
une 15	11 PM	Midnight
uly 1	10 PM	11 PM
uly 15	9 PM	10 PM

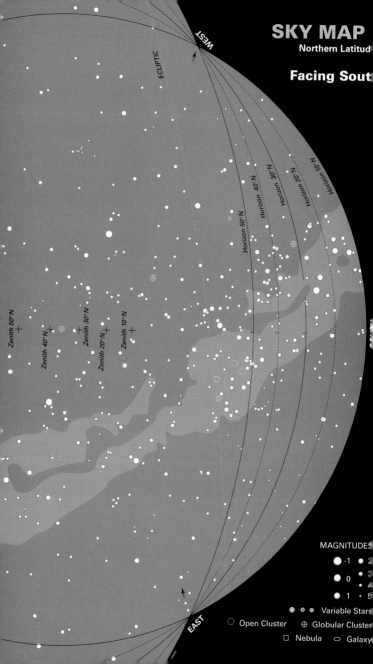

SKY MAP

Northern Latitud

Facing Sout

ECLIPTIC

WEST

EAST

Horizon 50° N
Horizon 40° N
Horizon 30° N
Horizon 20° N
Horizon 10° N

Zenith 50° N
Zenith 40° N
Zenith 30° N
Zenith 20° N
Zenith 10° N

MAGNITUDES

● -1   ● 2
● 0   ● 3
● 1   · 5

◉ ● · Variable Stars

○ Open Cluster   ⊕ Globular Cluster

□ Nebula   ○ Galaxy

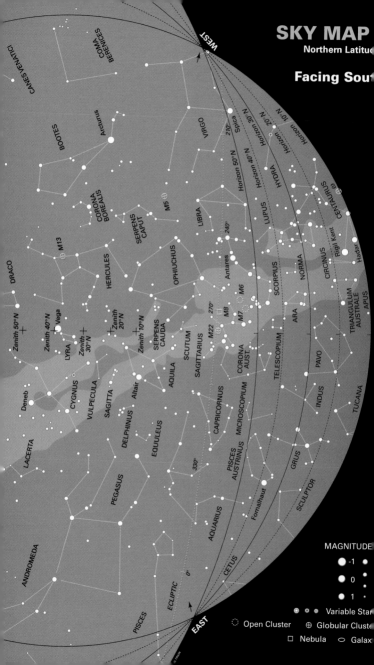

SKY MAP

Northern Latitu

Facing Sout

MAGNITUDE

- ● -1
- ● 0
- ● 1

◎ ◎ ● Variable Star
◌ Open Cluster ⊕ Globular Cluste
□ Nebula ◯ Galax

WEST

EAST

N 10° N
Horizon 30° N
Horizon 20° N
Horizon 40° N
Horizon 50° N

210°
240°
270°
330°
0°

Zenith 50° N
Zenith 40° N
Zenith 30° N
Zenith 20° N
Zenith 10° N

CANES VENATICI
COMA BERENICES
BOOTES
Arcturus
VIRGO
Spica
HYDRA
CENTAURUS
CORONA BOREALIS
M5
SERPENS CAPUT
LIBRA
LUPUS
Rigil Kent
DRACO
M13
HERCULES
OPHIUCHUS
Antares
SCORPIUS
NORMA
CIRCINUS
Hadar
Vega
LYRA
SERPENS CAUDA
M6
ARA
TRIANGULUM AUSTRALE
APUS
Zenith
SCUTUM
M8
M7
Deneb
CYGNUS
VULPECULA
SAGITTARIUS
M22
CORONA AUST
TELESCOPIUM
PAVO
SAGITTA
Altair
AQUILA
SAGITTA
DELPHINUS
CAPRICORNUS
MICROSCOPIUM
INDUS
TUCANA
LACERTA
EQUULEUS
PISCES AUSTRINUS
GRUS
PEGASUS
Fomalhaut
SCULPTOR
AQUARIUS
ANDROMEDA
CETUS
PISCES
ECLIPTIC

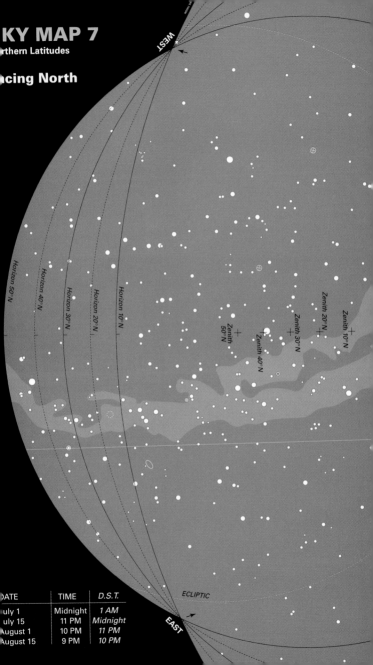

SKY MAP 7
rthern Latitudes

cing North

WEST

Horizon 50°N
Horizon 40°N
Horizon 30°N
Horizon 20°N
Horizon 10°N

Zenith 50°N
Zenith 40°N
Zenith 30°N
Zenith 20°N
Zenith 10°N

ECLIPTIC

EAST

DATE	TIME	D.S.T.
uly 1	Midnight	1 AM
uly 15	11 PM	Midnight
ugust 1	10 PM	11 PM
ugust 15	9 PM	10 PM

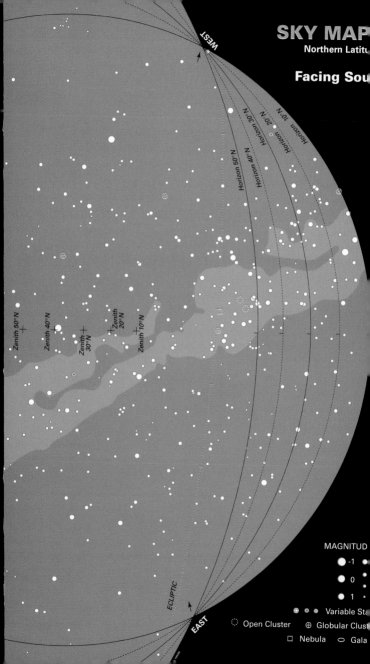

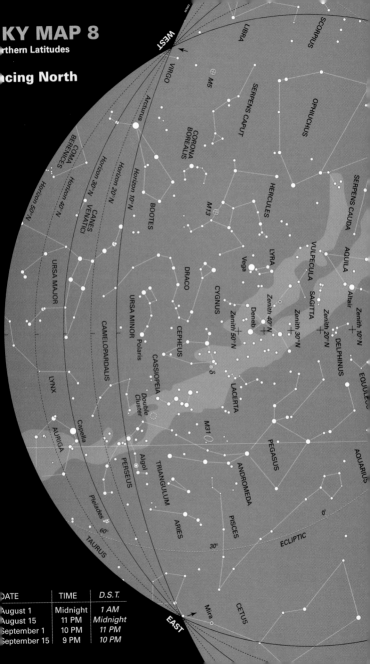

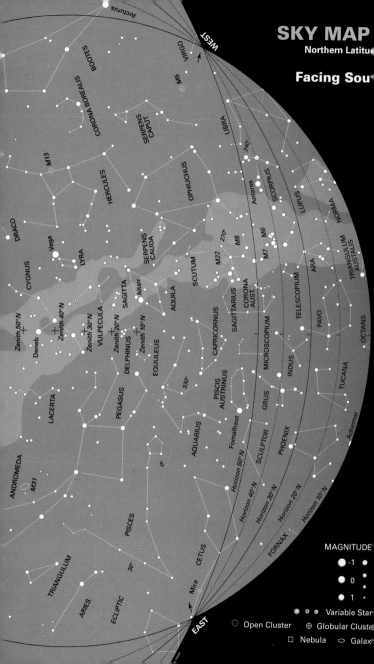

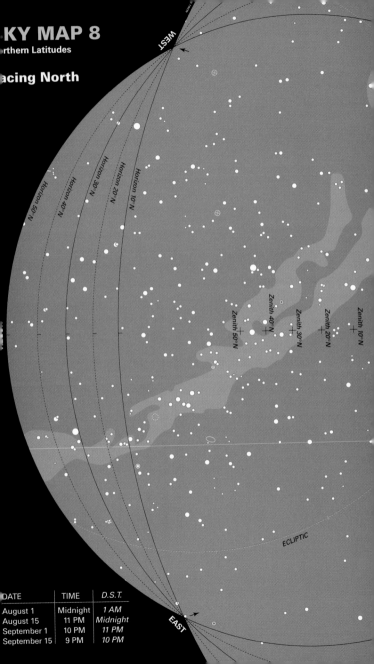

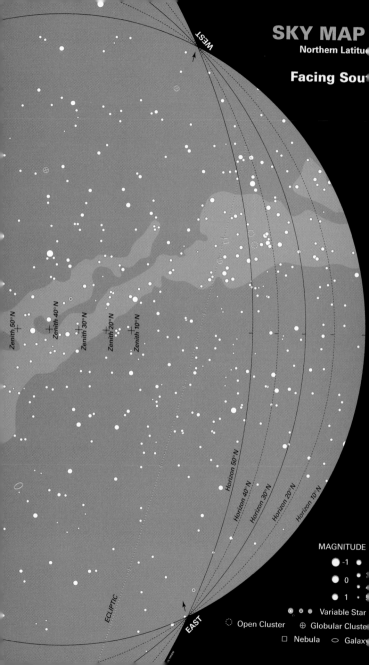

SKY MAP

Northern Latitu

Facing Sou

WEST

EAST

Zenith 50° N
Zenith 40° N
Zenith 30° N
Zenith 20° N
Zenith 10° N

Horizon 50° N
Horizon 40° N
Horizon 30° N
Horizon 20° N
Horizon 10° N

ECLIPTIC

MAGNITUDE

-1
0
1

Variable Star

Open Cluster ⊕ Globular Cluste

☐ Nebula ○ Galaxy

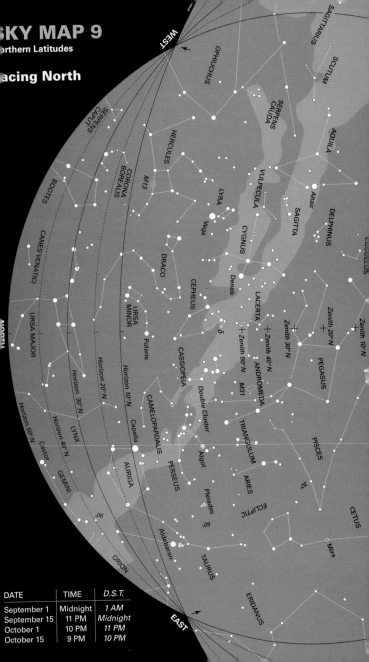

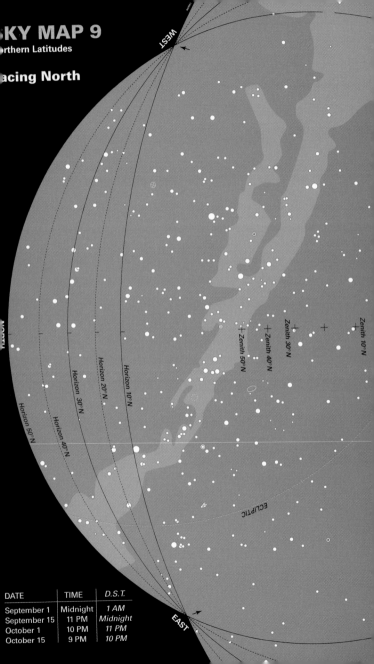

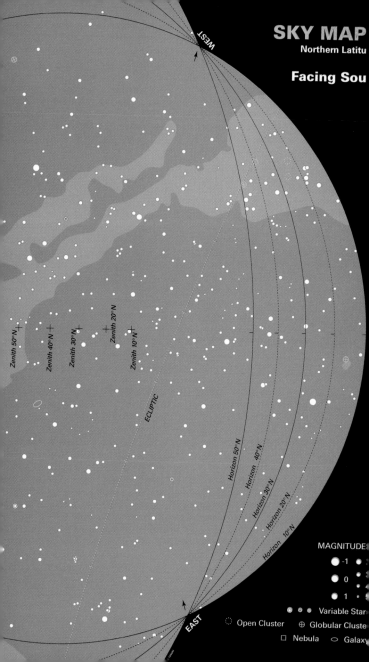

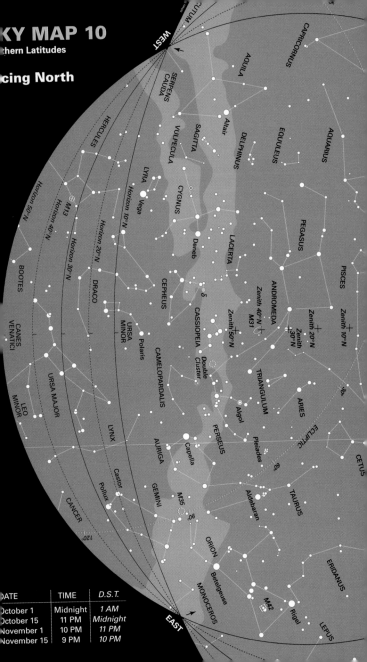

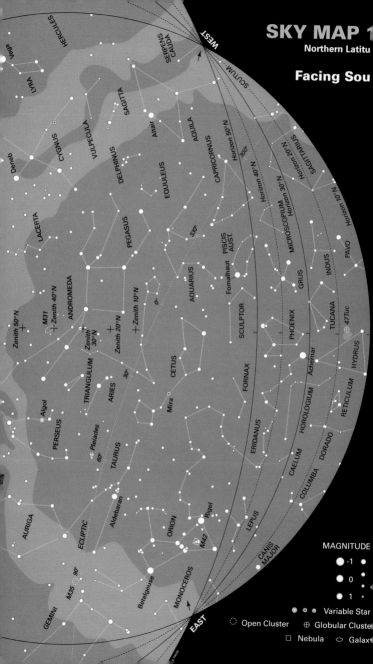

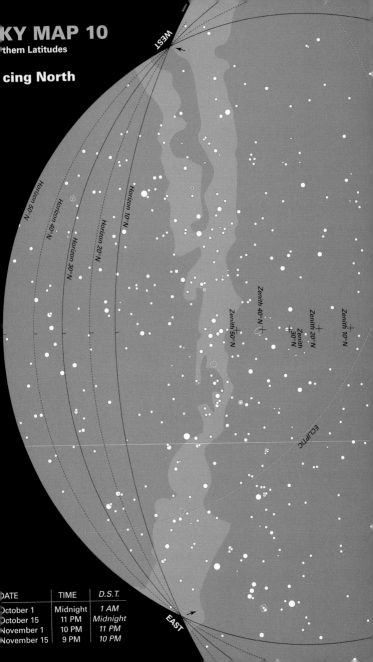

SKY MAP 10

Northern Latitudes

Facing North

WEST

NORTH

Horizon 50° N
Horizon 40° N
Horizon 30° N
Horizon 20° N
Horizon 10° N

Zenith 50° N
Zenith 40° N
Zenith 30° N
Zenith 20° N
Zenith 10° N

ECLIPTIC

EAST

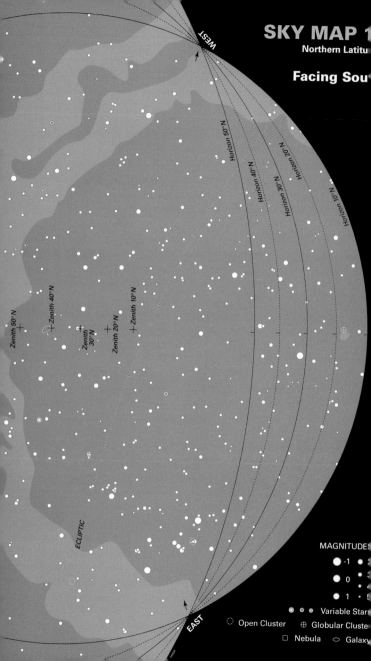

Horizon 50° N

Horizon 40° N

Horizon 30° N

Horizon 20° N

Horizon 10° N

WEST

EAST

Zenith 50° N

Zenith 40° N

Zenith 30° N

Zenith 20° N

Zenith 10° N

ECLIPTIC

MAGNITUDES

● -1

● 0

● 1

⊙ ● ● Variable Stars

○ Open Cluster ⊕ Globular Cluste

□ Nebula ○ Galaxy

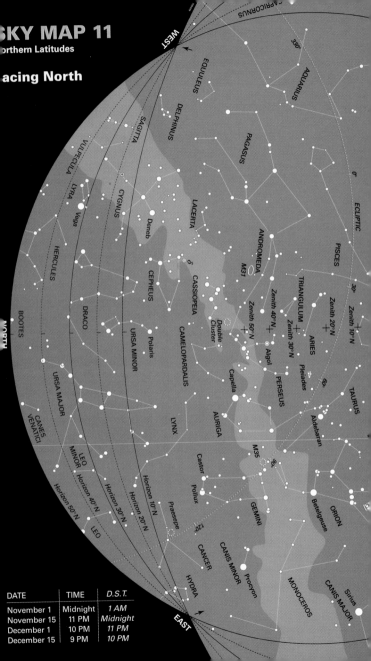

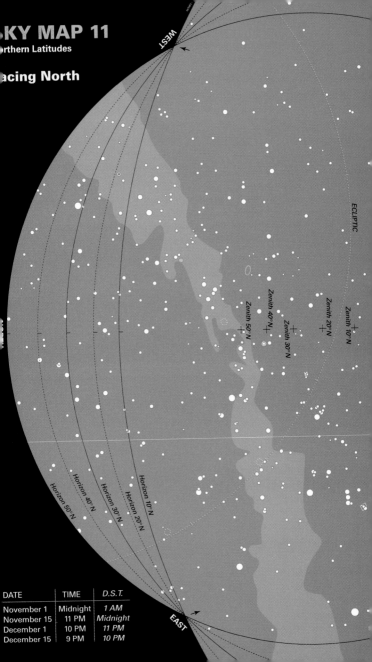

KY MAP 11

rthern Latitudes

acing North

WEST

EAST

ECLIPTIC

Zenith 50° N
Zenith 40° N
Zenith 30° N
Zenith 20° N
Zenith 10° N

Horizon 50° N
Horizon 40° N
Horizon 30° N
Horizon 20° N
Horizon 10° N

DATE	TIME	D.S.T.
November 1	Midnight	1 AM
November 15	11 PM	Midnight
December 1	10 PM	11 PM
December 15	9 PM	10 PM

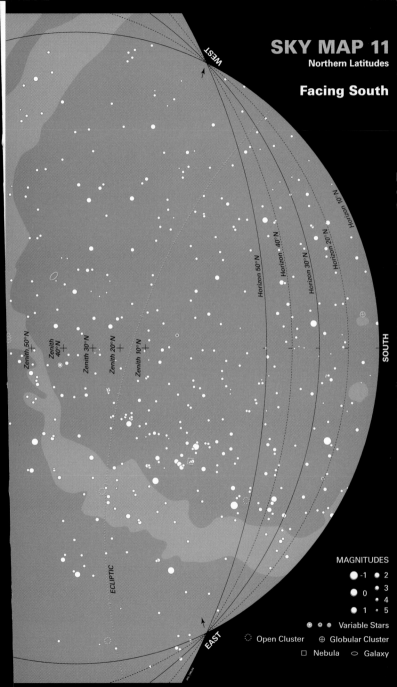

SKY MAP 11
Northern Latitudes

Facing South

WEST

Horizon 10° N

Horizon 20° N

Horizon 30° N

Horizon 40° N

Horizon 50° N

SOUTH

Zenith 50° N

Zenith 40° N

Zenith 30° N

Zenith 20° N

Zenith 10° N

ECLIPTIC

EAST

MAGNITUDES

● -1   ● 2

● 0   ● 3

● 1   · 4

       · 5

●●● Variable Stars

⊙ Open Cluster   ⊕ Globular Cluster

□ Nebula   ◇ Galaxy

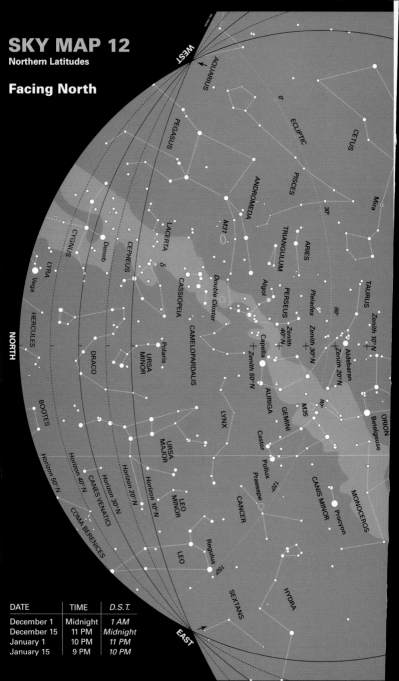

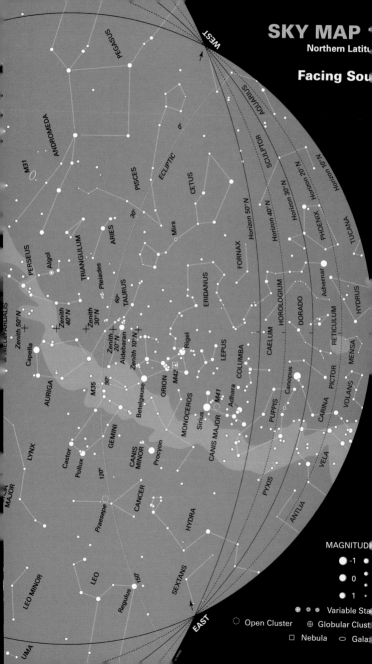

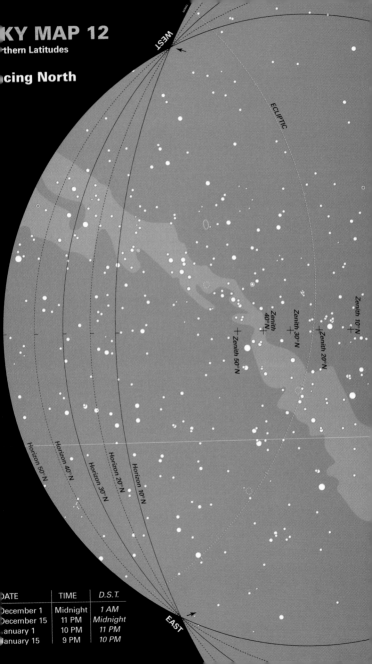

SKY MAP 12
thern Latitudes

cing North

ECLIPTIC

WEST

NORTH

Zenith 10° N

Zenith 20° N

Zenith 30° N

Zenith 40° N

Zenith 50° N

Horizon 50° N

Horizon 40° N

Horizon 30° N

Horizon 20° N

Horizon 10° N

EAST

DATE	TIME	D.S.T.
December 1	Midnight	1 AM
December 15	11 PM	Midnight
January 1	10 PM	11 PM
January 15	9 PM	10 PM

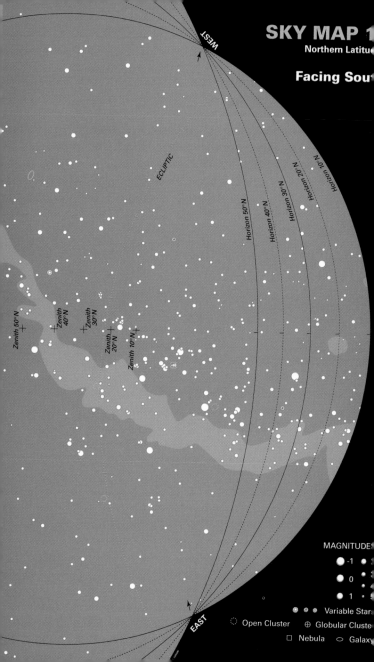

SKY MAP 1

Northern Latitud

Facing Sout

ECLIPTIC

Horizon 50° N
Horizon 40° N
Horizon 30° N
Horizon 20° N
Horizon 10° N

Zenith 50° N
Zenith 40° N
Zenith 30° N
Zenith 20° N
Zenith 10° N

WEST

EAST

MAGNITUDE
● -1
● 0
● 1

● ● ● Variable Star
◌ Open Cluster ⊕ Globular Cluste
□ Nebula ○ Galaxy

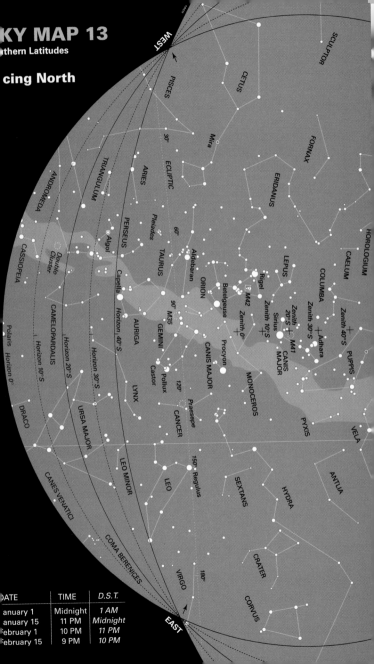

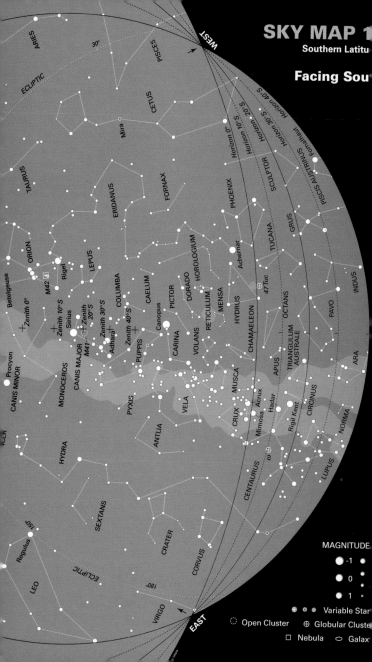

SKY MAP 1

Southern Latitu

Facing Sou

CONSTELLATIONS AND STARS

ARIES
PISCES
CETUS
TAURUS
ECLIPTIC
Mira
PISCES AUSTRINUS
Fomalhaut
Horizon 0°S
Horizon 10°S
Horizon 20°S
Horizon 30°S
Horizon 40°S
SCULPTOR
ERIDANUS
ORION
Betelgeuse
Rigel
M42
FORNAX
PHOENIX
GRUS
INDUS
LEPUS
COLUMBA
CAELUM
PICTOR
DORADO
HOROLOGIUM
Achernar
TUCANA
47 Tuc
Zenith 0°S
Sirius
Zenith 10°S
Zenith 20°S
Zenith 30°S
Canopus
Zenith 40°S
RETICULUM
MENSA
HYDRUS
OCTANS
PAVO
Adhara
Canis Major
M41
PUPPIS
CARINA
VOLANS
CHAMAELEON
APUS
ARA
Procyon
CANIS MINOR
MONOCEROS
PYXIS
VELA
CRUX
MUSCA
TRIANGULUM AUSTRALE
Acrux
Mimosa
Hadar
Rigil Kent
CIRCINUS
NORMA
ANTLIA
CENTAURUS
HYDRA
ω
LUPUS
SEXTANS
CRATER
CORVUS
Regulus
LEO
ECLIPTIC
180°
VIRGO

WEST
EAST

MAGNITUDE

- -1
- 0
- 1

Variable Star
Open Cluster   Globular Cluste
Nebula   Galax

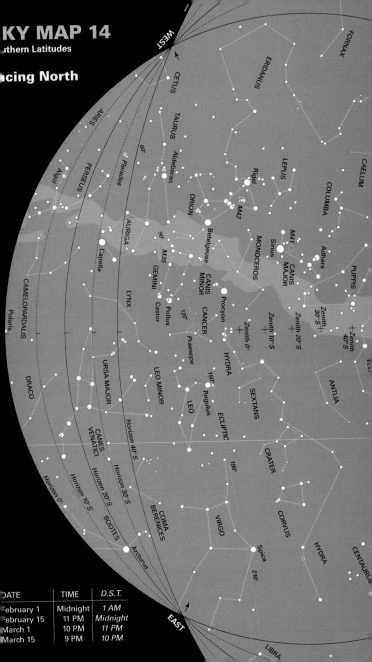

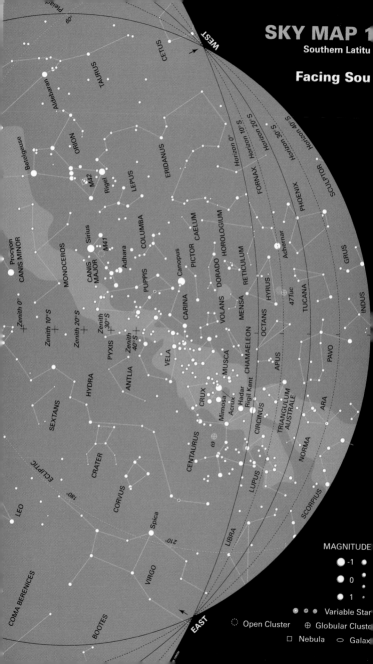

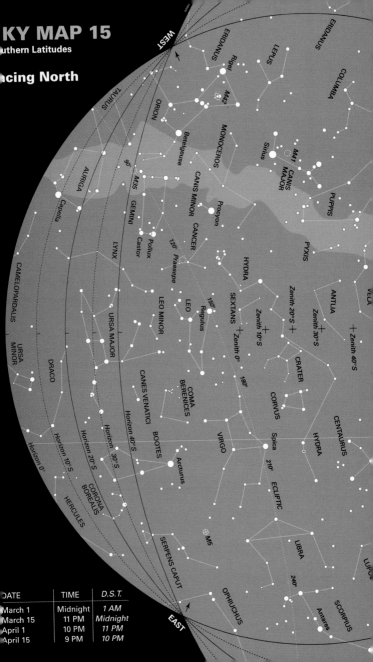

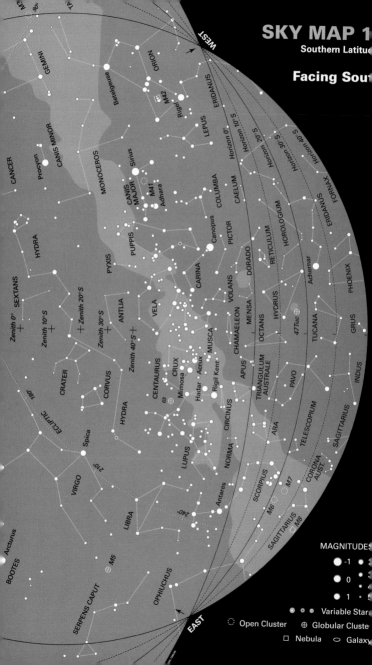

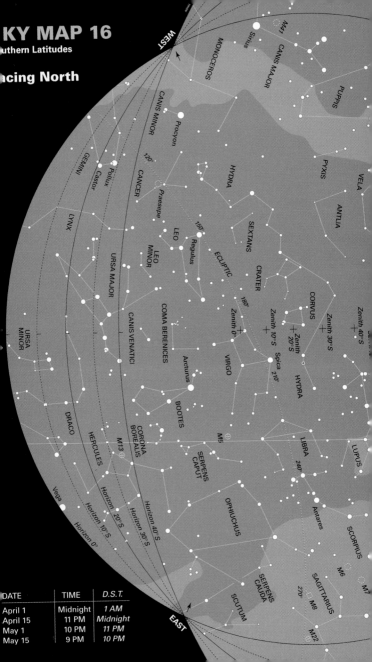

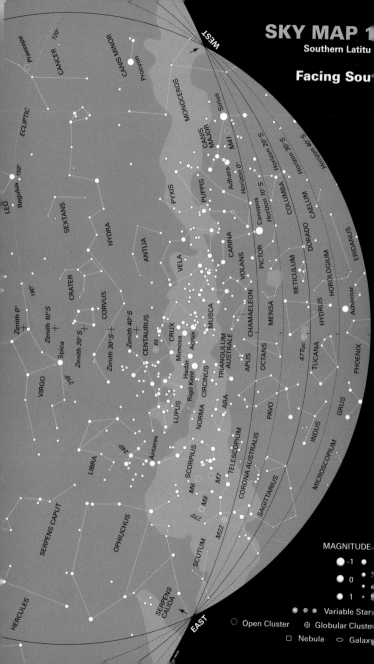

WEST

EAST

Praesepe

120°

CANCER

CANIS MINOR

Procyon

MONOCEROS

Sirius

ECLIPTIC

LEO

Regulus

150°

SEXTANS

CANIS MAJOR

M41

Adhara

Horizon 0°

PUPPIS

Horizon 10° S

Horizon 20° S

Horizon 30° S

Horizon 40° S

HYDRA

PYXIS

Canopus

COLUMBA

CAELUM

DORADO

ERIDANUS

CRATER

ANTLIA

VELA

CARINA

VOLANS

PICTOR

RETICULUM

HOROLOGIUM

Achernar

Zenith 0°

Zenith 10° S

CORVUS

Zenith 20° S

Zenith 30° S

Zenith 40° S

CENTAURUS

ω

CRUX

MUSCA

CHAMAELEON

MENSA

HYDRUS

PHOENIX

180°

Spica

Mimosa

Acrux

TRIANGULUM
AUSTRALE

47 Tuc

TUCANA

210°

VIRGO

Hadar

Rigil Kent

CIRCINUS

APUS

OCTANS

LUPUS

NORMA

ARA

PAVO

INDUS

GRUS

240°

LIBRA

Antares

SCORPIUS

TELESCOPIUM

MICROSCOPIUM

SERPENS CAPUT

M6

M7

CORONA AUSTRALIS

M8

270°

SAGITTARIUS

OPHIUCHUS

M22

SCUTUM

HERCULES

SERPENS
CAUDA

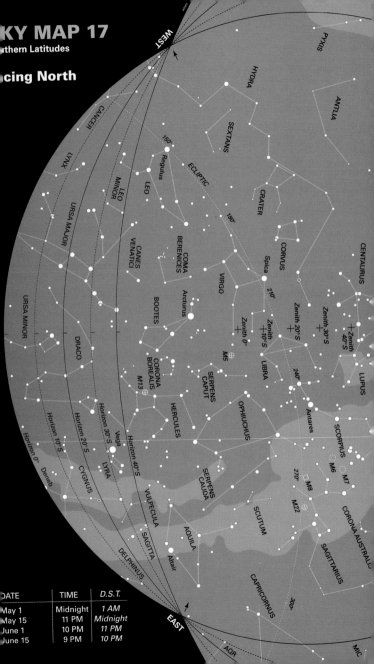

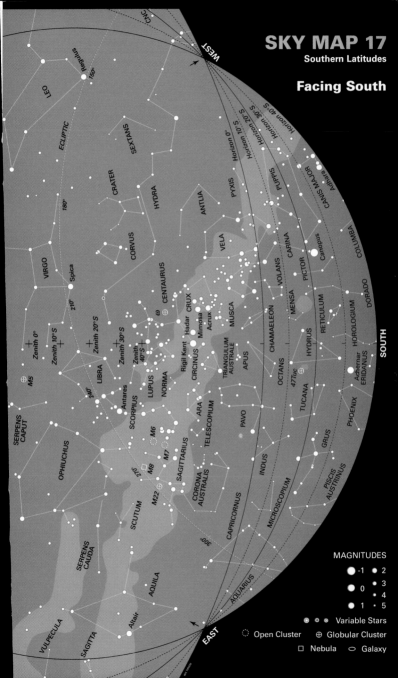

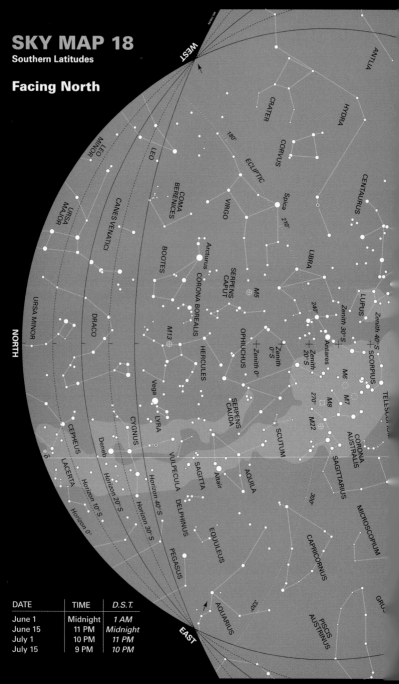

SKY MAP 18
Southern Latitudes

Facing North

DATE	TIME	D.S.T.
June 1	Midnight	1 AM
June 15	11 PM	Midnight
July 1	10 PM	11 PM
July 15	9 PM	10 PM

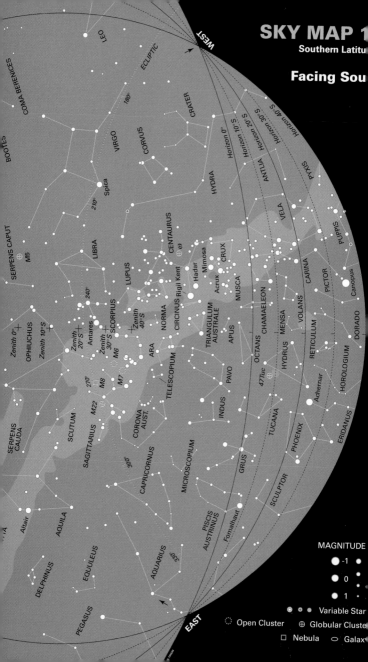

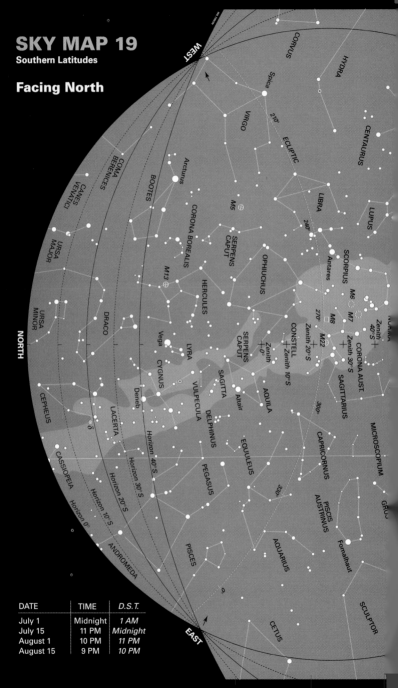

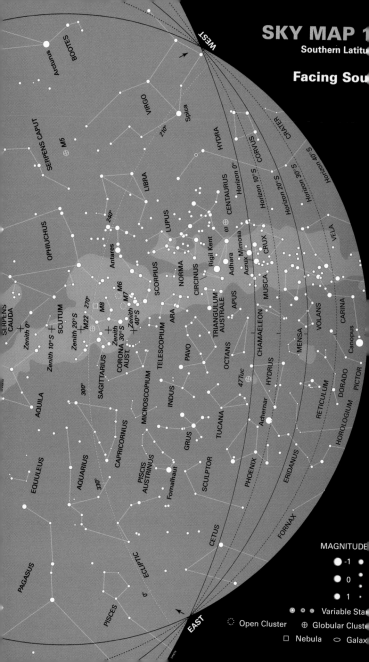

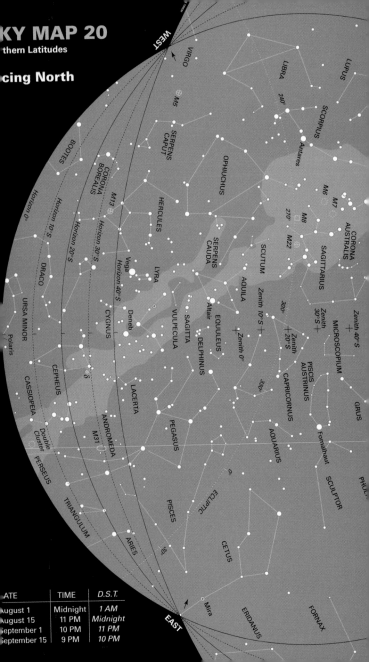

SKY MAP 2

Southern Latitu

Facing Sou

MAGNITUDES

- ● -1
- ● 0
- ● 1

● ● ● Variable Stars
○ Open Cluster   ⊕ Globular Cluste
□ Nebula   ○ Galaxy

WEST

HERCULES

SERPENS CAPUT

M5

VIRGO

OPHIUCHUS

LIBRA

240°

SERPENS CAUDA

Antares

SCORPIUS

LUPUS

CENTAURUS

SCUTUM

M8
M22
M6
M7
270°

NORMA

ARA

CIRCINUS

Rigil Kent
Hadar

CRUX

Acrux Mimosa
MUSCA

AQUILA

300°

SAGITTARIUS

CORONA AUST

TELESCOPIUM

TRIANGULUM AUSTRALE

APUS

OCTANS

CHAMAELEON

VELA

Zenith 0°

Zenith 10°S

Zenith 20°S

Zenith 30°S

Zenith 40°S

PAVO

EQUULEUS

CAPRICORNUS

330°

MICROSCOPIUM

INDUS

TUCANA

HYDRUS

MENSA

VOLANS

CARINA

PISCIS AUSTRINUS

GRUS

47Tuc

Achernar

RETICULUM

DORADO

Canopus

PUPPIS

AQUARIUS

Fomalhaut

SCULPTOR

PHOENIX

HOROLOGIUM

CAELUM

COLUMBA

0°

FORNAX

ERIDANUS

ECLIPTIC

CETUS

Mira

PISCES

30°

EAST

Horizon 0°
Horizon 10° S
Horizon 20° S
Horizon 30° S
Horizon 40° S

HYDRA

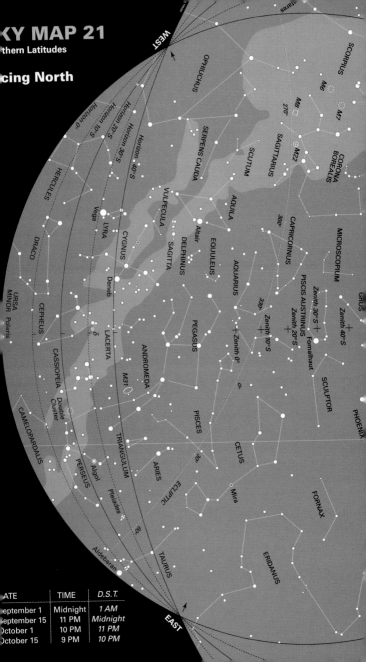

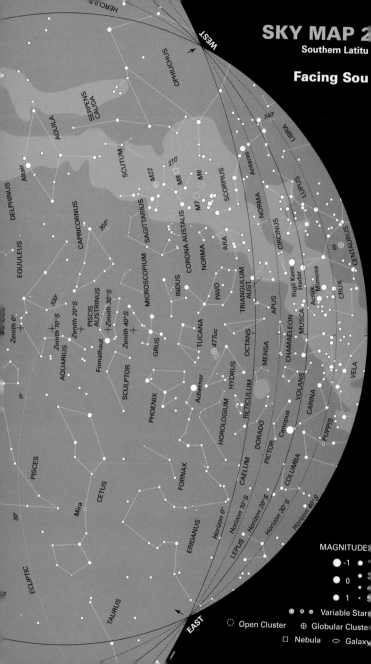

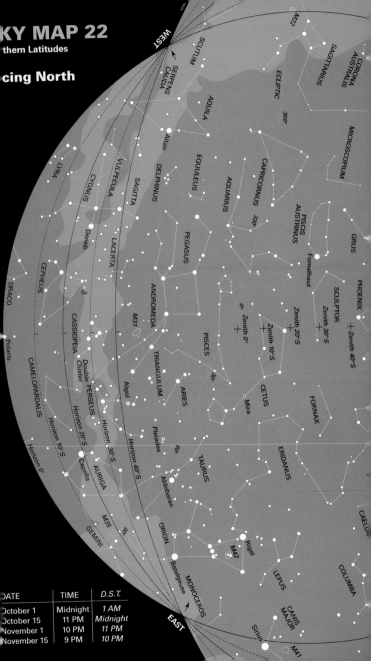

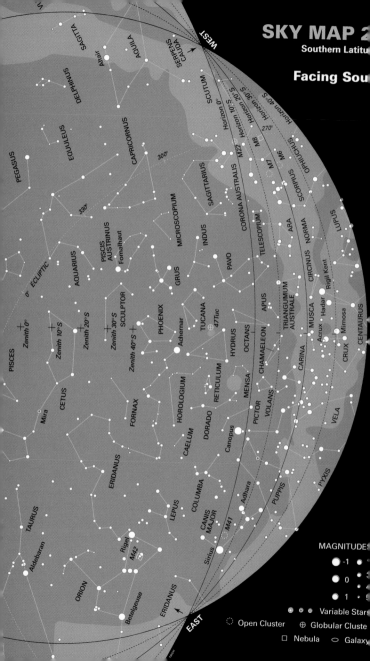

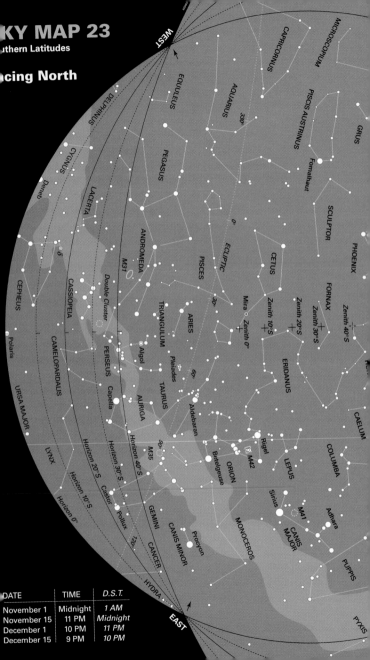

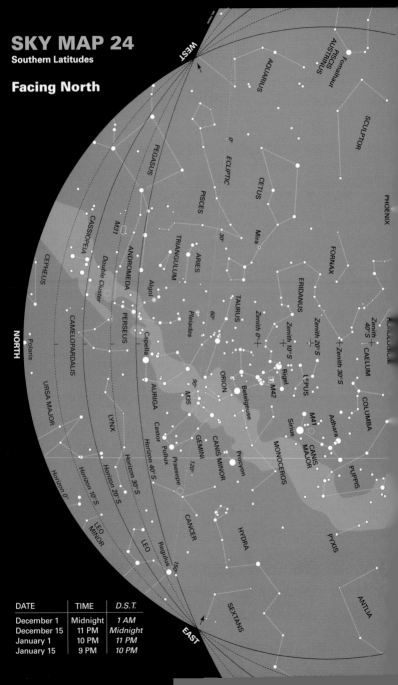

SKY MAP 24
Southern Latitudes

Facing North

DATE	TIME	*D.S.T.*
December 1	Midnight	*1 AM*
December 15	11 PM	*Midnight*
January 1	10 PM	*11 PM*
January 15	9 PM	*10 PM*

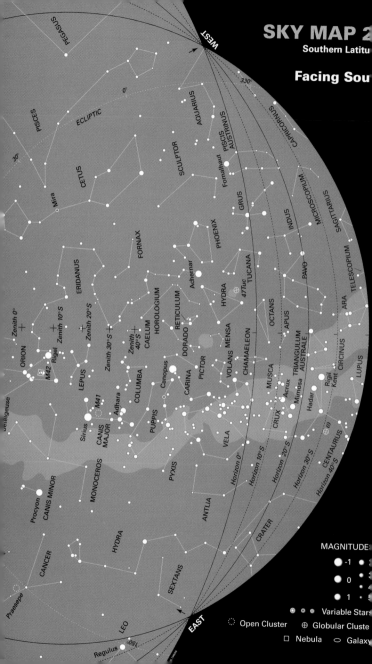

THE CONSTELLATIONS

Today we know that the stars in any given constellation do not necessarily have any physical relationship to one another. Some stars within a constellation may be relatively close to earth, while others may be relatively far away. All we know is that the stars are in roughly the same direction as seen from earth. Yet, just as it was convenient for the people of ancient civilizations to divide up the heavens into constellations or groups of seemingly related stars, it remains convenient for us to associate each star with one and only one constellation.

HISTORY OF THE CONSTELLATIONS

Exactly when and where the first system of constellations was devised is not known. Cuneiform texts and artifacts from the civilizations of the Euphrates Valley suggest that the lion, the bull, and the scorpion were already associated with constellations by 4000 B.C.E. Many scholars have been intrigued by the fact that there is some similarity between the names given to constellations by civilizations separated by vast distances. Perhaps a very ancient common tradition for naming a few groups of stars will ultimately be found. For the most part, however, the constellations of different civilizations appear to have developed independently of one another.

Of the 88 constellations listed by the International Astronomical Union in the definitive compilation of 1930, more than half were known to the ancients. Early records of the Greek constellations are found in the poetry of Homer, from about the ninth century B.C.E., and of Aratus, from about the third century B.C.E. Sometime between the lives of these two poets, probably in the mid- to late fifth century B.C.E., the ecliptic—the path the sun

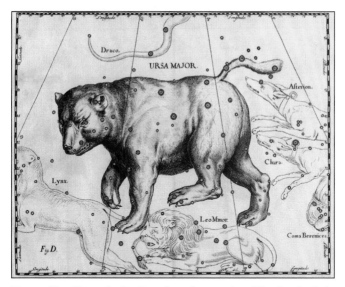

Fig. 4-1 Ursa Major, the Big Bear, from the star atlas of Hevelius (1687). The view is of the celestial sphere from outside, backward from the view we get from the ground. (Jay M. Pasachoff)

appears to follow across the celestial sphere in the course of a year—was identified in Babylon and perhaps in Greece as well. The Babylonians divided the ecliptic into the 12 parts of the zodiac—the band of constellations through which the sun, moon, and planets move (as seen from earth) in the course of the year.

During the second century, the astronomer Ptolemy cataloged information about 1,022 stars, grouped into 48 constellations. Not surprisingly, Ptolemy's catalog comprises only stars visible from the latitude of Alexandria, where he lived and wrote.

The *Almagest,* Ptolemy's chief work, remained the last word on the constellations until the sixteenth century, when European voyages of discovery took navigators into southern latitudes. The first star atlas, published by Johann Bayer in 1603, included 12 new constellations visible in southern-hemisphere skies. In 1624, the German astronomer Jakob Bartsch added three new constellations in the spaces that existed between previously named constellations. Bartsch also listed as a separate constellation the grouping we know as Crux, the Southern Cross, whose four chief stars Ptolemy had noted as part of the constellation Centaurus.

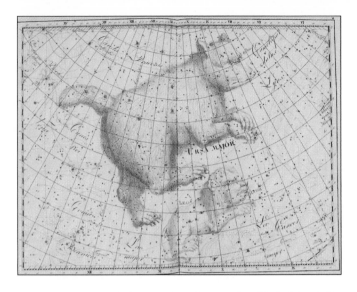

Fig. 4-2 Ursa Major, the Big Bear, from the star atlas of Bode (1 801). (Jay M. Pasachoff)

(The name Crux also reveals an attempt to depaganize the heavens that was typical of this period.) Earlier in the seventeenth century, Tycho Brahe likewise elevated to the status of constellation the asterism Coma Berenices, Berenice's Hair, which the ancients had thought of as part of Leo or Virgo.

In 1 687, the German astronomer Johannes Hevelius produced a major star atlas (fig. 4-1), in which he added seven more constellations visible from midnorthern latitudes. The visit of Nicolas Louis de Lacaille to the Cape of Good Hope in 1 750 resulted in 1 4 additional southern constellations. There have been other attempts since that time to invent new constellations, but official acceptance has been withheld. Johan Elert Bode's magnificent star atlas from 1 801 (fig. 4-2) shows the variety of constellations then accepted. However, since the mid-1 800s it has become traditional to break up Ptolemy's largest constellation, Argo Navis (Argo the Ship), into three constellations representing the ship's keel (Carina), stern (Puppis), and sails (Vela), in addition to the compass (Pyxis) invented by Lacaille.

The current list of constellations was adopted by the International Astronomical Union in 1 928 and codified in a list two years later. The IAU defined a constellation as one of 88 regions into which they divided the entire sky; each area of the sky belongs to one and only one of these regions. The boundaries zigzag so that the lines that separate constellations do not wreak havoc with ancient figures. A few stars that had been previously thought of as part of another constellation wound up in a new one — for example, one of the four stars of the Great Square of Pegasus is now officially in Andromeda. But on the whole, the IAU division provided a great simplification.

The lines dividing constellations were originally drawn along lines of right ascension and declination for the year 1875.0. (Astronomers use decimals to indicate parts of a year; 1875.0 means the beginning of the year 1875.) But because of *precession* (the drifting of the direction of the earth's axis among the stars), the lines between constellations have drifted slightly; the Atlas Charts in chapter 7 show that the lines dividing the constellations are no longer aligned so neatly with lines marking the coordinate scales that are now drawn for the date 2000.0.

Precession has also changed the dates when the sun appears to drift through each constellation of the zodiac, so that the sun is not actually in the sign listed in the newspapers' daily horoscopes. The sun actually passes through 13, not 12, constellations in the course of a year. Further, all or part of 24 constellations are actually in the zodiac, if we define it as the region within about 8° of the ecliptic — the band in which we find the first eight planets. (If we include Pluto, unknown to the ancients, the band would be wider still.)

The list below divides the constellations into three groups. First come the constellations of the traditional zodiac, listed in order around the sky: Aries, Taurus, Gemini, Cancer, Leo, Virgo, Libra, Scorpius, Sagittarius, Capricornus, Aquarius, and Pisces. Then come, alphabetically, the other constellations cataloged by Ptolemy, arranged here alphabetically. Finally, we list the constellations added since 1600, also alphabetically. Since in many cases more than one mythological story is associated with the same constellation, the list does not aim to provide full coverage of all the myths. For each constellation, we list the numbers of the Atlas Charts on which the constellation is found in chapter 7. A list of standard abbreviations for the constellations and the genitive form of their names used to make star names (e.g. alpha Centauri, in Centaurus) appears in appendix 1.

The old star atlases of Bayer (1603), Hevelius (1687), and Bode (1801) contain beautiful drawings of some of the constellations. Selected drawings from these atlases appear throughout this Field Guide.

ARIES, THE RAM. The golden fleece of this ram was the prize ultimately carried off by Jason, leader of the Argonauts. (Atlas Charts 10, 22, and 23)

TAURUS, THE BULL. Zeus disguised himself as a snow-white bull in order to attract Europa, Princess of Phoenicia. Drawn to the animal by its beauty, she climbed onto its back. Zeus then swam with his passenger to Crete, where he revealed his identity to her and won her. (Atlas Charts 10, 11, 23, and 24) See figure 4-3.

GEMINI, THE TWINS. These are the two devoted twins and, some say, half-brothers. Gemini is the Roman word for the Dioscuri, who in Greek mythology were the sons of Leda, the wife of Tyndarus, the king of Sparta. Though Castor was the son of Tyndarus, Pollux (Polydeuces in Greek) may have been the son of Zeus, which would have made him immortal. After Castor's death, Pollux was overwhelmed with grief and wanted to share his immortality with his twin. Finally Zeus reunited them by placing them together in the heavens. (Atlas Charts 12, 24, and 25)

CANCER, THE CRAB. When Hercules was struggling with Hydra (see p. 136), Juno sent this crab to attack him. The crab did not succeed in its mission, and instead was crushed. But Juno rewarded the crab for

Fig. 4-3 Taurus, the Bull, from the star atlas of Bayer (1603). (Jay M. Pasachoff)

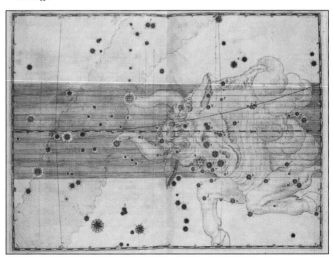

ARGO NAVIS, THE SHIP ARGO. This ship carried the Argonauts from Thessaly to Colchis in search of the Golden Fleece. No longer considered a single constellation, this grouping has been subdivided into four constellations: Carina, Puppis, Pyxis, and Vela.

ARIES, THE RAM. Zodiacal; see above.

AURIGA, THE CHARIOTEER. No story definitively explains the figure that this constellation's stars supposedly define—a charioteer (minus chariot and horse), holding the reins in his right hand, a goat on his left shoulder, and two small kids in his left arm. One myth associates Auriga with Erecthonius, the lame son of Vulcan and Minerva. Erecthonius invented the chariot, which not only enabled him to get around but also won him a place in the heavens. (Atlas Charts 3, 11, and 12)

BOÖTES, THE HERDSMAN. No single myth is definitively associated with this constellation. According to one story, Boötes was rewarded with a place in the heavens for his invention of the plow. (Atlas Charts 15, 16, 28, and 29)

CANCER, THE CRAB. Zodiacal; see above.

CANIS MAJOR, THE GREATER DOG. This constellation has been associated with several mythical dogs, including the hound of Actaeon, but it has also been known as the Dog of Orion. Orion loved to hunt wild animals, such as Lepus, the Hare; Canis Major, at Orion's heel in the sky, seems ready to pounce on the Hare. (Atlas Charts 24, 25, and 36)

CANIS MINOR, THE LESSER DOG. Among the dogs with which this constellation has been associated is Helen's favorite, which allowed Paris to abduct her without resistance. (Atlas Chart 25)

CAPRICORNUS, THE SEA GOAT. Zodiacal; see above.

CASSIOPEIA. See also Andromeda (above). When Cassiopeia objected to the wedding of Perseus and Andromeda, Perseus exhibited the head of Medusa. As a result, his enemies, including Cassiopeia, were turned into stone. Neptune placed Cassiopeia in the heavens, but in order to humiliate her, he arranged it so that at certain times of the year she would appear to be hanging upside down. (Atlas Charts 1, 2, and 9)

CENTAURUS, THE CENTAUR. This constellation is often identified with Chiron (see under Sagittarius). When Hercules accidentally wounded Chiron with one of his poisoned arrows, Chiron suffered greatly but, as an immortal, could not find a release in death. He resolved his problem by offering himself as a substitute for Prometheus, a Titan who was suffering because he had stolen fire from heaven for the benefit of humanity. The gods had punished Prometheus by chaining him to a rock, where each day a vulture devoured his liver, which was restored each night. At the request of Hercules, Zeus agreed to release Prometheus if a substitute could be found for him. Chiron offered his immortality to Prometheus and went to Tartarus, where Prometheus had been imprisoned by Zeus. Zeus placed Chiron among the stars. See also Corona Australis (below). (Atlas Charts 38–40, 49, and 50)

CEPHEUS, A KING OF ETHIOPIA. See Andromeda (above). (Atlas Charts 1, 2, and 8)

CETUS, THE WHALE. A sea monster; see Andromeda (above). (Atlas Charts 21–23, 33, and 34) See figure 4-4.

CORONA AUSTRALIS, THE SOUTHERN CROWN. Also called Corona Austrina. The Southern Crown is said to represent a crown of laurel or olive, placed on victors in games and on those who perform great service to their peers. According to one story, this constellation symbolizes a laurel wreath ready to be placed on the brow of Chiron in acknowledgment of his service to Prometheus. See Centaurus (above). (Atlas Chart 42)

CORONA BOREALIS, THE NORTHERN CROWN. This constellation is generally associated with Ariadne, daughter of Minos, king of Crete. Each year, as a tribute to Crete, 14 young people from Athens were fed to the Minotaur, the monster that Minos kept in a labyrinth. When Theseus arrived from Athens as one of the prospective victims, Ariadne fell in love with him. She offered to help him escape the Minotaur if he promised to take her back to Athens as his bride. Theseus agreed. Ariadne kept her part of the bargain, and after Theseus killed the Minotaur, they sailed off. They stopped on the island of Naxos, where Theseus and the Athenians abandoned the sleeping Ariadne. When she awoke, clamoring for vengeance, she found Dionysus, god of the vine, who married her immediately. As a wedding gift, Dionysus gave her a crown studded with gems. When Ariadne died, Dionysus set the crown among the stars. (Atlas Charts 16, 17)

CORVUS, THE RAVEN. According to some, this is the raven that Apollo sent to guard his beloved, Coronis, during his absences. During one of the god's absences, Coronis fell in love with another and was unfaithful to Apollo. When the raven told Apollo what had happened, Apollo rewarded the bird by placing it in the sky. (Atlas Charts 27 and 39)

CRATER, THE CUP. This constellation, which is supposed to resemble a goblet, has been associated with various gods and heroes, including Apollo, Bacchus, Hercules, and Achilles. (Atlas Charts 27 and 38)

CYGNUS, THE SWAN. According to one legend, this constellation is related to the story of Phaethon, a mortal who learned that his father was Helius, the sun god. Helius rashly promised to let Phaethon drive the chariot of the sun across the sky. Phaethon soon lost control, and his reckless driving threatened to destroy the earth with the sun's heat. Zeus intervened by hurling a thunderbolt at Phaethon, who fell into the Eridanus River (see below). In his grief, Phaethon's devoted friend Cygnus dived into the water in search of the body. Apollo took pity on Cygnus, changed him into a swan, and placed the swan in the sky. (Atlas Charts 7, 8, 1 8, and 1 9)

PHINUS, THE DOLPHIN. In one story, the dolphin successfully convinced the sea goddess Amphitrite to marry Poseidon, from whom she had been fleeing. As a reward, Poseidon placed Delphinus among the stars. (Atlas Charts 1 9 and 31)

ACO, THE DRAGON. Among the monsters with whom this constellation has been associated is the dragon slain by Cadmus, brother of Europa, on the site of the city that was later to become Thebes. After Cadmus accomplished this feat, Athena instructed him to plant some of the dragon's teeth. Armed men grew up from the soil; when Cadmus threw stones into their midst, they began to fight among themselves. All but five died, and the survivors helped Cadmus build the city of Thebes. (Atlas Charts 4–8)

UULEUS, THE LITTLE HORSE. This group of stars is associated with Celeris, the brother of Pegasus. Mercury gave Celeris to the hero Castor. (Atlas Chart 32)

DANUS, THE RIVER. The waters of the River Eridanus are said to steam perpetually as a result of Phaethon's fiery fall (see under Cygnus, above). (Atlas Charts 22–24, 34, 35, and 46)

MINI, THE TWINS. Zodiacal; see above.

RCULES. The most famous of all the Greek heroes, Hercules was worshiped throughout the Mediterranean world. Best known for his twelve labors, he performed many incredible deeds throughout his life. His death provided an equally dramatic end to his career. After accidentally killing a young man, Hercules decided to go into exile with his wife, Deianira. When they came to a river, Hercules swam on ahead, leaving Deianira in the charge of Nessus, a centaur who offered to carry her across the water on his back. When Nessus tried to rape Deianira, Hercules shot the centaur with a poisoned arrow. Before Nessus died, he suggested to Deianira that she keep his blood as a love charm, should she ever need to revive Hercules' interest in her. Some time later, learning of her husband's interest in another woman, Deianira rubbed the centaur's dried blood on one of Hercules' robes. The hero's body burned as soon as he put the robe on, and when he tried to remove the robe, his skin came off with it. When Deianira learned of the effects of the robe, she hanged herself. Hercules built and mounted a funeral pyre, which only the compassionate Philoctetes had the courage to light. Instantly, a flash of lightning was seen. The pyre burned out completely, leaving no bones behind. It was assumed that Hercules' body had been carried to Olympus. (Atlas Charts 1 7, 1 8, 29, and 30)

HYDRA, THE WATER SNAKE. Successfully eliminating this many-headed monster was Hercules' second labor—a difficult task because whenever one head was cut off, two new ones grew in its place. Hercules solved this problem by having his nephew, Iolaus, burn the stump of each head as soon as Hercules cut it off, thus preventing new heads from sprouting. Because Iolaus assisted him in this labor, Hercules was required to substitute another labor in its place. Do not confuse this constellation with Hydrus, a southern constellation. (Atlas Charts 25, 26, 37–39, and 40)

LEO, THE LION. Zodiacal; see above.

LEPUS, THE HARE. According to some, the Hare was placed in the heavens to be near its hunter, Orion. (Atlas Charts 24, 35, and 36)

LIBRA, THE SCALES. Zodiacal; see above.

LUPUS, THE WOLF. One story associates this constellation with the impious Lycaon, who doubted Zeus's claim to divinity. In order to test Zeus, Lycaon served the king of the gods the flesh of a child. To punish Lycaon for this impious act, Zeus transformed him into a wolf. (Atlas Charts 40 and 50)

LYRA, THE LYRE. This is the instrument given by Apollo to Orpheus, the most famous poet and musician in Greek legend. When his wife Eurydice died, Orpheus was told he could bring her back from the underworld on condition that he not look back at her until she was in the light of the sun. Orpheus led her out, playing on the lyre, but when he reached the light of the sun, he looked back. Because Eurydice had not yet reached the sunlight, he lost her forever. He remained inconsolable and refused the advances of all the women who tried to win his love. One day a group of women whom he had scorned attacked him; drowning out his music, they tore him apart and threw his head and lyre into the Hebrus River. Apollo intervened, however; Orpheus' head was placed in a cave, his limbs were buried at the foot of Mount Parnassus, and his lyre was placed among the stars. (Atlas Chart 18)

OPHIUCHUS, THE SERPENT-BEARER. This group is generally identified with Asclepius, the first physician and surgeon, who accompanied the Argonauts. He was so successful in curing the ill and wounded that Pluto began worrying about the declining immigration into the underworld. Pluto convinced Zeus to strike Asclepius with a thunderbolt and put him among the constellations. The snake twined around a staff remains the symbol of medicine today, perhaps because of the association of the snake's periodic sloughing off of its skin with the renewal of life. (Atlas Charts 29, 30, and 41)

ORION, THE HUNTER. When this giant hunter met Artemis, goddess of the hunt and of the moon, her brother Apollo feared for her virginity. Apollo sent Scorpius, the Scorpion (see p. 131), to attack Orion, who leaped into the sea to escape. Apollo then tricked his unsus-

Fig. 4-5 Perseus, Andromeda, and Cassiopeia, from the star atlas of Bode (1801). (Jay M. Pasachoff)

pecting sister into shooting at a dark spot on the waves that in actuality was Orion. The goddess tried to have Asclepius revive Orion, her hunting companion, but the physician had been killed by Zeus's thunderbolt (see under Ophiuchus, above). Artemis then placed Orion in the heavens, where he would be pursued eternally by the scorpion. (Atlas Charts 11, 12, 23, and 24)

EGASUS. This is the winged horse that sprang up from the blood of Medusa after Perseus killed her (see below). Pegasus was tamed by Bellerophon, whose successes against monsters and enemies had gone to his head. When Bellerophon decided to ride Pegasus up to Olympus, the gods were offended, so Zeus sent a gadfly to sting Pegasus. Pegasus reared in pain, causing Bellerophon to topple off. Bellerophon was lamed and blinded, but Pegasus continued to climb Olympus and earned a place among the stars as a constellation. (Atlas Charts 9, 19–21, and 32)

ERSEUS. Armed with a polished shield given to him by Athena, Perseus killed Medusa, the only one of the Gorgons who was mortal. The three Gorgons were winged monsters so horrible to behold that all who looked upon them were turned into stone. Athena told Perseus to use the shield as a mirror. He could thus avoid looking

Fig. 4-6 Argo Navis, the Ship, from the star atlas of Bayer (1603). It is now broken up into Vela (the Sails), Puppis (the Stern), Carina (the Keel), and Pyxis (the Compass). These constellations are not visible from midnorthern latitudes. (Jay M. Pasachoff)

directly at the Gorgons. After he cut off Medusa's head, the winged horse Pegasus sprang from her blood. The head of Medusa enabled Perseus to overcome many enemies, as well as to kill the monster Cetus (see under Andromeda, p. 132). (Atlas Charts 2, 3, 9–11) See Fig. 4-5.

PISCES, THE FISH. Zodiacal; see above.

PISCIS AUSTRINUS, THE SOUTHERN FISH. Also known as Piscis Australis, this constellation is linked by some to the story of the monster Typhon. After Zeus and the Olympians overthrew the Titans, sons of Gaea, Gaea gave birth to another son, Typhon. She incited Typhon to attack the Olympians, who assumed various animal forms to evade him—Venus, for example, assumed the form of a fish. (See also under Pisces, p. 132.) (Atlas Charts 43 and 44)

SAGITTA, THE ARROW. This constellation has been associated with several different arrows, including the one that Apollo used to slay the three one-eyed giants called Cyclops. (Atlas Charts 18, 19, and 31)

SERPENS, THE SERPENT. This constellation is associated with Ophiuchus (see above). (Serpens Caput, the head, is on Atlas Charts 16 and 29; Serpens Cauda, the tail, is on Atlas Chart 30)

TAURUS, THE BULL. Zodiacal; see above.

TRIANGULUM, THE TRIANGLE. This minor constellation has been associated with different geographical locations. Not surprisingly, because of its similarity in shape to the Greek letter Δ (delta), it was sometimes called Delta and thus was associated with Egypt and the Nile, whose delta provided fertile land. It was also associated with the island of Sicily, whose three promontories give it a triangular shape. (Atlas Charts 9 and 10)

URSA MAJOR, THE GREAT BEAR. Zeus fell in love with Callisto, daughter of Lycaon (see under Lupus, p. 136). Together they had a son, Arcas. Callisto was changed into a bear by one of the gods—some say by Artemis, who was angry with Callisto, her formerly chaste companion; some say by Zeus, who was anxious to protect Callisto from the jealousy of his wife, Hera; and some say by Hera. (Atlas Charts 4–6 and 13–15)

URSA MINOR, THE LESSER BEAR. When Arcas reached manhood, he was about to shoot a bear, unaware that it was his mother, Callisto. To protect Callisto, Zeus changed Arcas into a bear as well, and carried them both off by their tails to the heavens, where they became constellations. Annoyed at this honor, Hera took her revenge by convincing Poseidon not to allow the bears to bathe in the sea. For this reason, Ursa Major and Ursa Minor are circumpolar constellations, never sinking below the horizon. (Atlas Charts 2 and 6)

CONSTELLATIONS ADDED SINCE 1600

Note that the names of constellations reflect the times in which they were named; we find here many names of machines in addition to allusions to classical myths.

ANTLIA, THE PUMP. Lacaille called this constellation the *Machine Pneumatique,* and in Germany it is called the *Luftpumpe,* or air pump. (Atlas Charts 37 and 38)

APUS, THE BIRD OF PARADISE. The bird for which this constellation was named was originally found in Papua New Guinea. (Atlas Charts 50 and 51)

CAELUM, THE CHISEL. Lacaille formed this constellation from stars between Columba and Eridanus. (Atlas Chart 35)

CAMELOPARDALIS, THE GIRAFFE. Bartsch first outlined this constellation, claiming that it represented the camel that brought Rebecca to Isaac. The constellation lies in the large space between Perseus, Auriga, and the Bears. (Atlas Charts 2 and 3)

CANES VENATICI, THE HUNTING DOGS. Hevelius formed this constellation out of stars between Ursa Major and Boötes, to represent the two greyhounds held in leash by Boötes (see p. 133). (Atlas Chart 15)

CARINA, THE KEEL [OF ARGO]. See Argo Navis, p. 133. (Atlas Charts 47-49) See figure 4-6.

CHAMAELEON, THE CHAMELEON. This small constellation lies below Carina, separated from the south celestial pole by Octans. (Atlas Charts 48 and 49)

CIRCINUS, THE DRAWING COMPASSES. This constellation, south of Lupus, was added by Lacaille. (Atlas Chart 50)

COLUMBA, THE DOVE. Petrus Plancius, a sixteenth-century Dutch theologian and mapmaker, formed this constellation south of Lepus, to represent the dove that Noah sent out from the ark. (Atlas Charts 35 and 36)

COMA BERENICES, BERENICE'S HAIR. Long recognized as an asterism, Coma Berenices was first listed as a separate constellation in 1602 by Tycho Brahe. Berenice was the wife of Ptolemy Euergetes of Egypt in the middle of the third century B.C.E. She vowed to sacrifice her hair to Venus if her husband returned safely from war. After his safe return she honored her vow, but her hair disappeared overnight. A Greek astronomer pointed to the constellation, claiming it to be the missing hair, which Venus must have placed in the sky. Coma Berenices contains the north galactic pole (the north pole with respect to the orientation of our Milky Way Galaxy). (Atlas Charts 15, 27, and 28)

CRUX, THE CROSS. The ancient Greeks considered the four chief stars of Crux as part of the Centaur, which surrounds Crux on three sides. There is no central star to the cross, so it looks more like a kite. (Atlas Chart 49)

DORADO, THE SWORDFISH. The Large Magellanic Cloud lies within this constellation. (Atlas Chart 47)

FORNAX, THE FURNACE. Lacaille formed this group from stars within the southern bend of the River Eridanus. (Atlas Chart 34)

GRUS, THE CRANE. A southern constellation. (Grus americana is the scientific name for the whooping crane.) (Atlas Charts 43–45 and 52)

HOROLOGIUM, THE CLOCK. One of Lacaille's constellations. (Atlas Charts 34 and 46)

HYDRUS, THE WATER SNAKE. Not to be confused with the Ptolemaic constellation Hydra, Hydrus lies between the Large and Small Magellanic Clouds. (Atlas Chart 46)

INDUS, THE INDIAN. This constellation is supposed to represent a Native American, with arrows in both hands. (Atlas Charts 43 and 52)

LACERTA, THE LIZARD. Hevelius formed this constellation from stars lying between Cygnus and Andromeda. The shape supposedly was determined by the available space between the older constellations. (Atlas Charts 1, 8, and 20)

LEO MINOR, THE LESSER LION. Hevelius formed this constellation from stars lying between the zodiacal constellations Leo and Ursa Major. The name reflects his belief that this star grouping was similar in nature to the two other groups. (Atlas Charts 13 and 14)

«X. Hevelius chose this name, explaining that only those with the eyes of a lynx could discern this star group. (Atlas Charts 3, 4, 12, and 13)

NSA, THE TABLE. Lacaille named this constellation Mons Mensae, after Table Mountain, south of Cape Town, South Africa, where he did much of his work. (Atlas Chart 47)

ROSCOPIUM, THE MICROSCOPE. Lacaille formed this constellation from the stars south of Capricornus and west of Piscis Austrinus. (Atlas Chart 43)

NOCEROS, THE UNICORN. The Unicorn is generally attributed to Bartsch, but some claim that it is older. (Atlas Charts 24 and 25)

SCA, THE FLY. In early catalogs, this group, south of the Southern Cross and northeast of Chamaeleon, was sometimes also called the Bee. (Atlas Chart 49)

RMA, THE SQUARE. One of Lacaille's constellations, Norma lies immediately to the north of the Southern Triangle, Triangulum Australe (see below). (Atlas Charts 40 and 50)

TANS, THE OCTANT. Lacaille named this constellation to recognize John Hadley's invention of the octant in 1730. It includes the south celestial pole. (Atlas Charts 45–52)

VO, THE PEACOCK. According to myth, Argos, the builder of the ship Argo, was changed into a peacock by Hera after she brought his ship to the heavens. So it was fitting that one of the 12 new constellations in the southern skies should be named after the peacock, thus reuniting Argos and his ship. (Atlas Charts 51 and 52)

OENIX. Another one of the 12 new constellations in the southern skies, named after the mythical bird. The phoenix was said to live 500 or 600 years in the Arabian wilderness, to burn itself on a funeral pyre, and to rise from its ashes to live again. In ancient China, Egypt, India, and Persia, the phoenix was an astronomical symbol representing natural cycles. (Atlas Charts 33, 45, and 46)

TOR, THE EASEL. Lacaille named this grouping Equuleus Pictoris, or Painter's Easel. (Equuleus also means "small horse.") The name has been shortened to Pictor. The constellation lies south of Columba. (Atlas Charts 35 and 47)

PPIS, THE STERN [OF ARGO]. See Argo Navis, p. 133. (Atlas Charts 25, 36–37) See figure 4-6.

XIS, THE COMPASS [OF ARGO]. See Argo Navis, p. 133. (Atlas Chart 37) See figure 4-6.

TICULUM, THE RETICLE. This constellation was named by Lacaille in honor of his reticle, the network of fine lines he placed in the focus of his telescope in order to make his observations in the southern

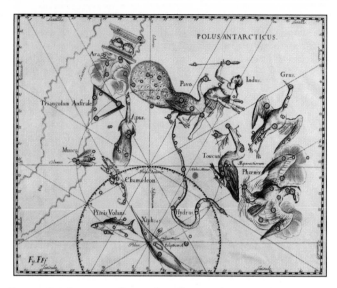

Fig. 4-7 Southern constellations from the star atlas of Hevelius (1687). The constellations are drawn backward from the way they appear in the sky because Hevelius drew the celestial sphere as it would appear from the outside. (Jay M. Pasachoff)

hemisphere. The constellation, however, had been drawn earlier by Isaak Habrecht of Strasbourg. (Atlas Chart 46)

SCULPTOR. Lacaille named this grouping l'Atelier du Sculpteur, the Sculptor's Studio, but the name has been shortened. The constellation lies between Cetus and Phoenix and contains the south galactic pole. (Atlas Charts 33 and 44)

SCUTUM, THE SHIELD. Hevelius formed this constellation from several stars in the Milky Way, between the tail of Serpens and the head of Sagittarius. The constellation is supposed to represent the coat of arms of John Sobieski, king of Poland, in honor of his successful resistance against the Turkish attack on Vienna in 1683. (Atlas Chart 30)

SEXTANS, THE SEXTANT. Hevelius formed this constellation, between Leo and Hydra, to recognize the importance of the sextant to his measurements of the stars. (Atlas Chart 26)

TELESCOPIUM, THE TELESCOPE. Lacaille's original formation of this constellation, between Ara and Sagittarius, encroached upon several older constellations. For example, the telescope's stand was in Sagittarius. Later catalogers redrew the outlines to avoid this kind of overlapping. (Atlas Charts 42, 51, and 52)

ANGULUM AUSTRALE, THE SOUTHERN TRIANGLE. Much more prominent than its northern counterpart, Triangulum, this group lies south of Norma. (Atlas Chart 50)

ANA, THE TOUCAN. Another one of the southern constellations named for exotic birds, this constellation is home to the Small Magellanic Cloud. (Atlas Chart 45)

A, THE SAILS [OF ARGO]. See Argo Navis, p. 133. (Atlas Charts 37, 38, 48, and 49). See figure 4-6.

ANS, THE FLYING [FISH]. Shortened from its original name, Piscis Volans. (Atlas Chart 48)

PECULA, THE FOX. A constellation added by Hevelius, Vulpecula is home to the Dumbbell Nebula (pp. 151 and 263). Shortened from Vulpecula cum Ansere, the Fox with the Goose. (Atlas Charts 18 and 19)

Fig. 4-8 *Star trails around the south celestial pole, including the Large Magellanic Cloud and the Small Magellanic Cloud.* (Akira Fujii)

STARS, NEBULAE,
AND GALAXIES

The universe contains many types of objects, and most types are represented on the Atlas Charts in chapter 7. In this chapter, I will give a brief description of the different types of objects whose positions in the sky are fixed to the celestial sphere, a fictitious bowl that appears to rotate around the celestial poles. (The sun, moon, and planets, on the other hand, change positions with respect to the stars.) At the end of the chapter are lists and graphic displays of a selection of the most interesting objects in the sky and when they are visible. For your convenience, I have chosen the objects so that some of them will be visible at each time of the year.

STARS

You have already seen a list of the brightest stars on the Monthly Sky Maps in chapter 3; they are also listed in appendix 2. Astronomers measure the brightness of stars and other celestial objects on the magnitude scale, which is described in chapter 3 (p. 48).

Both the Monthly Sky Maps and the detailed Atlas Charts in chapter 7 use circles of different sizes to represent the magnitudes of the stars, with smaller symbols for fainter stars. Remember that the higher the magnitude, the fainter the star. Stars as faint as 6th magnitude can be seen with the unaided ("naked") eye under perfect conditions, but depending on your eyesight and the brightness of the night sky from streetlights and other sources of light pollution, the visual limit is usually much lower in magnitude. It would not be uncommon if your sky were bright enough to prevent you from seeing stars any fainter than 3rd magnitude. Indeed, in cities, the sky itself may be so bright that you cannot see any stars at all!

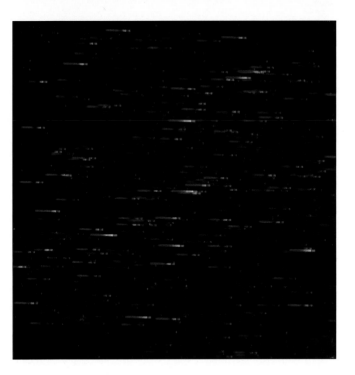

Fig. 5-1 A thin prism placed in front of a wide-field telescope shows all the stars as bands of color. These different spectra reveal the stars' temperatures. (Richard Hill)

ʀAR TEMPERATURES

All stars are balls of gas, so hot that they are glowing. Stars have different temperatures, ranging from about 2,100°C (about 3,800°F) up to about 50,000°C (about 90,000°F). Just as an iron poker glows more and more as it is heated, starting with red hot and then becoming white hot as it grows even hotter, a star's reddish color also indicates its temperature. The coolest stars are reddish balls of glowing gas; the hottest stars glow blue-white (fig. 5-1). These colors can even be seen with the naked eye. In the constellation Orion, for example, it is not hard to see that the star Betelgeuse, in Orion's shoulder, looks much redder and is therefore much cooler than the star Rigel, in Orion's heel.

Astronomers always give star temperatures using the kelvin scale, which uses Celsius degrees but begins at absolute zero, the coldest temperature possible. Absolute zero, 0 kelvin (0 K), is $-273\,°C = -459.4\,°F$.

SPECTRAL TYPES

Depending on its temperature, each star has its own *spectral type*, which is determined by analyzing its spectrum—the band of colors into which the light from the star can be broken down. The rainbow displays the visible part of the spectrum: red, orange, yellow, green, blue, indigo, and violet; a mnemonic for the order of colors is the name of an imaginary fellow, ROY G. BIV.

When a narrow band of light from the sun or another star is spread out into its spectrum, with blue, say, at the left and red at the right, the spectrum turns out to be crossed by dark lines, each indicating the absence of some particular color. These *spectral lines* are the signatures of different chemical elements at different temperatures. Each element has unique sets of spectral lines for various temperature ranges, so the spectrum of a star reveals the chemical composition of the star's atmosphere. These spectral lines are dark because they are absorbing some of the energy from deeper layers of the sun or the star; they are thus called *absorption lines*. All stars have spectra that are continuous bands of radiation crossed by absorption lines.

Astronomers determine a star's spectral type by seeing which spectral lines are darkest on its spectrum (fig. 5-1). Originally, when spectral types were assigned in the early twentieth century, the stars with the strongest (darkest) hydrogen lines (called Hα, Hβ, Hγ, and so on, using the Greek alphabet) were spectral type A, the next strongest spectral type B, and so on, alphabetically. But when theoretical interpretation allowed temperatures to be assigned, the spectral types in order of temperature, from hottest to coolest, turned out to be O B A F G K M (appendix 3). These types are then subdivided into tenths; for example, B2 is hotter than B5. You can see in the figure that the spectral lines that are darkest in the hottest stars, near the top, are mostly lines of hydrogen (H) and helium (He). The hydrogen line Hα represents a transition from energy level 2 to 3 of the hydrogen atom, Hβ is a transition from level 2 to level 4, and so on, until Hε is a transition from level 2 to 7, which explains why the following hydrogen lines are labeled H_8 and H_9 at upper left, given the apparent determination to stop using the less-known Greek letters. The Roman numeral I stands for the basic, un-ionized state of the complete atom, and the Roman numeral II stands for the first ion, the state in which one of the electrons has been stripped off because of the temperature. So HeI is "helium one," neutral helium, and HeII is "helium two," ionized helium, which occurs only in the hottest stellar atmospheres. The coolest stars, near the bottom of the figure, show not only some absorption lines, such as those labeled NaI for neutral sodium, but also some broader absorption bands of molecules like TiO, titanium oxide. Whenever you see those bands, you know you are observing a cool star.

When hot gas shines by itself, it may give off *emission lines,* specific colors that are brighter than adjacent colors. A few stars have some emission lines in addition to their absorption lines; the presence of emission lines is noteworthy because it usually shows that there is hot gas surrounding the star. Although emission lines are rare in the spectra of stars, the hot gas in most nebulae is so tenuous that these nebulae have emission-line spectra. They are, therefore, called *emission nebulae,* and appear reddish because the red emission line from heated hydrogen gas, Ha, is the strongest contributor to the visible part of their spectra.

In this book, we discuss mainly the visible part of the spectrum because our discussions are meant for optical observers. But professional astronomers are increasingly studying other parts of the spectrum. The whole spectrum is a set of waves of different wavelengths. Gamma rays, x-rays, and ultraviolet have wavelengths shorter than those of visible light. Infrared and radio waves have wavelengths longer than those of visible light. Of those other parts of the spectrum, only some radio studies can be carried out by amateur observers; along with a few parts of the infrared, radio waves are the only part of the spectrum besides the visible part that comes through the earth's atmosphere.

Over the past decades, a set of space missions has joined ground-based observing to cover the entire range of the spectrum. Three of NASA's Great Observatories—the Compton Gamma-Ray Observatory, the Hubble Space Telescope, and the Chandra X-ray Observatory—cover gamma rays, x-rays, the ultraviolet, and the visible, and the part of the infrared nearest to the visible. The Space InfraRed Telescope Facility is the remaining planned Great Observatory. New optical telescopes with mirrors up to 10 meters in diameter (the twin Keck Telescopes in Hawaii), and large new radio telescopes have also extended our ability to observe. Advances in electronics are enabling us to use even existing telescopes to get more sensitivity than was previously possible.

In the original determination of spectral types of stars, capital letters were assigned in alphabetical order to indicate the strength of hydrogen's spectral lines relative to the spectral lines from other elements (fig. 5-2), with A representing the greatest strength. But astronomers later realized that hydrogen radiation is strongest for stars of intermediate temperature rather than for the hottest or coolest stars. So the modern list in order of temperature is out of alphabetical order. From the hottest to the coolest, the order is: O B A F G K M, remembered by generations from the mnemonic "Oh, be a fine girl, kiss me" (now often also "Oh, be a fine guy, kiss me"). By comparing a star's spectrum with those in the figure, you can find the best match and therefore the star's temperature. The temperatures and other properties corresponding to the spectral types are given in appendix 3. Appendix 2 gives spectral types and other data for the 287 brightest stars in the sky.

Double and Variable Stars

Most stars in the sky that appear to be single are really made up of two or more objects revolving around each other. Sirius, for example, is really a *double* or *binary star*; the bright, bluish white star we see with the naked eye has a much fainter companion (known as Sirius B) revolving around it. Double stars (which are often called "double" even when three or more stars are present) can be interesting to observe, especially when the stars in a pair or multiple system present a beautiful contrast of colors in binoculars or a

Fig. 5-2 *A computer simulation of spectral types in order of temperature. The dark vertical lines are spectral absorption lines. The series of absorption lines from hydrogen are strongest in spectral type A, but the alphabetical system was rearranged to a system in order of descending temperature. (Roger A. Bell, University of Maryland; and Michael Briley, University of Wisconsin at Oshkosh)*

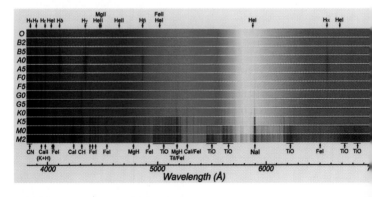

STARS, NEBULAE, AND GALAXIES

telescope (see table 7, p. 189). In some double-star systems, the stars periodically block each other as they orbit, making the total brightness we see vary. These particular binary star systems are examples of *eclipsing binaries.*

Some stars vary in brightness all by themselves. Sometimes a star actually periodically changes in size and therefore brightness. These stars are called *variable stars.*

Since double stars and variable stars are particularly interesting for casual star observers and amateur astronomers to observe, a whole chapter (chapter 6) is devoted to them. Two short tables of the double and variable stars that are easiest to observe appear at the end of this chapter; the times when they are visible above the horizon are shown in a Graphic Timetable (p. 191). You will find more charts that will help you observe interesting doubles and variables in chapter 6.

THE LIFE HISTORIES OF STARS

Much of the astronomical research over the last few decades has been devoted to studying the life cycles of stars. Stars begin as collapsing clouds of gas in interstellar space and become stars when nuclear fusion begins in their interiors, fusing hydrogen into helium. At that point, stars join the *main sequence.* During this period of their lives, energy from the fusion creates a pressure that pushes outward, balancing the inward force of gravity. Depending on how hot and bright a star is, it spends more or less time as a main-sequence star. The hottest stars, though more massive than the sun, use their fuel at such a furious pace that they have relatively short lifetimes—perhaps only 100,000 years. Stars like the sun live on the main sequence for about 10 billion years; the sun is now about halfway through that lifetime. Stars cooler and fainter than the sun can live much longer, 50 billion years or more.

In the type of nuclear fusion that goes on inside the sun and most other stars, four hydrogen nuclei—the central parts of hydrogen atoms—fuse to become one helium nucleus. The resulting helium has slightly less mass than the four hydrogens that went into forming it. The difference is converted into energy in the amount $E = mc^2$, a fact and formula discovered by Albert Einstein in 1905.

When a star begins to die, its fate depends on how much mass it has. Let us first consider stars with about as much mass as the sun, ranging from the least massive stars (with 0.07—7 percent —of the mass of the sun) to stars with about twice as much mass as the sun. As the stars use up all the hydrogen in their interiors, and so can no longer fuse hydrogen into helium, they can no longer fight off gravity, and their interiors begin to contract. During the contraction, energy is released and their outer layers are pushed out. These outer layers become large and cool, and the stars become *red giants*.

Eventually, the outer layers of some of the red giants become disconnected and float outward, carrying perhaps 20 percent of the star's mass out into space. The expanding shell of gas is known as a *planetary nebula*. Planetary nebulae are marked with a special symbol on the Atlas Charts in chapter 7. The escaped nebula bares the inner layers of the star, so the remaining "central stars" appear blue because they are so hot, hotter even than the hottest normal stars. The most commonly observed planetary nebula is the Ring Nebula in Lyra (fig. 5-3), which appears to the eye as a hazy white smoke ring through even a telescope but which reveals its colors to photography (fig. 5-4). Other planetary nebulae are shown in figures 5-5 to 5-10.

Throughout this chapter, we show almost entirely objects that can be photographed with small telescopes. They are often identified by their numbers in the Messier Catalogue (M) or the later and more extensive (and thus less exclusive) New General Cata-

150 Fig. 5-3 *The Ring Nebula in Lyra, seen through a small telescope. It is not an impressive object to the eye. Atlas Chart 18. (Richard Hill)*

Fig. 5-4 *The Hubble Space Telescope reveals more of the Ring Nebula's beauty. Atlas Chart 18. (Hubble Heritage Program)*

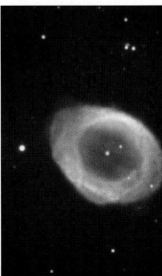

Fig. 5-5 M27, the Dumbbell Nebula in Vulpecula. Atlas Charts 18, 19. (Daniel Good)

Fig. 5-6 NGC 7009, the Saturn Nebula. Atlas Chart 32. See also fig. 7-33. (Tim Hunter and James McGaha) The Owl Nebula, M97, a faint planetary nebula in Ursa Major (Atlas Chart 5), appears on pp. 176 and 236.

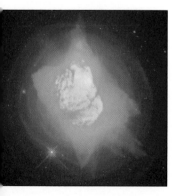

Fig. 5-7 NGC 7027, a planetary nebula, observed with the Hubble Space Telescope, showing several shells of ejected gas (blue), dust clouds (red), and a central white dwarf. Only the very high resolution of Hubble allows these shells to be seen. Atlas Chart 19. (H. Bond, STScI; and NASA/ESA)

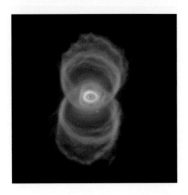

Fig. 5-8. The Hourglass Nebula, a planetary nebula, observed with the Hubble Space Telescope. We are seeing a cone, an hourglass-shape, of ejected material, with etched walls. Only the very high resolution of Hubble allows these shells to be seen. It is too faint to be marked on Atlas Chart 50 (13^h $39^m -67° 22^m$ in Musca). (Raghvendra Sahai and John Trauger, JPL; the WFPC2 science team; and NASA)

logue (NGC) and its Index Catalogue (IC). A few of the objects are from the new Caldwell Catalogue (C). These catalogs of non-stellar objects are described in chapter 7. The photographs come from a variety of telescopes, mostly by amateur astronomers with a few from professional observatories and a handful from the Hubble Space Telescope, mostly for comparison.

Fig. 5-9 The Helix Nebula, NGC 7293, in Aquarius. It is 700 light years away, making it the nearest planetary nebula to us. Its diameter in the sky is about half that of the moon's, but it is too faint to be seen with the unaided eye. Atlas Chart 44. (© Anglo-Australian Observatory)

Fig. 5-10 A small detail of the Helix Nebula, viewed with the Hubble Space Telescope. Hubble found thousands of the knots shown; each is twice the size of our solar system, with tails even longer. They result from hot gas from the star colliding with cooler gas ejected 10,000 years earlier. Atlas Chart 44. (Robert O'Dell and Kerry P. Handron, Rice University, Houston, Texas; and NASA)

After only 50,000 years or so—a twinkling of an eye in the lifetime of a star—the planetary nebula drifts away, and the central star cools. Eventually the central star contracts until it is about the size of the earth. If a mass less than 1.4 times the mass of the sun is left, the star remains at this stage indefinitely. It is then a *white dwarf.* White dwarfs are faint because they are so small. Thus we see that objects such as red giants, planetary nebulae, and white dwarfs, originally discovered as independent classes of objects, are really different stages in the lifetimes of ordinary stars.

Fig. 5-11 *A supernova shining brightly in the galaxy NGC 3877. Atlas Chart 14. You must get current briefings if you want to photograph a supernova—unless you find one for yourself. (© Tim Hunter and James McGaha)*

Fig. 5-12 *The Crab Nebula, M1, in Taurus, the remnant of the 1054 supernova. Atlas Chart 11. (From Palomar Observatory plates, California Institute of Technology, © David Malin/Jay M. Pasachoff/Caltech.)*

Sometimes a white dwarf is part of a double-star system, and its gravity attracts material from the other star in the system. When this material hits the surface of a white dwarf, it may begin nuclear fusion briefly, and the system brightens considerably. We see it as a *nova*. Novae were originally thought of as new stars, but are now known to be stars that suddenly become brighter or visible because of this effect. A few famous novae are marked on the charts.

A star that contains more than about twice as much mass as the sun comes to a more spectacular end. After it becomes a red giant, it grows bigger yet and becomes a supergiant. At that point it may explode completely, becoming a *supernova*. A supernova may rival in brightness the whole galaxy it is in (fig. 5-11). It was only in the 1920s that it was realized that supernovae and novae are completely different phenomena. A supernova completely devastates the star, forming heavy elements because of the high temperature and then spewing these elements out into space. The heavier elements in the human body (all those heavier than iron) were formed in supernovae that exploded before our solar system formed. When our sun and planets formed, they incorporated these heavy elements. Also, it seems increasingly likely that our solar system formed when a nearby supernova caused a cloud of gas and dust to collapse. So supernovae were important to the existence of humanity.

In a few cases, the remnants of a supernova can still be seen in space, as in the Crab Nebula (fig. 5-12), which exploded in the year 1054. The Veil Nebula is part of the Cygnus Loop (figs. 5-13 and 5-14), the visible remnant of a supernova that exploded even longer ago. Astronomers detect still other supernovae from their radio or x-ray emissions.

In 1987, the first supernova visible to the naked eye since 1604 erupted in the Large Magellanic Cloud (figs. 5-15 and 5-16). Unfortunately for northern-hemisphere denizens, it was visible only fairly far south on the earth; it was not visible in midnorthern latitudes. Astronomers have been able to make detailed observations as the supernova brightened and faded. They have verified that heavy elements such as iron, cobalt, and nickel were formed and are learning many details about stellar evolution. Their basic idea of how massive stars explode to become supernovae was verified, though there were many surprises. For example, the star that exploded was a blue supergiant, and it had been thought that only red supergiants would make supernovae.

During a supernova explosion, the center of the star can be extremely compressed and can collapse until it is quite small. Stars that have 1.4 to 4 times the mass of the sun collapse until they

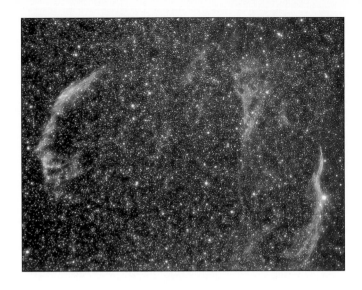

Fig. 5-13 *The Veil Nebula in Cygnus. Atlas Chart 19. (© Jerry Lodriguss)*

Fig. 5-14 *Part of the Veil Nebula in Cygnus. Atlas Chart 19. (Tim Puck-ett)*

Fig. 5-15 The Large Magellanic Cloud (Atlas Chart 47), with the reddish Tarantula Nebula and, near it in the sky, the supernova that exploded in 1987. It faded below naked-eye visibility but continues to be widely studied. It is predicted to brighten again, perhaps to naked-eye visibility, so if you are in the southern hemisphere, check its status. (Akira Fujii)

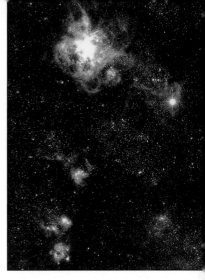

Fig. 5-16 Supernova 1987A in the Large Magellanic Cloud, in the upper right middle of the photo, to the lower right of the Tarantula Nebula. Before the supernova, only a faint star, like hundreds of others on the picture, was detectable. (© 1989 UK Astronomy Technology Center, Royal Observatory, Edinburgh)

are only about 12 miles (20 km) across. At that point, the neutrons in the star cannot be pushed closer together. Most of the star is then a gas containing neutrons only, or perhaps the still smaller particles called quarks that make up neutrons. The star is now a *neutron star*.

Neutron stars are too small and faint to be seen directly in visible light. But some neutron stars give off beams of radio radiation. As the star rotates very rapidly—once every second or so—in some cases this beam of radiation sweeps by the earth, the way a lighthouse beacon apparently gives pulses of light as it turns. We detect pulses of radio waves from the neutron star, and so call the object a *pulsar*. Even though we can't see pulsars, a few prominent pulsars are marked on the Atlas Charts in chapter 7 for inspirational purposes. More than 1,000 pulsars are known.

Other neutron stars exist in binary systems (double stars). The gravity of the neutron star sometimes pulls mass off the other star. This mass falls on the neutron star with such force that it heats up enough to give off x-rays. X-ray telescopes in orbit have detected signs of a number of these *binary x-ray sources*.

In a few cases, a mass greater than about four times that of the sun remains after a supernova explosion. Then nothing can stop the collapse of the star. Because of its gravity, it becomes more and more compressed. In cases where the gravity is so high, scientists must use the theory of gravity developed by Einstein in 1917 to explain the situation. Einstein's theory is called "the general theory of relativity." The theory predicts that even light will be bent so much by the strong gravity that the light will no longer be able to escape. Neither could any object escape. The star has become a *black hole*.

Black holes are even fainter than neutron stars, so we must look for signs of their presence other than their optical appearance. They are too small to hope to see them block out light from behind. Our best chance of detecting a black hole is if it is located in a binary system. Then the black hole should be attracting material from its companion star. The matter will swirl around the black hole before it enters and heat up until it gives off x-rays. So x-ray sources are candidates we must study in order to see if other evidence tells us that they are black holes. We conclude that they are black holes if we find that they have too much mass to be neutron stars. We measure the mass of the invisible star in a binary by studying the motions of its visible companion star. The effect is similar to deducing that an invisible dancer is present by watching the movements of the companion. One of the longest studied black-hole candidates in our galaxy, known as Cygnus X-1, is in the constellation Cygnus; its location is marked on Atlas Chart 19 in chapter 7.

STAR CLUSTERS

Sometimes stars form in groups. With binoculars or a small telescope, we might see a dozen or so stars in an irregular grouping in a certain area, though the cluster may actually contain hundreds of stars. The stars are contained within a region about 30 light-years across. Usually, using a larger telescope or making a longer photographic exposure will reveal more stars. The Pleiades (figs. 5-17 and 5-18) and the Hyades in Taurus are among the most famous examples of this type of cluster, which is often called an

open cluster. Such groupings are also called *galactic clusters,* because they lie in the plane of our galaxy. Open clusters are relatively young on a cosmic scale; the Pleiades formed only 100 million years ago. Long-exposure photos of this cluster show that
some of the gas and dust from which the stars formed is still present nearby. Only about 400 light-years away from us, the
Pleiades is one of the closest open clusters. Many other clusters
can be observed in the sky, some with the naked eye, some with
binoculars, and others with telescopes (figs. 5-17 to 5-23).

In other places in the sky, thousands or hundreds of thousands
of stars of a common origin may be located within 300 light-years
or so, forming a huge spherical ball. These groupings are called
globular clusters. In the northern sky, the globular clusters M3 in
Canes Venatici (fig. 5-24), M13 in Hercules, and M15 in Pegasus
are among the brightest and thus the easiest to see. A globular
cluster looks like a hazy mothball to the naked eye or when
viewed through a small telescope; larger telescopes are necessary
to see individual stars in this type of cluster. The nearest globular
cluster to us is M4 in Scorpius, near the bright star Antares in the
sky (fig. 7-41). Photographs of the region reveal gas and dust
around several of the stars in the region, including Antares and
(text continues on p. 164)

*Fig. 5-17 The Pleiades, showing many more stars (named on p. 246)
than the mythical seven sisters. Atlas Chart 10. (Ronald Royer)*

Fig. 5-18 *The Pleiades, in a longer exposure that reveals the dust surrounding the stars. Atlas Chart 10. (© Jerry Lodriguss)*

Fig. 5-19 *The double cluster in Perseus, h and χ (chi) Persei. Atlas Chart 2. The crossed spikes, as always in astronomical photographs, are an artifact resulting from the mounting of the telescope's secondary mirror. (© Jerry Lodriguss)*

Fig. 5-20 M25, an open cluster in Sagittarius. Atlas Chart 30. (Richard Hill)

Fig. 5-21 M34, an open cluster in Perseus. Atlas Chart 10. (Richard Hill)

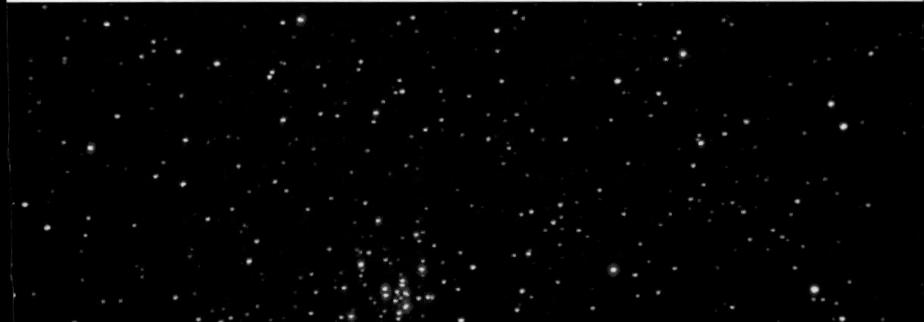

Fig. 5-22 The open cluster NGC 6520 in Sagittarius is near the dark dust cloud Barnard 86 (Atlas Chart 42). The dust shields molecules from ultraviolet radiation that breaks it apart. Radio telescopes and infrared telescopes allow us to penetrate the dust to detect stars forming inside clouds like this one. (© Anglo-Australian Observatory)

Fig. 5-23 The open cluster M11 in Scutum. Atlas Chart 30. (Daniel Good)

Fig. 5-24 M3, a globular cluster in Canes Venatici. Atlas Chart 15. (Akira Fujii)

Fig. 5-25 M15, a globular cluster in Pegasus. Atlas Chart 32. (Richard Hill)

Fig. 5-26 M22, a globular cluster in Sagittarius. Atlas Chart 42. (Tim Hunter and James McGaha)

Fig. 5-27 M71, a globular cluster in Sagitta. Atlas Chart 42. (Daniel Good)

rho Ophiuchi. The globular cluster is at a different distance, however, and is free of dust.

Scientists have found that all the globular clusters are very old —perhaps 12 billion years old. Thus they are free of dust. Open clusters, on the other hand, have a wide range of ages. Some may even be forming now. Often, we see around them the dust and gas from which they have formed. A number of interesting open and globular clusters that can be seen at various times of year are listed in table 7 at the end of this chapter; the times when the clusters are visible are plotted on a Graphic Timetable.

NEBULAE

Nebulae—clouds of gas and dust that appear hazy to our view—are some of the most beautiful objects to observe in space. The word nebulae (singular, nebula) comes from the Greek word for cloud. A few particularly interesting nebulae are listed in table 8 at the end of this chapter; the times when they can be observed are shown on a Graphic Timetable there.

Some nebulae represent shells of gas thrown off by dying stars. Planetary nebulae (pp. 151–153) and the remnants of supernovae (pp. 154–156) are examples. Planetary nebulae—often called planetaries, for short—got their name because they often appear in small telescopes as small greenish disks; their telescopic images resemble those of the planets Uranus and Neptune, but they have no other similarity to planets. Planetary nebulae often appear greenish because hot oxygen in a certain condition emits a lot of green radiation. Two planetary nebulae are included in table 10 (p. 190); others are listed in appendix 5.

Other nebulae represent gas and dust still surrounding young stars. The dust reflects the starlight to us, making a reflection nebula. The nebulae surrounding the stars in the Pleiades (fig. 5-18, p. 160) are reflection nebulae, as is dust around other stars (fig. 7-41, p. 320).

Some nebulae glow; we call them emission nebulae. Other nebulae are dark and absorb radiation from behind. We call them dark nebulae or absorption nebulae. The Lagoon Nebula, M8, glows with the red light of hydrogen and has a dark absorbing band across it (fig. 5-28).

Some nebulae contain small dark nebulae in which stars may be forming. The nebula M16 in Serpens (fig. 5-29) is a good example. Since radio waves and infrared rays penetrate the nebulae better than visible light does, scientists interested in the birth of stars are now studying nebulae not only in visible light (figs. 5-28 to 5-35) but also in other parts of the spectrum.

Fig. 5-28 M8, the Lagoon Nebula in Sagittarius. Atlas Chart 42. (Daniel Good)

Fig. 5-29 (left) The open cluster and nebula M16, often known as the Eagle Nebula, in Serpens. Atlas Chart 30. (Tim Hunter and James McGaha)

Fig. 5-30 (right) A Hubble Space Telescope false-color view of part of the Eagle Nebula, M16. This image has become iconographic for Hubble. The pillars are dense clouds of molecular hydrogen gas and dust that are being evaporated by ultraviolet light. They will leave newborn stars behind. (Jeff Hester and Paul Scowen, Arizona State University; and NASA)

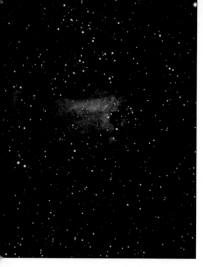

Fig. 5-31 M17, the Omega Nebula, in Sagittarius. Atlas Chart 30.
(Daniel Good)

Fig. 5-32 M20, the Trifid Nebula,
in Sagittarius. Atlas Charts 41,
42. (Tom Montemayor, McDonald
Observatory, University of Texas)

Fig. 5-33 The California Nebula,
NGC 1499), named because of its
shape. Atlas Chart 10.
(Ronald Royer)

Fig. 5-34 The Rosette Nebula,
NGC 2237-39, in Monoceros,
around the open cluster NGC
2244. Atlas Chart 24. (R. Royer)

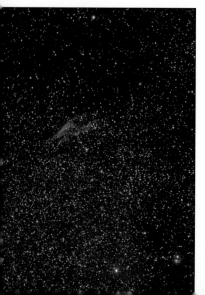

Fig. 5-35 The Cone Nebula, NGC 2264. Atlas Chart 24. (© Anglo-Australian Observatory)

The North America Nebula (fig. 7-20) in the constellation Cygnus is an example of an emission nebula with an absorption nebula that defines the boundaries we see. The part of the emission nebula we see is roughly the same shape as the continent of North America. Notice how few stars are visible in what would be the Gulf of Mexico because of the dark absorption nebula there. (Notice also that the name is "America," not "American.")

The Great Nebula in Orion (Figs. 2-8 to 2-10) turns out to be a particularly exciting place. The emission nebula we see is a blister sticking out toward us from a large cloud of dust and gas in which stars are actually forming now. Though we can't see through the emission nebula with visible light, infrared and radio waves penetrate the dark cloud behind it. The absorbing gas and dust in that cloud prevent radiation from the emission nebula from entering and breaking up the gas clouds from which stars will soon form.

A *galaxy* is an island of matter in space—a giant collection of gas, dust, and millions or billions of stars. The galaxy we live in, which includes about a trillion (a thousand billion) stars, is called the *Milky Way Galaxy*. Most of the Milky Way Galaxy is in the shape of a disk; the earth is approximately halfway out from the center. Since the earth is located inside the Milky Way Galaxy, various parts of the galaxy are always visible in our sky. The disk of the Milky Way Galaxy is thin. When we look toward or away from the center or in any other direction in the plane of that disk, we see many stars and much gas and dust; in those directions we see a

Fig. 5-36 *This drawing of the entire sky shows 7,000 stars plus the Milky Way. (Lund Observatory, Sweden)*

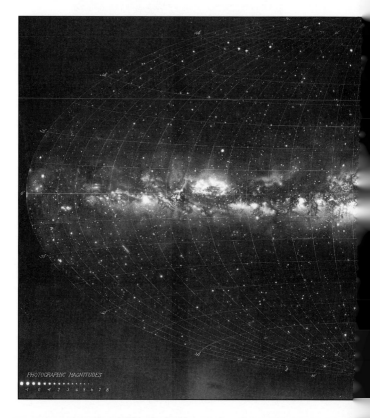

PHOTOGRAPHIC MAGNITUDES

band of white across the sky, mottled with some dark material. People have long called the band of light that appears to cross the sky by the name the *Milky Way* because of its appearance (fig. 5-36); it is the Milky Way from which our galaxy draws its name.

If we look in directions other than that of the plane of our galaxy, our line of sight quickly passes out of the galaxy, and we do not see many stars or much gas or dust; in those directions we see only some of the nearer stars against a dark background. Imagine living deep in a pit of a compact disk: when you look toward the center or toward the outside edge of the CD, you look through a lot of material; when you look upward or downward, though, you quickly see out of the CD.

Most of the open clusters lie in the disk of our galaxy, so we see them in or near the Milky Way. Most of the nebulae lie in the disk as well. The globular clusters, though, form a large spherical halo

surrounding the center of our galaxy. The halo extends far above and below the center of our galaxy; thus these globular clusters do not, for the most part, lie in or near the Milky Way. Indeed, it was the realization around 1920 that the globular clusters seemed to be forming a spherical halo around something that led Harlow Shapley to deduce that whatever was at the center of that halo was the center of our galaxy. That discovery was the proof that the sun does not lie in the center of the galaxy or of the universe.

When we look away from the Milky Way, we are able to look out of our galaxy and see more distant objects, such as other galaxies. (Galaxies and nebulae are known as *deep-sky objects*.) Still, when we look at photographs of these distant objects, the individual points of light we see are usually stars in the foreground, stars that are relatively close to us—in our galaxy, rather than in distant galaxies.

From 1989 to 1993 NASA's Cosmic Background Explorer spacecraft studied the universe in the infrared and on the border of the radio spectrum. The view it sent back in the infrared of our galaxy (fig. 5-37) gave us a remarkable view of the plane of the Milky Way and the bulge at our galaxy's center.

Fig. 5-37 Our Milky Way Galaxy, as observed in three infrared wavelengths by the Cosmic Background Explorer and reproduced here in three visible colors. The cooler dust, shown as yellow, extends through the plane of our galaxy. (COBE Science Team, NASA)

GALAXIES

The Milky Way is only one of millions of galaxies in the universe. Most galaxies outside our own are too faint and distant to be seen with the naked eye or binoculars, but it is fascinating to study their shapes in a telescope. A few interesting examples of galaxies are listed in table 10 at the end of this chapter; the times of the year when they are visible are shown on a Graphic Timetable (p. 193).

CLASSES OF GALAXIES

A basic scheme for classifying galaxies was worked out in the 1920s by Edwin Hubble. More recent and elaborate methods, such as those of Gerard de Vaucouleurs, have also been developed. The Milky Way Galaxy is an example of a *spiral galaxy*—a galaxy in which several "arms" seem to unwind from a central region. Hubble classified the spiral galaxies with an S (for *spiral*) followed by an *a*, *b*, or *c*, depending on the increasing looseness of the arms of the galaxy. Thus a galaxy with tightly wound arms is

Fig. 5-38 *The Andromeda Galaxy, M31 (Sb), with the elliptical galaxies NGC 205 (next to it) and M32. Fig. 7-9 and Atlas Chart 9. (Richard Hill)*

Fig. 5-39 *The inner part of the spiral galaxy M33 (Sc) in Triangulum, seen face-on. Atlas Chart 9. (Tom Montemayor, McDonald Observatory, University of Texas)*

Fig. 5-40 M51, the Whirlpool
Galaxy (Sc) in Canes Venatici. Atlas Chart 14. (Daniel Good)

Fig. 5-41 The spiral galaxy M63
(Sb) in Canes Venatici. Atlas
Chart 15. (Tim Hunter and James
McGaha)

classified as an Sa; one with less tightly wound arms, like our galaxy or the Andromeda Galaxy (fig. 5-38), is an Sb; and one with loosely wound arms is an Sc (figs. 5-39 and 5-40).

We cannot move around to different sides of a galaxy to see it from different perspectives. A telescope reveals so many galaxies, though, that we can see examples of each class from different angles—broadside (face-on), somewhat tilted, or edge-on. From studying all these galaxies we know, for example, that an Sa galaxy has a smaller bulge at its center than an Sc galaxy does.

The youngest stars and most of the gas and dust are located in the arms of a spiral galaxy. Since the hottest stars burn out quickly, whenever we see a hot star (which we can recognize by its blue color), it must be relatively young. Color photographs of galaxies show that the central regions are relatively yellow, indicating that older stars are dominant there, while the arms are relatively blue and therefore contain relatively young stars.

Some spiral galaxies have a central bar from which the arms unwind. They are classified with a *B* (for *bar*): SBa, SBb, and SBc (fig. 5-47, for example).

Most galaxies do not have spiral arms, but have an elliptical shape instead. They are classified with an *E* (for *elliptical*), with a number from 0–7, showing how far out of round they appear. E0

Fig. 5-42 *M65 (Sa, right of center), M66 (Sb, lower left of center), and NGC 3628 (Sb, upper left), three galaxies in Leo. Atlas Chart 27. (Daniel Good)*

Fig. 5-43 *M74, a spiral galaxy (Sc) in Pisces. Atlas Charts 9, 22. (Tim Hunter and James McGaha)*

galaxies appear spherical (fig. 5-46), while E7 galaxies appear quite oblong. Elliptical galaxies are quite old, contain only old stars, and have no gas or dust. They come in a wide range of sizes, from huge to relatively small on a galactic scale.

In between the most elliptical of the elliptical galaxies and the spiral galaxies with the most tightly wound arms is a transitional class of galaxy: S0. An S0 galaxy has a disk but no arms. It is currently thought that the shapes of galaxies result from the different conditions under which they formed, and that galaxies do not evolve from one type to another.

Some galaxies of each type have a noticeable peculiarity in addition to one of the basic shapes described above. A galaxy might have a jet of gas sticking out, for example, or be wrapped with extra dust (fig. 5-60). These galaxies are labeled (pec), for *peculiar,* as in "E7(pec)."

Other galaxies have no special shape and are known as Irregular (Irr). Some irregular galaxies seem to have a trace of spiral structure and might really be the equivalent of Sd spirals (that is, spirals with very loose arms). Others are truly irregular. The nearest type-Irr galaxies to us are the two small satellite galaxies of the Milky Way Galaxy. Since they were first seen by Magellan's crew as they sailed far enough south, these satellite galaxies are known as the Magellanic Clouds. They are best visible from southern latitudes and cannot be seen from the continental United States. In a telescope, the Large Magellanic Cloud (LMC) seems to show traces of spiral structure (fig. 5-58), while the Small Magellanic Cloud (SMC) seems completely irregular (fig. 5-59).

All galaxies except the very closest ones are moving away from us. We can tell this by studying their spectra, which are shifted toward the red end of the spectrum. A *redshift*—a lengthening of the wavelength of any wave—results from motion away from you, just as the length of a sound wave increases as a motorcycle passes you and starts to recede from you. Edwin Hubble discovered that the farther away a galaxy is, the faster it is receding. We measure how fast galaxies are receding by studying the redshifts in their spectra. This method works no matter how far away the galaxy is, as long as we can take a long enough exposure to measure a redshift. Once we measure the redshift, we use *Hubble's Law,* which links redshift with distance, to find out how far away galaxies are. This is the only method we have for finding the distances to the farthest objects in the universe.

The enormous distances to galaxies give us an indication of the immense scale of the universe. The nearest galaxy that northern-hemisphere observers can see is the Great Galaxy in Andromeda, sometimes simply called the Andromeda Galaxy. The center of

(text continues on p. 182)

Fig. 5-44 M81 (Sb) and M82 (Ir-regular), galaxies in Ursa Major. Atlas Charts 4, 5. (© Jerry Lodriguss)

Fig. 5-45 The spiral galaxy M83 (Sc) in Hydra. Atlas Chart 39. (Tim Hunter and James McGaha)

Fig. 5-46 The elliptical galaxy M87 in Virgo. Atlas Chart 27A. (© Jerry Lodriguss)

Fig. 5-47 M95, a barred spiral galaxy (SBb) in Leo. Atlas Chart 26. (Tim Hunter and James McGaha)

Fig. 5-48 M96, a spiral galaxy (Sa) in Leo, with a supernova on its visible edge. Atlas Chart 26. (Tim Hunter and James McGaha)

Fig. 5-49 M97, the Owl Nebula, a planetary nebula, with M108, a spiral galaxy (Sb), both in Ursa Major. Atlas Chart 5. (Daniel Good)

Fig. 5-50 A high-resolution Hubble Space Telescope view of the inner part of M100, an Sc spiral galaxy in Coma.

Fig. 5-51 M104, the Sombrero Galaxy, an edge-on spiral galaxy (Sa) in Virgo. Atlas Chart 27. (European Southern Observatory)

Fig. 5-52 NGC 6946, Caldwell 12, an Sc spiral galaxy in Cepheus. Atlas Chart 8. (Tim Hunter and James McGaha)

Fig. 5-53 NGC 891, Caldwell 23, an Sb spiral galaxy in Andromeda. Atlas Chart 10. (Tim Puckett)

Fig. 5-54 NGC 4244, Caldwell 26, an Sc spiral galaxy in Canes Venatici. Atlas Chart 15. (Tim Hunter and James McGaha)

Fig. 5-55 NGC 4631, Caldwell 32, an Sc spiral galaxy in Canes Venatici. Atlas Chart 15. (Tim Puckett)

Fig. 5-56 NGC 1300, an Sb spiral galaxy. Atlas Chart 23. (Tim Hunter and James McGaha)

Fig. 5-57 NGC 1365, an SBb barred spiral galaxy in Fornax. Atlas Chart 34. (Tim Hunter and James McGaha)

Fig. 5-58 The Large Magellanic
Cloud, an irregular satellite galaxy
to our own. Atlas Chart 47. (Alan
Dyer)

Fig. 5-59 The Small Magellanic
Cloud, and the globular cluster 47
Tucanae. Atlas Chart 45.
(Ronald Royer)

Fig. 5-60 NGC 5128, the optical galaxy that corresponds to the radio
source Centaurus A. Atlas Chart 39. (© Jerry Lodriguss)

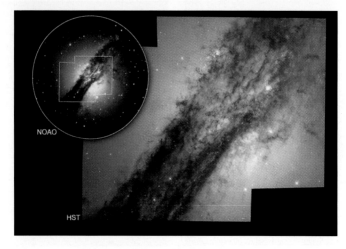

Fig. 5-61 A Hubble Space Telescope high-resolution view the part of NGC 5128 outlined in the ground-based image from the National Optical Astronomy Observatory. The giant elliptical galaxy, with a giant black hole at its center, is cannibalizing a smaller galaxy. (E. J. Schrier, A. Marconi, D. Axon, N. Caon, and D. Macchetto, STScI; and NASA)

Fig. 5-62 The Virgo Cluster of galaxies, about 60 million light-years away, including the elliptical galaxies M86 (upper center) and M84 (right). NGC 4438 is the distorted galaxy at upper left. The positions of these galaxies are shown on Chart 27A in chapter 7. (© Anglo-Australian Observatory/Royal Observatory, Edinburgh, photograph from UK Schmidt plates by David Malin)

Fig. 5-63 The Coma Cluster of galaxies, about 300 million light years away, from the Hubble Space Telescope. The giant elliptical M87 is at the zigzag. Atlas Chart 27A. (STScI, NASA)

Fig. 5-64 View of part of the Coma Cluster of galaxies, from a 24-inch telescope with CCD. NGC 4889 is at upper left and NGC 4874 at upper right. Atlas Chart 27A. (Tim Hunter and James McGaha)

Fig. 5-65 The Hercules Cluster of galaxies, about 500 million light years away, from a 24-inch telescope with CCD. The galaxies are too faint to be on Atlas Charts 17 and 29; they are located near κ Herculis. (Tim Hunter and James McGaha)

Fig. 5-66 *The Hubble Deep Field. Numbers show the redshifts as a fraction of the wavelength. A redshift of 5, fantastically far, means that the wavelengths are shifted by five times their original wavelength. Such galaxies are over 10 billion years old. (Robert Williams and STScI; and NASA; redshifts from M. Dickinson and Z. Levay)*

this galaxy is dimly visible to the naked eye, but it is so far away that this light has been traveling for 2.2 million years—at a speed of 300,000 kilometers per second (186,000 miles per second)—to reach us. When we observe the Andromeda Galaxy, we are looking 2.2 million years back in time.

All galaxies apparently come in groups. Our Milky Way Galaxy, the Andromeda Galaxy, and M33 are the three spirals in our Local Group of some two dozen galaxies, which also includes a couple of giant elliptical galaxies and many dwarf elliptical and irregular galaxies.

The Local Group is an outlying member of a large *cluster* of *galaxies* that covers much of the constellation Virgo and is the

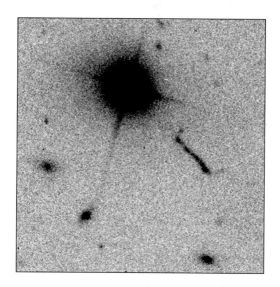

Fig. 5-67 The quasar 3C 273, in Virgo, is magnitude 13.6. The long jet of gas coming from it is a sign of violent activity there. The four spikes are telescope artifacts. Atlas Chart 27A. (H.-J. Röser et al., Max-Planck-Institut für Astronomie, Heidelberg/ ESO/NASA/ESA, courtesy of William Keel and the Astronomical Society of the Pacific)

known as the Virgo Cluster (fig. 5-62). Most of the galaxies in the Virgo Cluster are about 60 million light-years away. Other prominent clusters are in Coma Berenices, about 300 million light-years away, and in Hercules, about 500 million light-years away. (These values are computed using the measurement that Hubble's constant—the rate at which velocity increases with distance—is 65 km/sec/megaparsec, which is 13 km/sec per million light-years.) Clusters of galaxies have recently been studied particularly well from x-ray telescope observations, which have revealed the presence of gas in between the galaxies, and from optical mapping.

Many clusters of galaxies are linked together in *superclusters*. In between the superclusters are giant voids where no galaxies are found. 3D mapping has found chains of superclusters.

UASARS

It came as a surprise in 1960 when some of the sources of celestial radio waves appeared to be pointlike—"quasi-stellar"—instead of looking like galaxies. These "quasi-stellar radio sources" —*quasars,* for short—were discovered three years later to be extremely far away, when Maarten Schmidt discovered that the spectrum of one of them showed an extreme redshift. A class of radio-quiet quasars—quasi-stellar objects with huge redshifts—has since been discovered. We now know of thousands of quasars.

The redshift can be expressed as the fraction or percentage by which the wavelengths of light are shifted. For speeds much slower than the speed of light, this fraction is the same as the fraction that the recession speed of the quasar is of the speed of light. For example, a redshift (called z) of 0.2 means that the wavelengths are shifted toward the red by 0.2 times (20 percent of) the original wavelength. It is also true that the galaxy or quasar emitting such light is receding from us at 0.2 times the speed of light, or 60,000 km/sec. A quasar with a redshift of 1.0 has its wavelengths shifted by 1.0 times (100 percent of) the original wavelength; each wavelength is doubled. But Einstein's formulas from the special theory of relativity must be used to calculate how fast the quasars are receding; their velocities are always less than the speed of light.

The most distant known quasar is redshifted by about $z = 5$. Relatively few quasars with redshifts greater than 3 (300 percent) have been found, even though instruments exist that would be sensitive enough to do so. The farther out the quasar is, the farther back in time the light reaching us was emitted. That few quasars of the largest redshifts are found indicates that we are seeing back to the instant at which the first quasars were formed —somewhat over 10 billion years ago. Though for a long time, quasars all seemed farther than the farthest galaxies, new methods of detecting galaxies have found them still farther out.

Since quasars are so far away yet still send us some visible light and relatively strong radio radiation, the quasars must be astonishingly bright. Astronomers now agree that giant black holes are probably present in the centers of quasars. These black holes would contain millions of times the mass of the sun. As gas is sucked into the black holes, it heats up and gives off energy. Quasars may be a stage in the evolution of galaxies, for there is evidence that some quasars have structure around them not unlike that of galaxies. Since we see most quasars quite far away, we are seeing them far back in time, and we can conclude that the epoch at which quasars were brightest took place long ago.

Only one quasar—3C 273, in the constellation Virgo (fig. 5-67)—is bright enough to be accessible for visual observing with amateur telescopes, but we have plotted a handful of quasars on the Atlas Charts in chapter 7 since they are such astonishing objects that it is exciting even to look in their directions and since electronic detectors allow even amateur astronomers to observe them.

OSMOLOGY

Hubble's observations that the most distant galaxies are receding from us faster than closer galaxies can be explained if the universe is expanding in a way similar to the way a giant raisin cake rises (fig. 5-68). If you picture yourself as sitting on a raisin, all the other raisins will recede from you as the cake rises. Since there is more dough between you and the more distant raisins, the dough will expand more and the distant raisins will recede more rapidly than closer ones. Similarly, the universe is expanding. We have the same view no matter which raisin or galaxy we are on, so the fact that all raisins and galaxies seem to be receding doesn't say that we are at the center of the universe. Indeed, the universe has no center. (We would have to picture a raisin cake extending infinitely in all directions to get a more accurate analogy.)

Since the universe is uniformly expanding now, we can ask what happened in the past. As we go back in time, the universe must have been more compressed, until it was at quite a high density 15 billion or so years ago. Most models of the origin of the universe say that there then was a *big bang* that started the expansion. A widely accepted model—the *inflationary universe*—holds that the early universe grew larger rapidly for a short time before it settled down to its current rate of expansion. The big bang itself may not have occurred; the first matter could have formed as a chance fluctuation in the nothingness of space.

Fig. 5-68 A raisin cake rising is analogous to the expansion of the universe, since from any raisin, farther raisins appear to be separating at greater speeds than nearer ones, just as galaxies do not change in size but appear to be separating at greater speeds from farther galaxies than from nearer ones. (Jay M. Pasachoff)

The temperatures were so high in the first microsecond of the universe that not even the chemical elements had formed. But as the universe cooled, first basic atomic particles, such as protons (hydrogen nuclei), and then heavier atoms formed. Still, the universe was opaque until it cooled down to 3,000°C. When it reached that temperature, protons and electrons combined to make hydrogen atoms, and the universe became transparent. The hot radiation that permeated the universe at that time has cooled as the universe has continued to expand, and it is now detectable as a faint glow in sensitive radio telescopes. The glow has a temperature of only 3 kelvins (3°C above absolute zero) and is our best evidence that the early universe was hot and dense.

Astronomers ask what will happen to the universe in the future. The evidence is not all in. One possibility is that the universe is *open*—it will continue to expand forever. Another possibility is that the universe will eventually stop its expansion and begin to contract. We know that this cannot happen for at least 50 billion years more—at least three times longer than the current age of the universe—because we can observe the rate at which the universe is now expanding. Still, if the universe does contract in the long run, then we will have a *closed universe* that will wind up in a *big crunch*. The inflationary model indicates that the universe will expand forever. Evidence from studies of distant supernovae released in 1998, seem to indicate that the universe is accelerating its expansion, rather than slowing up. The interpretation as of this writing is that the universe contains energy in a form characterized by a cosmological constant, a term in the equations of Albert Einstein used to describe the universe.

Observations from new telescopes on the ground, from the Hubble Space Telescope and the Chandra X-ray Observatory in orbit, from the Very Long Baseline Array of radio telescopes stretching over the whole United States, and from other instruments are expected to help us improve the accuracy of our forecast of the future of the universe.

LIGHT POLLUTION

Hundreds or thousands of years ago, people were more familiar with the sky. Today's outdoor lighting brings many benefits, but it does tend to brighten the night sky, preventing most people from seeing it at its best. A view from space demonstrates how many sources of lighting there are (fig. 5-69). Note that we are seeing the light that escapes upward, yet most of that light is intended to light the ground. Light that goes in unwanted directions is known as *light pollution*.

Fig. 5-69 *A satellite view of the world, showing the light that escapes upward, plus an aurora borealis over Greenland. Most of the light is from urban sources. Some roads are visible, such as Interstate 5 in California. Natural gas burning in oil fields in northwestern South America, the Persian Gulf, and elsewhere also shows, as do fires from slash-and-burn agriculture in the tropics. (Defense Mapping Satellite Program, © Dr. Y. Nakayama, Remote Sensing Technology Center of Tokyo, courtesy of Syuzo Isobe)*

Sometimes limiting light pollution is as easy as shutting off outside lights when you do not expect nighttime guests or anyone else coming home. Sometimes it merely involves shutting off parking-lot lighting when stores close. And sometimes it merely means choosing lampposts with shields on their tops, so that the light they give off hits only the ground, where it is needed, and doesn't escape outward to illuminate satellites and the moon!

Other forms of light-pollution control are more subtle. Some forms of streetlights use sodium gas under low pressure. Most of their emission occurs in a couple of yellow wavelengths, which can be filtered out by professional or amateur astronomers taking pictures. Some light-pollution filters even exist for visual observing when such lighting is used. Convincing your local government to control lighting, with regulations against lights escaping upward and with the best kinds of streetlights, can help make the sky available for view.

Astronomers have long been plagued by airplanes and satellites passing through their fields of view. Between 1998 and 2000, a set of 66 working satellites plus 8 spares were aloft in the Iridium system to provide worldwide hand-held telephones. (Iridium is element number 77, the original number of satellites planned for the system; dysprosium, element 66, is a less pleasant-sounding name.) The system went bankrupt in 2000, and the satellites are being brought down, burning up in the atmosphere. The solar panels on the ones that are still aloft catch the sun from time to time. When they do, they gleam as brightly as −7th magnitude, 100 times brighter than the brightest star and brighter than any planet. The times of these Iridium flares are very predictable and can be found on Web sites, including some available through this book's Web site. It is fun to look in the sky at the appointed time and see something brighten for about 20 seconds.

Fig. 5-70 An Iridium flare. We see the spacecraft apparently brighten as its solar panels reflect the sun especially well during its passage across the sky. This image shows the spacecraft's trail, brightest in the center of its 20 seconds of visibility. (Pekka Parviainen)

BLE 7. SELECTED DOUBLE STARS

NAME	ATLAS CHART	MAGNITUDES	COLOR	SEPARATION OF COMPONENTS (IN ARC SEC)
gamma Andromedae	9	2.3 & 5.5	orange & blue	10"
beta Monocerotis	24	5.2, 6.1, 4.7	blue	2.8" & 7.3"
iota Cancri	13	4.2 & 6.6	yellow & blue	31"
gamma Virginis	27	3.5 & 3.5	yellow	4"
beta Scorpii	29	2.6 & 4.9	blue & blue	14"
epsilon Lyrae	18	5.0, 6.0, 5.2, 5.4	uniform	2.5", 180",2.4" ("double double")
beta Cygni (Albireo)	18	3.1 & 5.1	yellow & green	34"

NOTE: The times when these doubles are visible are shown in the Graphic Timetable of Double and Variable Stars (p. 191). The positions of these and other double stars are given in appendix 6 (pp. 530–531) and are plotted on the Atlas Charts in chapter 7.

BLE 8. SELECTED VARIABLE STARS

NAME	ATLAS CHART	TYPE (SEE P. 198)	MAGNITUDE (RANGE)	PERIOD (DAYS)
Mira (omicron Ceti)	22	long-period	2.6–10.1	332
Algol (beta Persei)	10	eclipsing binary	2.1–3.4	2.9
delta Cephei	8	Cepheid variable	3.5–4.4	5.4
zeta Geminorum	12	Cepheid variable	3.7–4.2	10.2
beta Lyrae	18	eclipsing binary	3.3–4.3	12.9
RR Lyrae	18	RR Lyrae variable	7.0–8.1	0.6

NOTE: The times when these variables are visible are shown in the Graphic Timetable of Double and Variable Stars (p. 191). The positions of these and other variable stars are given in appendix 7 (p. 532) and appendix 8 (p. 533); they are also plotted on the Atlas Charts in chapter 7. Special finding charts for Mira, Algol, and beta Lyrae appear at the end of chapter 6 (pp. 204–207).

TABLE 9. SELECTED OPEN AND GLOBULAR CLUSTERS

NAME	CONSTELLATION	TYPE	MAG.	DIAM.	R.A. (2000.0)	DEC.
M103	Cassiopeia	open	7	6'	01^h33^m	+60°42'
h and χ Persei	Perseus	open*	4	30', 30'	02^h22^m	+57°07'
M45 (Pleiades)	Taurus	open	1	110'	03^h47^m	+24°07'
M79	Lepus	globular	8	9'	05^h25^m	-24°33'
M35	Gemini	open	5	28'	06^h09^m	+24°20'
M44 (Praesepe)	Cancer	open	3	95'	08^h40^m	+19°59'
M3	Canes Venatici	globular	6	16'	13^h42^m	+28°23'
M5	Serpens	globular	6	17'	15^h19^m	+02°05'
M13	Hercules	globular	6	17'	16^h42^m	+36°28'
M92	Hercules	globular	7	11'	17^h17^m	+43°08'
M23	Sagittarius	open	6	27'	17^h57^m	−19°01'
M24	Sagittarius†	open	6	5'	18^h17^m	−18°29'
M11	Scutum	open	6	14'	18^h51^m	−06°16'
M15	Pegasus	globular	6	12'	21^h30^m	+12°10'
M39	Cygnus	open	5	32'	21^h32^m	+48°26'
M22	Sagittarius	globular	5	24'	18^h36^m	−23°54'

The following objects are visible only from southern latitudes:

NAME	CONSTELLATION	TYPE	MAG.	DIAM.	R.A. (2000.0)	DEC.
ω Cen	Centaurus	globular	4	36'	13^h27^m	−47°29'
47 Tuc	Tucana	globular	4	31'	0^h24^m	−72°05'

*The double cluster; a 1° field is necessary to include both.
†M24 is the star cloud in the direction of the galactic center.

TABLE 10. SELECTED NEBULAE AND GALAXIES

NAME	CONSTELLATION	TYPE	DIAM.	R.A. (2000.0)	DEC.
M31	Andromeda	spiral	2°×40'	00^h43^m	+41°16'
M1 (Crab)	Taurus	supernova	6'×4'	05^h35^m	+22°01'
M42 (Orion)	Orion	emission	60'	05^h35^m	−05°27'
M81	Ursa Major	spiral	16'×10'	09^h56^m	+69°04'
M49	Virgo	elliptical	4'	12^h30^m	+08°00'
M51 (Whirlpool)	Canes Venatici	spiral	12'×6'	13^h30^m	+47°12'
M20 (Trifid)	Sagittarius	emission	29'	18^h03^m	−23°02'
M57 (Ring)	Lyra	planetary	1'	18^h54^m	+33°02'
M27 (Dumbbell)	Vulpecula	planetary	8'×4'	20^h00^m	+22°43'

NOTES: Diameters are expressed in degrees (°) and arc minutes ('); the dimensions are guide to the size of the object when seen through a 6–8" reflector in average conditions Abbreviations: spiral = spiral galaxy; supernova = supernova remnant; emission = emission nebula; elliptical = elliptical galaxy; planetary = planetary nebula.

GRAPHIC TIMETABLE OF THE HEAVENS
DOUBLE AND VARIABLE STARS

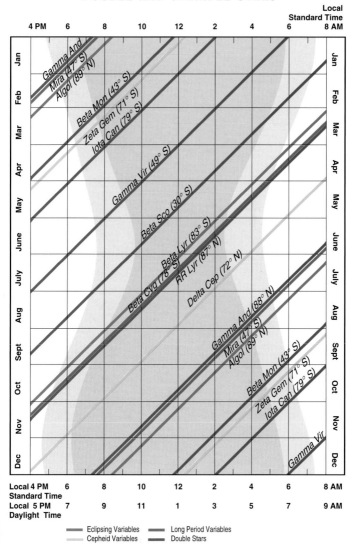

The times when selected double and variable stars transit.

(© 2000 Scientia, Inc.)

GRAPHIC TIMETABLE OF THE HEAVENS
OPEN AND GLOBULAR CLUSTERS

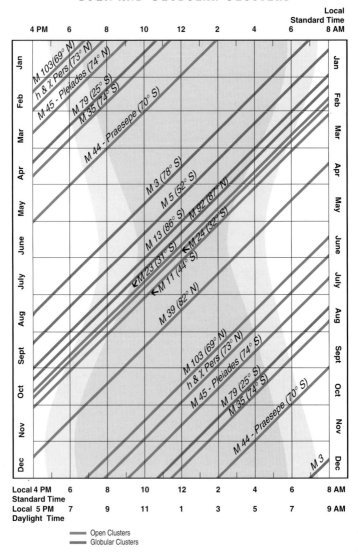

The times when selected star clusters transit. (© 2000 Scientia, Inc.)

GRAPHIC TIMETABLE OF THE HEAVENS
GALAXIES AND NEBULAE

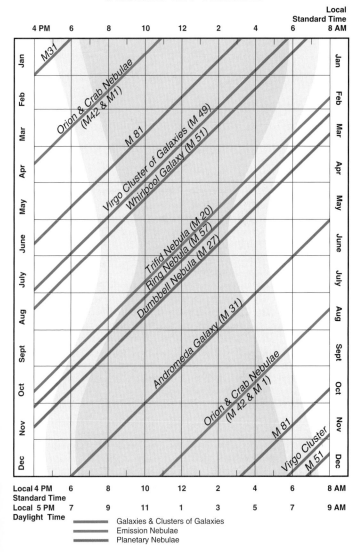

The times when selected galaxies and nebulae transit. (© 2000 Scientia, Inc.)

6

DOUBLE AND VARIABLE STARS

DOUBLE STARS

When we look at the sky, most of the stars we see seem to be single points of light. But more detailed investigation has revealed that most of the stars in our galaxy are actually in *double star* systems—systems containing two or more stars. The periods with which the components of double star systems orbit each other range from hours up through centuries.

In some cases, the stars are distant enough from each other and the star system is close enough to us that we can actually see that the star system contains more than one object. A system of this type is called a *visual binary*. In other cases, even though we may not be able to actually see more than one star, we can tell that the system is a double star because the component stars sometimes block each other as they orbit, making the total brightness of the system vary. This type of variable is an *eclipsing binary star* (fig. 6-1). Sometimes the fact that a star is double can be told only by examination of its spectrum, the breakup of its light into its component colors. The spectrum might show distinct contributions from each of the member stars or might show signs of the stars' motions. A double star that has been identified as such through variations of its spectrum is called a *spectroscopic binary*. Still another possibility is that we can see one star appear to wobble in the sky slightly from side to side over the years, as a result of the gravity of an otherwise invisible companion. This type of double star is called an *astrometric binary*, since it is the delicate measurements of the science of astrometry ("star measuring") that reveals it to us as a double.

Sometimes we see something that looks like a double star, but that is really two stars that happen to be almost in the same line of sight, even though they are at quite different distances from us

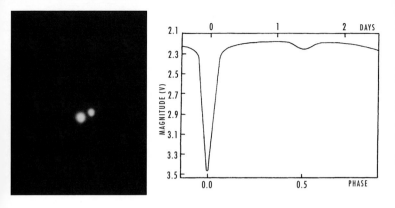

Fig. 6-1 (left) *The double star Albireo (β Cygni), notable because of the contrasting colors of the components. With magnitudes 3.2 and 5.4, it is easily visible in binoculars or a small telescope. See table 7. (Photo by S. Martin, Hopkins Observatory, Williams College)*

Fig. 6-2 (right) *The light curve of Algol (β Persei), an eclipsing binary. Algol is popular with variable-star observers because of its wide range in brightness over a conveniently short period. (Anthony D. Mallama)*

and therefore are not close to each other. These stars, which are not physically linked, are called *optical doubles,* but are not considered true double stars.

Double stars can be recognized on Atlas Charts in the next chapter by the horizontal lines drawn through the dots that indicate their brightnesses. Many interesting doubles are listed in appendix 6. A few double stars, including a selection of doubles that are visible at different times around the year, are listed in table 7, chapter 5 (p. 189); the times when they are highest in the sky are shown in the Graphic Timetable on p. 191. Appendix 6 includes information about the *position angle*—the angle measured eastward (counterclockwise), with the brighter star at the center of the "clock," for convenience, from the direction north around to the fainter star. The position angle and the apparent separation (in angular units—minutes or seconds of arc) between components of double systems change over time.

Some double star systems are particularly beautiful to observe with a telescope, even a small one, because the component stars are of different colors. Whenever a contrast of colors is seen in the members of a double, it results from the different temperatures of the individual stars.

One eclipsing binary system, Algol, is particularly easy to observe. It is also known as beta Persei because it is the second brightest star in Perseus. The name Algol, meaning "the demon star," comes from the Arabic words *ras-al-ghul,* meaning "the head of the demon." Figure 6-2 shows Algol's variations in brightness over a 2.9-day period. Algol is such a popular object to observe that we have listed some of the times of its minima in table 11 (p. 197). A finding chart for Algol (fig. 6-12) appears at the end of this chapter. On this and other finding charts there, decimal points are omitted from the magnitudes listed, since the decimal points might be confused with stars.

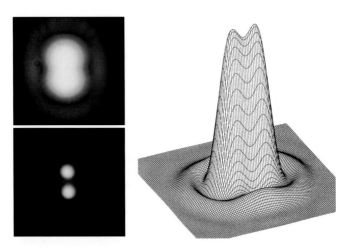

Fig. 6-3 (left) Dawes's limit, the point at which we begin to be able to distinguish that two objects are present, corresponds approximately to the bottom photo at left. The stars are even better distinguished in the top photo. (Chris Jones, Union College)

Fig. 6-4 (right) Dawes's limit corresponds to the brightness distribution at right. (Raymond G. Wilson and Diane Bootz, Illinois Wesleyan University)

TABLE 11. MINIMA OF ALGOL

2000 January 1, 19:41; February 2, 8:43; March 2, 0:56; April 2, 13:58; May 1, 6:09; June 1, 19:07; July 3, 8:03; August 1, 0:10; September 1, 13:04; October 3, 1:58; November 3, 14:55; December 2, 7:04.

2001 January 2, 20:04; February 3, 9:06; March 1, 4:30; April 1, 17:32; May 3, 6:32; June 3, 19:31; July 2, 11:38; August 3, 0:33; September 3, 13:27; October 2, 5:33; November 2, 18:29; December 1, 10:39.

2002 January 1, 23:39; February 2, 12:41; March 3, 4:54; April 3, 17:56; May 2, 10:07; June 2, 23:05; July 1, 15:13; August 2, 4:08; September 2, 17:02; October 1, 9:08; November 1, 22:04; December 3, 11:02

2003 January 1, 3:13; February 1, 16:15; March 2, 8:28; April 2, 21:30; May 1, 13:41; June 2, 2:40; July 3, 15:36; August 1, 7:42; September 1, 20:36; October 3, 9:31; November 1, 1:39; December 2, 14:37

2004 January 3, 3:37; February 3, 16:39; March 3, 8:52; April 1, 1:05; May 2, 14:05; June 3, 3:03; July 1, 19:11; August 2, 8:06; September 2, 21:00; October 1, 13:06; November 2, 2:02; December 3, 15:00.

2005 January 1, 7:12; February 1, 20:13; March 2, 12:27; April 3, 1:28; May 1, 17:40; June 2, 6:38; July 3, 19:34; August 1, 11:40; September 2, 0:34; October 3, 13:29; November 1, 5:37; December 2, 18:35.

2006 January 3, 7:35; February 3, 20:37; March 1, 16:01; April 2, 5:03; May 3, 18:03; June 1, 10:12; July 2, 23:09; August 3, 12:04; September 1, 4:09; October 2, 17:04; November 3, 6:00; December 1, 22:09.

2007 January 2, 11:10; February 3, 0:12; March 3, 16:25; April 1, 8:37; May 2, 21:38; June 3, 10:36; July 2, 2:43; August 2, 15:38; September 3, 4:32; October 1, 20:38; November 2, 9:35; December 1, 1:44.

2008 January 1, 14:44; February 2, 3:46; March 1, 19:59; April 2, 9:01; May 1, 1:12; June 1, 14:11; July 3, 3:07; August 3, 16:02; September 1, 8:07; October 2, 21:02; November 3, 9:58; December 2, 2:07.

2009 January 2, 15:08; February 3, 4:10; March 3, 20:23; April 1, 12:35; May 3, 1:36; June 3, 14:34; July 2, 6:41; August 2, 19:36; September 3, 8:30; October 2, 0:36; November 2, 13:33; December 1, 5:42.

2010 January 1, 18:42; February 2, 7:44; March 2, 23:57; April 3, 12:59; May 2, 5:10; June 2, 18:09; July 1, 10:16; August 1, 23:11; September 2, 12:05; October 1, 4:11; November 1, 17:07; December 3, 6:05.

NOTE: The first minimum of each month is given; add 2 days 20 hours 49 minutes successively to find other minima in a given month. (Calculations by Roger W. Sinnott, *Sky & Telescope*)

A nineteenth-century English astronomer, William Dawes, worked out a rule of thumb to give the telescope aperture (the diameter of the lens or mirror) necessary to separate (resolve) double stars into their components, that is, to tell that two stars are present in a system (figs. 6-3 and 6-4). His rule is that an aperture of I inches should make barely detectable the presence of a pair of 6th-magnitude stars separated by $4.6/I$ seconds of arc; in the metric system, an aperture of C centimeters should resolve the pair of stars separated by $11.7/C$ seconds of arc. Dawes's rule assumes good sky conditions for viewing and that the stars are not too different in magnitude.

VARIABLE STARS

In addition to eclipsing binaries—stars that vary because they eclipse each other periodically—there are many stars that vary in brightness all by themselves. All such stars are known as *variable stars*.

The first variable star to be discovered in each constellation has usually been given the letter R, followed by the genitive (possessive) form in Latin of the constellation's name (see appendix 1). For example, R Andromedae, the first variable star discovered in Andromeda, is a long-period variable. The second variable star in each constellation was given the name S; S Andromedae was apparently a bright star that newly appeared in Andromeda in 1885, though we now know that it was a supernova. The next variable was T, and so on up to Z. Then the lettering scheme started over again with RR and continued to RZ, then SS (not SR) up to SZ, TT up to TZ, and so on up to YY, YZ, and ZZ. Following are AA to AZ, BB to BZ, and so on up to QZ, omitting the letter J (which might be confused with I). This lettering scheme provides names for up to 334 variable stars in each constellation. Additional variables in each constellation are numbered following a V, for variable: V1500 Cygni. Stars that have Bayer designations have retained them, even when they are variables.

Plotting a star's brightness over a period of time is called "finding its light curve." A star's light curve is a graph of its brightness as it varies over time. The light curve of SS Cygni (fig. 6-5), for example, shows that this variable star changes irregularly in brightness, increasing every 50 days or so from its normal level of 12th magnitude up to 8th magnitude. A finding chart for SS Cygni (fig. 6-9) appears at the end of this chapter. SS Cygni is an example of a *dwarf nova*. U Geminorum and AY Lyrae are other prominent members of the class; in fact, dwarf novae are often called *U Gem stars*.

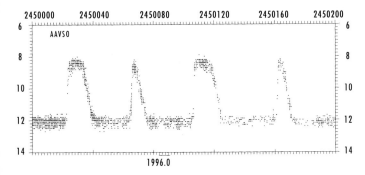

Fig. 6-5 The light curve of SS Cygni, a dwarf nova whose outbursts occur about every 50 days. The numbers at the top are Julian Days (p. 203). (AAVSO)

Sometimes stars are variable as a result of an actual change in a star's size and therefore its brightness. One type of variable, known as a *long-period variable,* takes months to complete a cycle of variation. This type of variable is also called a *Mira variable,* after the prototype star Mira (omicron Ceti) in the constellation Cetus, the Whale. Six cycles of variation for Mira are shown at the top of figure 6-6; a finding chart for this star (fig. 6-14) also appears at the end of this chapter.

Mira was the very first variable star (except for novae and supernovae) to be discovered. It was noticed as a variable with the naked eye in 1596, even before the invention of the telescope. Since it changed from 9th magnitude, which is invisible to the naked eye, to 3rd magnitude, this star seemed to appear anew in the sky from time to time. It was named "Mira," the Latin word for "wonderful," because of its wonderful changes in brightness. R Boötis, which ranges from 7th to 13th magnitude (fig. 6-6), is another example of this class.

RR Lyrae variables, named after the prototype star RR Lyrae in the constellation Lyra (Atlas Chart 18), on the other hand, have very short, regular periods of less than one day (fig. 6-7). Since RR Lyrae stars are found mostly in globular clusters, they are also known as *cluster variables.* The fact that all variables of this type have essentially the same absolute magnitude—about 0.5—was important for measuring the distance to the globular clusters in which RR Lyrae stars are located. This, in turn, led to Harlow Shapley's discovery in the 1920s that the sun is not at the center of our galaxy, since distances to these RR Lyraes showed that the clusters were centered at a point far off to the side of the sun.

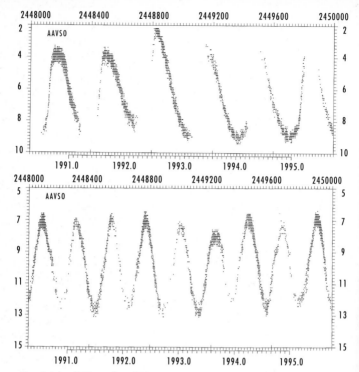

Fig. 6-6 (top) *Six cycles of Mira, omicron Ceti (o Ceti), a long-period variable. (AAVSO)* (bottom) *The light curve of R Boötis, a Mira variable with a period of 223.4 days. (AAVSO)*

Cepheid variables, named after the star δ (delta) Cephei, have a distinct pattern to their variations (fig. 6-8), with periods (cycles) that can range from about one day up to a few weeks. RR Lyrae stars and Cepheids are two types of variables that have been very important in helping astronomers determine the scale of distances in the universe. All RR Lyrae stars have about the same intrinsic brightness, so any differences in apparent brightness from one RR Lyrae star to the next result only from differences in their distances from us. Each Cepheid variable, on the other hand, can have a different intrinsic brightness, but it is fairly easy for astronomers to calculate the distance to a Cepheid once they know its period of variation. (The period of variation can be measured easily at a telescope; then the intrinsic brightness can be calculated from the star's period. From a comparison of this intrinsic

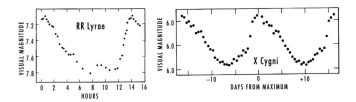

Fig. 6-7 (left) *RR Lyrae, the prototype of the cluster variables, has a period of 13.6 hours. It rises rapidly to maximum brightness and declines more slowly. This light curve is based on 330 visual observations by Erich Leiner, a German amateur astronomer.* (Sky & Telescope)

Fig. 6-8. (right) *The light curve of X Cygni, a Cepheid variable with a 17-day period.* (AAVSO)

brightness with the star's apparent brightness, astronomers can calculate the distance to the star.) Cepheid variables are giant stars, so large that we can detect them in some of the nearby galaxies; they provide the major method we have of measuring distances to these galaxies. All measurements to more distant galaxies depend, in turn, on the measurements of distances to nearby galaxies.

A star like R Coronae Borealis (abbreviated R CrB) varies in brightness because it occasionally gives off a cloud of opaque material—essentially soot—that masks the surface that we ordinarily see. The namesake for R CrB-class stars thus fades from 6th magnitude to 11th magnitude or even fainter (see the light curve facing Atlas Chart 16 in chapter 7, and the finding chart for this star at the end of this chapter). RY Sagittarii, which covers about the same range in magnitude, is another example of the class.

Flare stars, such as UV Ceti or YZ Canis Minoris, occasionally flare up within minutes by one to six magnitudes.

RV Tauri stars, such as R Scuti and V Vulpeculae, vary with a period of one to five months. During each period these stars have one deep minimum and one shallow minimum; at a deep minimum, the brightness of these yellow supergiants may drop by as much as three magnitudes.

For several periods, *semiregular variables* have periods lasting months or years; then they are irregular in their variations for a while.

Variable double stars such as RS Canum Venaticorum (abbreviated RS CVn and informally pronounced "RS Can Ven") are unusual eclipsing binaries that are especially interesting to study for those who have photoelectric equipment available. (Photoelectric equipment that can measure magnitudes accurately to 0.01 magnitude is needed.) RS CVn is a binary in which both stars are somewhat more massive than our sun, but generally one member is smaller and hotter and the other is a larger and cooler subgiant. The stars orbit each other very closely every 4.8 days. Every time the hotter star is eclipsed by the subgiant, the light drops by about 1 magnitude. One of the brightest RS CVn–type binaries is λ (lambda) Andromedae, which varies between visual magnitudes 3.70 and 4.05 every 54 days. Other RS CVn stars have periods between 0.6 and 80 days. The light variations arise as one star, with large starspots (comparable to the sunspots on our sun) darkening up to 40 percent of one hemisphere, rotates on its axis. Enormous bursts of radio waves coincide with the violent flare-ups of the star's extremely active outer atmosphere.

Variable stars are marked on the Atlas Charts in chapter 7 with both inner and outer circles whenever possible. The inner circles give their faintest magnitude and the outer circles give their brightest magnitude. A number of variables are listed in appendices 7 and 8; several also appear in table 8 and the Graphic Timetable for Double and Variable Stars in chapter 5. The stars listed in table 8 and appendix 7 have shorter periods—and are thus easier to observe as variables—than the other stars listed in appendix 8.

Novae—newly visible stars—can be considered as variables in one sense; they are marked with special symbols on the Atlas Charts, however. A nova brightens when gas from a large companion star falls on the white-dwarf member of a binary system, triggering nuclear fusion on the white dwarf's surface. Most of the 30 novae discovered in the 1990s were found by amateur astronomers. As professional automated search programs continue to expand, it is not clear whether amateur discoveries will continue. Amateur astronomers can, on the other hand, become still more useful in such areas as monitoring the brightnesses of faint variable stars, measuring accurate positions, and so on.

Supernovae—stars that explode—are also a sort of variable star. The supernovae that went off in the years 1572 (Atlas Chart 1), 1604 (Atlas Chart 41), and 1987 (Atlas Chart 47) are marked; the supernova of 1054 is M1, the Crab Nebula. More than 15 or so of the more than 100 supernovae discovered each year have been found by amateur astronomers in recent years. This number is a recent increase, attributable to the widespread availability and use of CCD (charge-coupled device) cameras by amateur astronomers. Since the late 1990s, professional astronomers using automated searches have greatly increased the rate of supernova discovery. The farthest of these supernovae have been used to

Fig. 6-9 The constellation Cygnus with Nova Cygni 1975 (N1975 on Chart 19 near 21ʰ10', +48°). It became about as bright as Deneb. The North America Nebula glows red. (George East)

probe the farthest scales in a study of the expansion of the universe.

Do not confuse *pulsars*—neutron stars that give off regular pulses of radio waves and which are not ordinarily detectable optically—with variable stars, whose visible brightness varies.

Variable-star observers often give dates in terms of *Julian days*, a system of calendar-keeping in which the days are numbered consecutively from January 1, 4713 B.C.E. The Julian-day system eliminates the need to calculate the number of days in a month or the effects of leap years. It also eliminates the need to worry about changes from the Julian to the Gregorian or other calendars. Also, Julian days begin at noon, so the Julian date does not change in the course of a night's observing. For example, noon on January 1, 2000, was the beginning of Julian Day 2,451,525. Every tenth Julian Day is listed in appendix 11. Portions of a day are customarily indicated by decimals in this system, instead of by hours, minutes, and seconds.

The American Association of Variable Star Observers (25 Birch St., Cambridge, Mass. 02138, aavso@aavso.org; www.aavso.org) is an international organization of amateur astronomers devoted to the study and cataloging of variable stars. They have a central registry of all variable star observations, and they calculate many light curves. They are always glad to have new members and to receive observations of variable stars from amateur astronomers. The AAVSO can provide finding charts for hundreds of variable stars. They provided suggestions for the variables included in table 8 as a set of first objects for amateurs who want to start making variable-star observations and reports.

The AAVSO is a major link between professional astronomers and amateurs who make visual observations of variable stars. The International Amateur-Professional Photoelectric Photometry association (IAPPP) also often provides important scientific data and is another amateur-professional link. Light curves and special observing projects are sometimes set up in collaboration with the AAVSO and IAPPP. For example, objects observed by professional astronomers in the x-ray region of the spectrum from satellites aloft often correspond to optical objects whose variations in brightness can be followed by amateurs on the ground. The address for the IAPPP is c/o Prof. Douglas Hall, Dyer Observatory, Vanderbilt University, Nashville, TN 37325. See the Web site at www.iappp.vanderbilt.edu/. (E-mail: Prof. Arnold Heiser, heiser@astro.dyer.vanderbilt.edu or to hall@astro.dyer.vanderbilt.edu.) The *IAPPP Communications* journal is edited by Prof. Terry D. Oswalt, oswalt@tycho.pss.fit.edu.

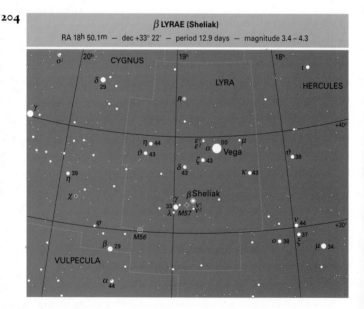

Fig. 6-1 o Beta Lyrae. (Wil Tirion) Note: Decimal points are omitted from magnitudes to avoid confusion with points representing stars.

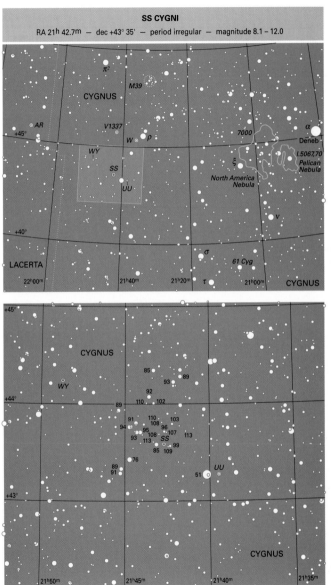

Fig. 6-11 SS Cygni. (Wil Tirion) Note: Decimal points are omitted from magnitudes to avoid confusion with points representing stars.

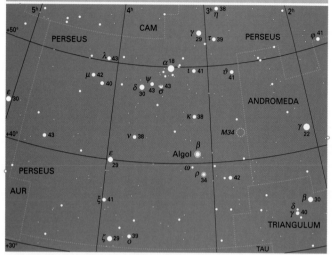

206

Fig. 6-12 (above) Algol (beta Persei). (Wil Tirion) Note: Decimal points are omitted from magnitudes to avoid confusion with points representing stars. Fig. 6-13 (below) Algol, seen at maximum through the blue ion tail of Comet C/1995 O1 (Hale-Bopp). (© Samuel E. Rhodes)

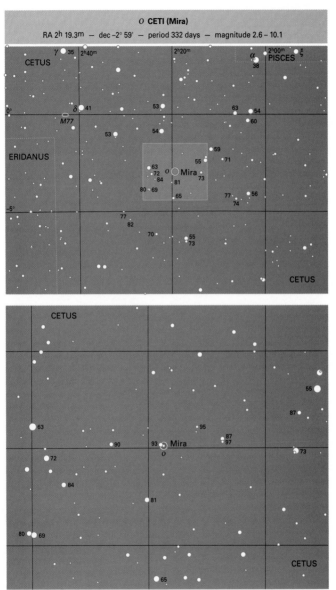

Fig. 6-14 Mira (omicron Ceti). (Wil Tirion) Note: Decimal points are omitted from magnitudes to avoid confusion with points representing stars.

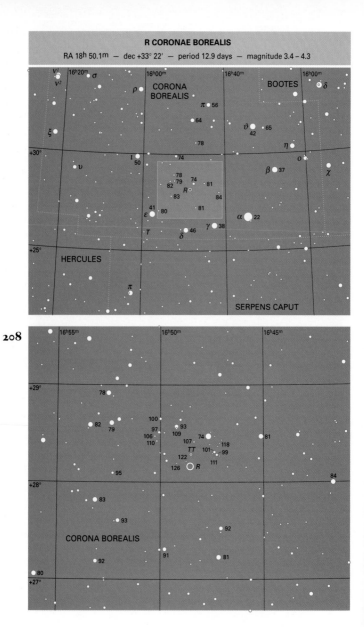

Fig. 6-1 5 R Coronae Borealis. (Wil Tirion) Note: Decimal points are omitted from magnitudes to avoid confusion with points representing stars.

7

ATLAS OF THE SKY

On the following pages are 52 Atlas Charts that together cover the entire sky, accompanied by 5 charts showing close-ups of areas of special interest. A visual key to the Atlas Charts appears on pp. 224–227; appendix 13 outlines the regions of sky each covers.

The charts have been drawn with high precision by Wil Tirion to show positions of celestial objects as they appear in epoch 2000.0, that is, at the beginning of the year 2000; objects deviate only very slowly from these positions. The charts contain 25,000 stars down to visual magnitude 7.5 and about 2,500 deep-sky objects (star clusters, nebulae, galaxies, and other objects among and beyond the stars). A key to the symbols for these objects and a list of lower-case Greek letters (which are used to label the brightest stars) appear below every chart. (Note that θ and ϑ are alternative italic forms of lowercase theta.)

The Greek letters were used to name the brightest stars in constellations by Johann Bayer in his sky atlas of 1603. On the whole, the stars were lettered in order of brightness, with α (alpha) usually being the brightest star in a constellation. This rule was not always followed; for example, the stars in the Big Dipper (which is part of the constellation Ursa Major) are lettered in order around the bowl and handle. Some of the fainter stars are labeled with lowercase Roman (regular) or italic letters or with numbers from the star catalog John Flamsteed compiled in 1725.

Astronomers mark positions in the sky using coordinates comparable to those we use to plot positions on earth. In astronomy, *right ascension* is celestial longitude, analogous to terrestrial longitude. *Declination* is celestial latitude, analogous to terrestrial latitude. The Atlas Charts in this chapter are marked in right ascension and declination. Right ascension is marked in hours, minutes, and seconds (abbreviated ^{h}, ^{m}, and ^{sec}) of time, with each 24 hours representing a full rotation of 360°. Declination is marked in degrees, minutes, and seconds (°, ', and ") north (+) or south (−) of the celestial equator.

The positions of celestial objects take account of *precession*— the drifting of the direction of the earth's axis of rotation and thus of the celestial coordinate system. Chapter 15 gives further details on right ascension, declination, precession, and matters of time and calendars.

CONSTELLATION figures are outlined, and the official constellation boundaries are marked with dotted lines. The line connecting stars stops short when a star that has traditionally been part of an asterism or constellation is no longer considered an official part of it; for example, one of the four stars marking the Great Square of Pegasus is really in the constellation Andromeda (see Atlas Charts 9 and 20).

STELLAR MAGNITUDES (see table 3) are represented by dots of different sizes. Each size represents one whole magnitude, including a range of ½ magnitude on each side of the integer (whole number) given, as shown in the key beneath each chart. The relative brightness of stars is shown as they appear to the eye rather than as they appear on a photographic plate, which is usually sensitive to a different range of colors (wavelengths).

DOUBLE STARS (chapter 6) are marked with a horizontal line through the dot (filled circle) that represents the total magnitude of the components. Only visual binaries—those systems in which two or more stars can be seen through a telescope (as opposed to stars that have been identified as binaries by studying their spectrum or in some other way)—are marked as doubles. The members of double-star systems are plotted as separate stars, though, if they are separated by more than one minute of arc. A few double stars are listed on the charts using the letter A (for Aitken) followed by their numbers in the Aitken Double Star (ADS) catalog, which is decreasingly used. Sometimes several members of a binary system share the same letter but each component is marked with a different superscript, as in ι^1 (iota-1) and ι^2 (iota-2).

VARIABLE STARS (chapter 6) whose maximum brightness is within the range of magnitude covered on these charts are indicated by a pair of concentric circles, with the outer circle showing the maximum brightness. Only variables whose brightness varies by more than 0.1 magnitude are marked.

DEEP-SKY OBJECTS. In 1784, Charles Messier compiled a list (table 12) of fuzzy objects in the sky, so that he would not be confused by them when searching for comets. A few additional objects were added later, to make the current total of 109. The list turns out to contain many of the most interesting objects in the sky that are accessible to amateur observers. Most deep-sky objects are marked with their numbers in the Messier list, in J. L. E. Dreyer's *New General Catalogue* (NGC) of 1888, or in the supplemental *Index Catalogues* (ICs) of 1895 and 1908. Messier numbers are preceded by an M, NGC numbers are written (on the charts) without prefixes, and IC numbers are preceded by the letter *I* and a period. A few objects are also labeled with their popular names.

In the 1990s, the noted British astronomy author and amateur astronomer Patrick Moore compiled a list of the 109 deep-sky objects that are best to observe but that are not on Messier's list of the same number of objects. His Caldwell Catalogue, from his rarely used official name, Patrick Caldwell-Moore, appears after the Messier Catalogue in this chapter. (He didn't want to use Moore for the catalog because it has the same initial M as Messier, which would be more confusing than having his list go from C1 to C109.)

THE MILKY WAY is shown in light blue on the charts.

OPEN STAR CLUSTERS (also known as *galactic clusters*) are marked with dotted circles, whose size signifies the size of each cluster. A special chart, following Atlas Chart 10, shows the stars of the Pleiades at a larger scale. Some star clusters are often referred to by their numbers in special catalogs, such as the Trumpler Catalogue (e.g., Tr 1) or the Melotte Catalogue (e.g., Mel 71).

GLOBULAR STAR CLUSTERS are marked with symbols showing three size ranges.

PLANETARY NEBULAE, shells of gas given off when stars such as the sun end their lives, are marked with circles bearing four external spikes.

PULSARS, tiny neutron stars that give off regular pulses of radio waves, representing the death of stars somewhat more massive than the sun, and **BLACK HOLES,** representing the death of extremely massive stars, also have their own symbols. One of the first likely black-hole candidates was Cygnus X-1 (Atlas Chart 19). The name Cygnus X-1 means that it was the first x-ray source to be discovered in the constellation Cygnus, a discovery that was not possible until observations could be made from x-ray telescopes orbiting above the earth's atmosphere. Even though pulsars and black holes cannot usually be seen in visible light (only two pulsars have been detected optically, even with large telescopes), these objects are so fascinating that it is interesting to know that one is nearby when you are observing the sky.

NOVAE —sudden brightenings of stars—are listed with the year in which they went off, such as Nova 1934 (for the nova visible in that location in the year 1934). We now know that novae are events that take place in binary star systems containing a white dwarf. Three **SUPERNOVAE** are also given the same symbol (a circle with a dot inside) to show that they have varied in brightness.

These supernovae—the one Tycho saw in 1572 (Atlas Chart 1, marked SN 1572), the one Kepler saw in 1604 (Atlas Chart 41, marked SN 1604), and the one in the Large Magellanic Cloud in 1987 (Atlas Chart 47, marked SN 1987A)—represented the explosive deaths of massive stars. These supernova remnants are not easy to see; the Crab Nebula and S1 47 in Taurus (Atlas Chart 11) and the Veil Nebula in Cygnus (Atlas Chart 19) are much more interesting supernova remnants to see with small telescopes.

NEBULAE, regions of gas and dust in space, are shown as shapes on the charts when they are more than 10 minutes of arc across; a square signifies a nebula smaller than 10 minutes of arc across. A special chart following Atlas Chart 24 shows at an enlarged scale the complex of stars and nebulae in Orion's sword (including the Orion Nebula) that are such beautiful objects for observers to see or photograph.

GALAXIES are drawn as ovals, in four size ranges. They are labeled down to magnitude 13.0 and fainter. The central part of the nearest cluster of galaxies, the Virgo Cluster, is featured on a larger-scale chart, 27A (p. 291). The Hubble Deep Field, the area of Ursa Major observed so carefully by the Hubble Space Telescope to reveal galaxies exceedingly faint and far back in time, is marked on its Chart (Atlas Chart 5), as is the Southern Hubble Deep Field (Atlas Charts 45 and 52).

A few selected **QUASARS** are marked with triangles. Quasars are farther from us than most galaxies marked on the charts, and all are optically fainter than the normal limit of our charts. In parts of the spectrum other than the visible, though, some of the quasars can be relatively bright. Some of the quasars are listed by their commonly used 3C *numbers,* that is, their numbers in the Third Cambridge Catalogue of radio sources. Others are named for their position in other catalogs.

Radio astronomers—astronomers using radio telescopes—originally named objects emitting radio waves with a capital letter and the name of the constellation, such as Taurus A for the first radio source discovered in Taurus. We now know that Taurus A is the Crab Nebula, named for its shape. It is also known as M1, because it was the first object in Messier's list (see table 12).

The Atlas Charts are in the following order: the north polar charts come first, then the midnorthern charts, equatorial charts, midsouthern charts, and south polar charts. In order to minimize distortion on these charts, Tirion has used a conic projection for the polar regions and intermediate declinations, and a cylindrical projection for the equatorial region.

Numbers set in triangles at the borders of each map direct you to adjacent charts. (The adjacent charts in each group are in order of right ascension.) North is up, west is to the right, south is down, and east is to the left on the charts.

Keep a life list of the Messier objects you have seen, recording the date on which you observed each one and sketching what it looked like. In March each year, it is possible (though difficult) to see all the Messier objects in a single night. The table below and the Atlas Charts will help you find these objects, whether you want to track them all down in a "Messier marathon" or pursue them at a more leisurely pace.

\.E 12. MESSIER CATALOGUE/LIFE LIST

NGC/IC	R.A. (2000.0) H M	DEC. ° '	CONST.	VISUAL MAG.	DESCRIPTION	ATLAS CHART
1952 ≥ seen_____	05 34.5	+22 01 initials____	Tau description_____	8.4	Crab Nebula	11
7089 ≥ seen_____	21 33.5	−00 49 initials____	Aqr description_____	6.5	Globular cluster	32
5272 ≥ seen_____	13 42.2	+28 23 initials____	CVn description_____	6.4	Globular cluster	15
6121 ≥ seen_____	16 23.6	−26 32 initials____	Sco description_____	5.9	Globular cluster	41
5904 ≥ seen_____	15 18.6	+02 05 initials____	Ser description_____	5.8	Globular cluster	29
6405 ≥ seen_____	17 40.1	−32 13 initials____	Sco description_____	4.2	Open cluster	41
6475 ≥ seen_____	17 53.9	−34 49 initials____	Sco description_____	3.3	Open cluster	41
6523 ≥ seen_____	18 03.8	−24 23 initials____	Sgr description_____	5.8	Lagoon Nebula	42
6333 ≥ seen_____	17 19.2	−18 31 initials____	Oph description_____	7.9	Globular cluster	30
6254 ≥ seen_____	16 57.1	−04 06 initials____	Oph description_____	6.6	Globular cluster	29, 30
6705 ≥ seen_____	18 51.1	−06 16 initials____	Sct description_____	5.8	Open cluster	30
6218 ≥ seen_____	16 47.2	−01 57 initials____	Oph description_____	6.6	Globular cluster	29
6205 ≥ seen_____	16 41.7	+36 28 initials____	Her description_____	5.9	Globular cluster	17
6402 ≥ seen_____	17 37.6	−03 15 initials____	Oph description_____	7.6	Globular cluster	30
7078 ≥ seen_____	21 30.0	+12 10 initials____	Peg description_____	6.4	Globular cluster	32
6611 ≥ seen_____	18 18.8	−13 47 initials____	Ser description_____	6.0	Open cl.; Eagle Nebula	30
6618 ≥ seen_____	18 20.8	−16 11 initials____	Sgr description_____	7	Omega Nebula	30

TABLE 12. MESSIER CATALOGUE/LIFE LIST (CONTINUED)

M	NGC/IC	R.A. (2000.0) H M	DEC. ° '	CONST.	VISUAL MAG.	DESCRIPTION	ATLAS CHAR
18	6613	18 199	−17 08	Sgr	6.9	Open cluster	30
date seen_____			initials____	description_____			
19	6273	17 02.6	−26 16	Oph	7.2	Globular cluster	41
date seen_____			initials____	description_____			
20	6514	18 02.6	−23 02	Sgr	8.5	Trifid Nebula	41,
date seen_____			initials____	description_____			
21	6531	18 04.6	−22 30	Sgr	5.9	Open cluster	41,
date seen_____			initials____	description_____			
22	6656	18 36.4	−23 54	Sgr	5.1	Globular cluster	42
date seen_____			initials____	description_____			
23	6494	17 56.8	−19 01	Sgr	5.5	Open cluster	30,
date seen_____			initials____	description_____			
24	6603	18 16.9	−18 29	Sgr	4.5	Open cluster	30
date seen_____			initials____	description_____			
25	IC4725	18 31.6	−19 15	Sgr	4.6	Open cluster	30
date seen_____			initials____	description_____			
26	6694	18 45.2	−09 24	Sct	8	Open cluster	30
date seen_____			initials____	description_____			
27	6853	19 59.6	+22 43	Vul	8.1	Dumbbell Nebula	18,
date seen_____			initials____	description_____			
28	6626	18 24.5	−24 52	Sgr	6.9	Globular cluster	42
date seen_____			initials____	description_____			
29	6913	20 23.9	+38 32	Cyg	6.6	Open cluster	18,
date seen_____			initials____	description_____			
30	7099	21 40.4	−23 11	Cap	7.5	Globular cluster	43
date seen_____			initials____	description_____			
31	224	00 42.7	+41 16	And	3.4	Andromeda Galaxy	9
date seen_____			initials____	description_____			
32	221	00 42.7	+40 52	And	8.2	Elliptical galaxy	9
date seen_____			initials____	description_____			
33	598	01 33.9	+30 39	Tri	5.7	Spiral galaxy (Sc)	9
date seen_____			initials____	description_____			
34	1039	02 42.0	+42 47	Per	5.2	Open cluster	10
date seen_____			initials____	description_____			
35	2168	06 08.9	+24 20	Gem	5.1	Open cluster	11,
date seen_____			initials____	description_____			
36	1960	05 36.1	+34 08	Aur	6.0	Open cluster	11
date seen_____			initials____	description_____			
37	2099	05 52.4	+32 33	Aur	5.6	Open cluster	11,
date seen_____			initials____	description_____			
38	1912	05 28.7	+35 50	Aur	6.4	Open cluster	11
date seen_____			initials____	description_____			
39	7092	21 32.2	+48 26	Cyg	4.6	Open cluster	19
date seen_____			initials____	description_____			
40	WNC4	12 22.4	+58 05	UMa	8	Double star	
date seen_____			initials____	description_____			
41	2287	06 46.0	−20 44	CMa	4.5	Open cluster	36
date seen_____			initials____	description_____			

NGC/IC	R.A. (2000.0) H M	DEC. (2000.0) ° '	CONST.	VISUAL MAG.	DESCRIPTION	ATLAS CHART
1976 te seen____	05 35.4	−05 27 initials____	Ori description____	4	Orion Nebula	24
1982 te seen____	05 35.6	−05 16 initials____	Ori description____		Orion Nebula; smaller part	24
2632 te seen____	08 40.1	+19 46 initials____	Cnc description____	3.1	Praesepe; open cluster	13, 25
— te seen____	03 47.0	+24 07 initials____	Tau description____	1.2	Pleiades; open cluster	10, 11
2437 te seen____	07 41.8	−14 49 initials____	Pup description____	6.1	Open cluster	25
2422 te seen____	07 36.6	−14 30 initials____	Pup description____	4.4	Open cluster	25
2548 te seen____	08 13.8	−05 48 initials____	Hya description____	5.8	Open cluster	25
4472 te seen____	12 29.8	+08 00 initials____	Vir description____	8.4	Elliptical galaxy	27A
2323 e seen____	07 02.8	−08 20 initials____	Mon description____	5.9	Open cluster	24, 25
5194-5 e seen____	13 29.9	+47 12 initials____	CVn description____	8.1	Whirlpool Galaxy	15
7654 te seen____	23 24.2	+61 35 initials____	Cas description____	6.9	Open cluster	1
5024 e seen____	13 12.9	+18 10 initials____	Com description____	7.7	Globular cluster	15
6715 e seen____	18 55.1	−30 29 initials____	Sgr description____	7.7	Globular cluster	42
6809 e seen____	19 40.0	−30 58 initials____	Sgr description____	7.0	Globular cluster	42
6779 e seen____	19 16.6	+30 11 initials____	Lyr description____	8.2	Globular cluster	18
6720 e seen____	18 53.6	+33 02 initials____	Lyr description____	9.0	Ring Nebula	18
4579 e seen____	12 37.7	+11 49 initials____	Vir description____	9.8	Spiral galaxy (SBb)	27A
4621 e seen____	12 42.0	+11 39 initials____	Vir description____	9.8	Elliptical galaxy	27A
4649 e seen____	12 43.7	+11 33 initials____	Vir description____	8.8	Elliptical galaxy	27A
4303 e seen____	12 21.9	+04 28 initials____	Vir description____	9.7	Spiral galaxy (Sc)	27A
6266 e seen____	17 01.2	−30 07 initials____	Oph description____	6.6	Globular cluster	41
5055 e seen____	13 15.8	+42 02 initials____	CVn description____	8.6	Spiral galaxy (Sb)	15
4826 e seen____	12 56.7	+21 41 initials____	Com description____	8.5	Spiral galaxy (Sb)	15
3623 e seen____	11 18.9	+13 05 initials____	Leo description____	9.3	Spiral galaxy (Sa)	27

TABLE 12. MESSIER CATALOGUE/LIFE LIST (CONTINUED)

M	NGC/IC	R.A. (2000.0) H M	DEC. (2000.0) ° '	CONST.	VISUAL MAG.	DESCRIPTION	ATLAS CHART
66 date seen___	3627	11 20.2 initials___	+12 59 description___	Leo	9.0	Spiral galaxy (Sb)	27
67 date seen___	2682	08 51.4 initials___	+11 49 description___	Cnc	6.9	Open cluster	25
68 date seen___	4590	12 39.5 initials___	−26 45 description___	Hya	8.2	Globular cluster	39
69 date seen___	6637	18 31.4 initials___	−32 21 description___	Sgr	7.7	Globular cluster	42
70 date seen___	6681	18 43.2 initials___	−32 18 description___	Sgr	8.1	Globular cluster	42
71 date seen___	6838	19 53.8 initials___	+18 47 description___	Sge	8.3	Globular cluster	19, 31
72 date seen___	6981	20 53.5 initials___	−12 32 description___	Aqr	9.4	Globular cluster	31, 32
73 date seen___	6994	20 58.9 initials___	−12 38 description___	Aqr		4 stars	31, 32
74 date seen___	628	01 36.7 initials___	+15 47 description___	Psc	9.2	Spiral galaxy (Sc)	9, 22
75 date seen___	6864	20 06.1 initials___	−21 55 description___	Sgr	8.6	Globular cluster	42, 43
76 date seen___	650-1	01 42.4 initials___	+51 34 description___	Per	11.5	Planetary nebula	1, 2, 9, 10
77 date seen___	1068	02 42.7 initials___	−00 01 description___	Cet	8.8	Spiral galaxy (Sb)	22
78 date seen___	2068	05 46.7 initials___	+00 03 description___	Ori	8	Emission nebula	24
79 date seen___	1904	05 24.5 initials___	−24 33 description___	Lep	8.0	Globular cluster	35
80 date seen___	6093	16 17.0 initials___	−22 59 description___	Sco	7.2	Globular cluster	41
81 date seen___	3031	09 55.6 initials___	+69 04 description___	UMa	6.8	Spiral galaxy (Sb)	4, 5
82 date seen___	3034	09 55.8 initials___	+69 41 description___	UMa	8.4	Irregular galaxy (Irr)	4, 5
83 date seen___	5236	13 37.0 initials___	−29 52 description___	Hya	7.6	Spiral galaxy (Sc)	39
84 date seen___	4374	12 25.1 initials___	+12 53 description___	Vir	9.3	Elliptical galaxy	27A
85 date seen___	4382	12 25.4 initials___	+18 11 description___	Com	9.2	Ellip./lenticular galaxy	15, 27
86 date seen___	4406	12 26.2 initials___	+12 57 description___	Vir	9.2	Elliptical galaxy	27A
87 date seen___	4486	12 30.8 initials___	+12 24 description___	Vir	8.6	Elliptical galaxy	27A
88 date seen___	4501	12 32.0 initials___	+14 25 description___	Com	9.5	Spiral galaxy (Sb)	27A

TABLE 12. MESSIER CATALOGUE/LIFE LIST (CONTINUED)

M	NGC/IC	R.A. (2000.0) H	M	DEC. °	'	CONST.	VISUAL MAG.	DESCRIPTION	ATLAS CHART
89 date seen____	4552	12	35.7	+12	33	Vir initials____	9.8	Elliptical galaxy description_____	27A
90 date seen____	4569	12	36.8	+13	10	Vir initials____	9.5	Spiral galaxy (Sb) description_____	27A
91 date seen____	4548	12	35.4	+14	30	Com initials____	10.2	Spiral galaxy or M58? description_____	27A
92 date seen____	6341	17	17.1	+43	08	Her initials____	6.5	Globular cluster description_____	17
93 date seen____	2447	07	44.6	−23	52	Pup initials____	6.2	Open cluster description_____	36
94 date seen____	4736	12	50.9	+41	07	CVn initials____	8.1	Spiral galaxy (Sb) description_____	15
95 date seen____	3351	10	44.0	+11	42	Leo initials____	9.7	Barred spiral galaxy (SBb) description_____	26
96 date seen____	3368	10	46.8	+11	49	Leo initials____	9.2	Spiral galaxy (Sa) description_____	26
97 date seen____	3587	11	14.8	+55	01	UMa initials____	11.2	Owl Nebula description_____	5, 14
98 date seen____	4192	12	13.8	+14	54	Com initials____	10.1	Spiral galaxy (Sb) description_____	27A
99 date seen____	4254	12	18.8	+14	25	Com initials____	9.8	Spiral galaxy (Sc) description_____	27A
100 date seen____	4321	12	22.9	+15	49	Com initials____	9.4	Spiral galaxy (Sc) description_____	15, 27A
101 date seen____	5457	14	03.2	+54	21	UMa initials____	7.7	Spiral galaxy (Sc) description_____	6, 15, 16
102 date seen____	—	—		—		— initials____		M101 description_____	
103 date seen____	581	01	33.2	+60	42	Cas initials____	7.4	Open cluster description_____	1, 2
104 date seen____	4594	12	40.0	−11	37	Vir initials____	8.3	Sombrero Galaxy description_____	27
105 date seen____	3379	10	47.8	+12	35	Leo initials____	9.3	Elliptical galaxy description_____	26
106 date seen____	4258	12	19.0	+47	18	CVn initials____	8.3	Spiral galaxy (Sb) description_____	14, 15
107 date seen____	6171	16	32.5	−13	03	Oph initials____	8.1	Globular cluster description_____	29
108 date seen____	3556	11	11.5	+55	40	UMa initials____	10.0	Spiral galaxy (Sb) description_____	5, 14
109 date seen____	3992	11	57.6	+53	23	UMa initials____	9.8	Barred spiral gal. (SBc) description_____	5, 14, 15

Note: Magnitudes based on a table in *Sky Catalogue* 2000.0 (vol. 2).

TABLE 13. CALDWELL CATALOGUE/LIFE LIST

C	NGC/IC	R.A. H M	DEC. ° '	CONST.	VISUAL MAG.	DESCRIPTION	ATLAS CHART
C1	188	00 44.4	+85 20	Cep	8.1	Open cluster	8
date seen_____			initials____ description_____				
C2	40	00 13.0	+72 32	Cep	11.6	Planetary nebula	8
date seen_____			initials____ description_____				
C3	4236	12 16.7	+69 28	Dra	9.7	Sb galaxy	7
date seen_____			initials____ description_____				
C4	7023	21 01.8	+68 12	Cep	6.8	Bright reflection nebula	8
date seen_____			initials____ description_____				
C5	IC 342	03 46.8	+68 06	Cam	9.2	SBc galaxy	3
date seen_____			initials____ description_____				
C6	6543	17 58.6	+66 38	Cas	8.8	PN: Cat's Eye Nebula	7
date seen_____			initials____ description_____				
C7	2403	07 36.9	+65 36	Cam	8.9	Sc Galaxy	3
date seen_____			initials____ description_____				
C8	559	01 29.5	+63 18	Cas	9.5	Open cluster	1
date seen_____			initials____ description_____				
C9	Sh2-155	22 56.8	+62 37	Cep	7.7	Cave Nebula	8
date seen_____			initials____ description_____				
C10	663	01 46.0	+61 15	Cas	7.1	Open cluster	1
date seen_____			initials____ description_____				
C11	7635	23 20.7	+61 12	Cas	7.0	Bubble Nebula	1
date seen_____			initials____ description_____				
C12	6946	20 34.8	+60 09	Cep	9.7	Sc galaxy	8
date seen_____			initials____ description_____				
C13	457	01 19.1	+58 20	Cas	6.4	Open cluster: Phi Cas	1
date seen_____			initials____ description_____				
C14	869/884	02 20.0	+57 08	Per	4.3	Double cluster	2
date seen_____			initials____ description_____				
C15	6826	19 44.8	+50 31	Cyg	9.8	Blinking Nebula	19
date seen_____			initials____ description_____				
C16	7243	22 15.3	+49 53	Lac	6.4	Open cluster	20
date seen_____			initials____ description_____				
C17	147	00 33.2	+48 30	Cas	9.3	dE4 galaxy	1
date seen_____			initials____ description_____				
C18	185	00 39.0	+48 20	Cas	9.2	dE0 galaxy	1
date seen_____			initials____ description_____				
C19	IC 5146	21 53.5	+47 16	Cyg	10.0	Cocoon Nebula	19
date seen_____			initials____ description_____				
C20	7000	20 58.8	+44 20	Cyg	6.0	North America Nebula	19
date seen_____			initials____ description_____				

TABLE 13. CALDWELL CATALOGUE/LIFE LIST (CONTINUED)

C	NGC/IC	R.A. H	M	DEC. °	′	CONST.	VISUAL MAG.	DESCRIPTION	ATLAS CHART
C21	4449	12	28.2	+44	06	CVn	9.4	Irregular galaxy	15
date seen_____				initials____		description_____			
C22	7662	23	25.9	+42	33	And	9.2	Planetary nebula	9
date seen_____				initials____		description_____			
C23	891	02	22.6	+42	21	And	9.9	Sb galaxy	9
date seen_____				initials____		description_____			
C24	1275	03	19.8	+41	31	Per	11.6	Seyfert galaxy in Perseus	2
date seen_____				initials____		description_____			
C25	2419	07	38.1	+38	53	Lyn	10.4	Globular cluster	13
date seen_____				initials____		description_____			
C26	4244	12	17.5	+37	49	CVn	10.6	Sc galaxy	15
date seen_____				initials____		description_____			
C27	6888	20	12.0	+38	21	Cyg	7.5	Crescent Nebula	19
date seen_____				initials____		description_____			
C28	752	01	57.8	+37	41	And	5.7	Open cluster	9
date seen_____				initials____		description_____			
C29	5005	13	10.9	+37	03	CVn	9.8	Sb galaxy	15
date seen_____				initials____		description_____			
C30	7331	22	37.1	+34	25	Peg	9.5	Sb galaxy	20
date seen_____				initials____		description_____			
C31	IC 405	05	16.2	+34	16	Aur	9.5	Flaming Star Nebula	11
date seen_____				initials____		description_____			
C32	4631	12	42.1	+32	32	CVn	9.3	Sc galaxy	15
date seen_____				initials____		description_____			
C33	6992/5	20	56.4	+31	43	Cyg		East Veil Nebula, SN	19
date seen_____				initials____		description_____			
C34	6960	20	45.7	+30	43	Cyg		West Veil Nebula, SN	19
date seen_____				initials____		description_____			
C35	4889	13	00.1	+27	59	Com	11.4	Brightest in Coma Cl.	15
date seen_____				initials____		description_____			
C36	4559	12	36.0	+27	58	Com	9.8	Sc galaxy	15
date seen_____				initials____		description_____			
C37	6885	20	12.0	+26	29	Vul	5.7	Open cluster	18
date seen_____				initials____		description_____			
C38	4565	12	36.3	+25	59	Com	9.6	Sb galaxy	15
date seen_____				initials____		description_____			
C39	2392	07	29.2	+20	55	Gem	9.9	PN: Eskimo Nebula	12
date seen_____				initials____		description_____			
C40	3626	11	20.1	+18	21	Leo	10.9	Sb galaxy	14
date seen_____				initials____		description_____			

TABLE 13. CALDWELL CATALOGUE/LIFE LIST (CONTINUED)

C	NGC/IC	R.A. H M	DEC. ° '	CONST.	VISUAL MAG.	DESCRIPTION	ATLAS CHART
C41 date seen_____		04 27.0	+16 00	Tau initials____	1.0	Open cluster: Hyades description_____	11
C42 date seen_____	7006	21 01.5	+16 11	Del initials____	10.6	Very distant globular cl. description_____	19
C43 date seen_____	7814	00 03.3	+16 09	Peg initials____	10.5	Sb galaxy description_____	20
C44 date seen_____	7479	23 04.9	+12 19	Peg initials____	11.0	SBb galaxy description_____	20
C45 date seen_____	5248	13 37.5	+08 53	Boo initials____	10.2	Sc galaxy description_____	28
C46 date seen_____	2261	06 39.2	+08 44	Mon initials____	10	Hubble's Variable Nebula description_____	24
C47 date seen_____	6934	20 34.2	+07 24	Del initials____	5.9	Globular cluster description_____	19
C48 date seen_____	2775	09 10.3	+07 02	Com initials____	10.3	Sa galaxy description_____	15
C49 date seen_____	2237-9	06 32.3	+05 03	Mon initials____		Rosette Nebula description_____	24
C50 date seen_____	2244	06 32.4	+04 52	Mon initials____	4.8	Open cluster description_____	24
C51 date seen_____	IC 1613	01 04.8	+02 07	Cet initials____	9.0	Irregular galaxy description_____	22
C52 date seen_____	4697	12 48.6	−05 48	Vir initials____	9.3	E4 galaxy description_____	27
C53 date seen_____	3115	10 05.2	−07 43	Sex initials____	9.1	Spindle Galaxy description_____	26
C54 date seen_____	2506	08 00.2	−10 47	Mon initials____	7.6	Open cluster description_____	25
C55 date seen_____	7009	21 04.2	−11 22	Aqr initials____	8.3	PN: Saturn Nebula description_____	32
C56 date seen_____	246	00 47.0	−11 53	Cet initials____	8.0	Planetary Nebula description_____	21
C57 date seen_____	6822	19 44.9	−14 48	Sgr initials____	9.3	Barnard's Galaxy, irr. description_____	31
C58 date seen_____	2360	07 17.8	−15 37	CMa initials____	7.2	Open cluster description_____	25
C59 date seen_____	3242	10 24.8	−18 38	Hya initials____	8.6	PN: Ghost of Jupiter description_____	26
C60 date seen_____	4038	12 01.9	−18 52	Crv initials____	10.7	Sc galaxy, The Antennae description_____	27
C61	4039	12 01.9	−18 53	Crv	10.7	Sc galaxy, The Antennae	27

TABLE 13. CALDWELL CATALOGUE/LIFE LIST (CONTINUED)

C	NGC/IC	R.A. H M	DEC. ° '	CONST.	VISUAL MAG.	DESCRIPTION	ATLAS CHART
date seen_____				initials____		description_____	
C62	247	00 47.1	−20 46	Cet	8.9	SBc galaxy	33
date seen_____				initials____		description_____	
C63	7293	22 29.6	−20 48	Aqr	6.5	PN: Helix Nebula	44
date seen_____				initials____		description_____	
C64	2362	07 17.8	−24 57	CMa	4.1	Tau CMa open cluster	36
date seen_____				initials____		description_____	
C65	253	00 47.6	−25 17	Scl	7.1	Sculptor Galaxy, Scp	33
date seen_____				initials____		description_____	
C66	5694	14 39.6	−26 32	Hya	10.2	Globular cluster	40
date seen_____				initials____		description_____	
C67	1097	02 46.3	−30 17	For	9.2	SBb galaxy	34
date seen_____				initials____		description_____	
C68	6729	19 01.9	−36 57	CrA	9.7	R CrA Nebula	42
date seen_____				initials____		description_____	
C69	6302	17 13.7	−37 06	Sco	12.8	PN: Bug Nebula	41
date seen_____				initials____		description_____	
C70	300	00 54.9	−37 41	Scl	8.1	Sd galaxy	33
date seen_____				initials____		description_____	
C71	2477	07 52.3	−38 33	Pup	5.8	Open cluster	36
date seen_____				initials____		description_____	
C72	55	00 14.9	−39 11	Scl	8.2	SB, brightest in Sculptor	33
date seen_____				initials____		description_____	
C73	1851	05 14.1	−40 03	Col	7.3	Globular cluster	35
date seen_____				initials____		description_____	
C74	3132	10 07.7	−40 26	Vel	8.2	Planetary nebula	37
date seen_____				initials____		description_____	
C75	6124	16 25.6	−40 40	Sco	5.8	Open cluster	41
date seen_____				initials____		description_____	
C76	6231	16 54.0	−41 48	Sco	2.6	Open cluster	41
date seen_____				initials____		description_____	
C77	5128	13 25.5	−43 01	Cen	7.0	Cen A	39
date seen_____				initials____		description_____	
C78	6541	18 08.0	−43 42	CrA	6.6	Globular cluster	42
date seen_____				initials____		description_____	
C79	3201	10 17.6	−46 25	Vel	6.7	Globular cluster	38
date seen_____				initials____		description_____	
C80	5139	13 26.8	−47 29	Cen	3.6	Globular cluster	39
date seen_____				initials____		description_____	
C81	6352	17 25.5	−48 25	Ara	8.1	Globular cluster	41
date seen_____				initials____		description_____	

TABLE 13. CALDWELL CATALOGUE/LIFE LIST (CONTINUED)

C	NGC/IC	R.A. H M	DEC. ° '	CONST.	VISUAL MAG.	DESCRIPTION	ATLAS CHART
C82 date seen_____	6193	16 41.3	−48 46	Ara initials____	5.2	Open cluster description_____	41
C83 date seen_____	4945	13 05.4	−49 28	Cen initials____	9.5	SBc galaxy description_____	39
C84 date seen_____	5286	13 46.4	−51 22	Cen initials____	7.6	Globular cluster description_____	39
C85 date seen_____	IC 2391	08 40.2	−53 04	Vel initials____	2.5	Open cluster: o Vel description_____	37
C86 date seen_____	6397	17 40.7	−53 40	Ara initials____	5.6	Globular cluster description_____	41
C87 date seen_____	1261	03 12.3	−55 13	Hor initials____	8.4	Globular cluster description_____	46
C88 date seen_____	5823	15 05.7	−55 36	Cir initials____	7.9	Open cluster description_____	50
C89 date seen_____	6087	16 18.9	−57 54	Nor initials____	5.4	Open cluster: S Nor description_____	50
C90 date seen_____	2867	09 21.4	−58 19	Car initials____	9.7	Planetary nebula description_____	48
C91 date seen_____	3532	11 06.4	−58 40	Car initials____	3.0	Open cluster description_____	49
C92 date seen_____	3372	10 43.8	−59 52	Car initials____	6.2	Eta Carinae Nebula description_____	49
C93 date seen_____	6752	19 10.9	−59 59	Pav initials____	5.4	Globular cluster description_____	51
C94 date seen_____	4755	12 53.6	−60 20	Cru initials____	4.2	Jewel Box description_____	49
C95 date seen_____	6025	16 03.7	−60 30	TrA initials____	5.1	Open cluster description_____	50
C96 date seen_____	2516	07 58.3	−60 52	Car initials____	3.8	Open cluster description_____	48
C97 date seen_____	3766	11 36.1	−61 37	Cen initials____	5.3	Open cluster description_____	49
C98 date seen_____	4609	12 42.3	−62 58	Cru initials____	6.9	Open cluster description_____	49
C99 date seen_____		12 53.0	−63 00	Cru initials____		Coal Sack, dark nebula description_____	49
C100 date seen_____	[IC 2944]	11 36.6	−63 02	Cru initials____	4.5	Open cl., λ Cen, Collinder 249 description_____	49
C101 date seen_____	6744	19 09.8	−63 51	Pav initials____	9.0	SBb galaxy description_____	51

TABLE 13. CALDWELL CATALOGUE/LIFE LIST (CONTINUED)

NGC/IC	R.A. H M	DEC. ° ′	CONST.	VISUAL MAG.	DESCRIPTION	ATLAS CHART
102 IC 2602	10 43.2	−64 24	Car	1.9	Open cluster, θ Car	48
date seen_____		initials____		description_____		
103 2070	05 38.7	−69 06	Dor	1.0	Tarantula Nebula in LMC	47
date seen_____		initials____		description_____		
104 362	01 03.2	−70 51	Tuc	6.6	Globular cluster	45
date seen_____		initials____		description_____		
105 4833	12 59.6	−70 53	Mus	7.3	Globular cluster	49
date seen_____		initials____		description_____		
106 104	00 24.1	−72 05	Tuc	4.0	47 Tuc	45
date seen_____		initials____		description_____		
107 6101	16 25.8	−72 12	Aps	9.3	Globular cluster	50
date seen_____		initials____		description_____		
108 4372	12 25.8	−72 40	Mus	7.8	Globular cluster	49
date seen_____		initials____		description_____		
109 3195	10 09.5	−80 52	Cha	8.4	Planetary nebula	48
date seen_____		initials____		description_____		

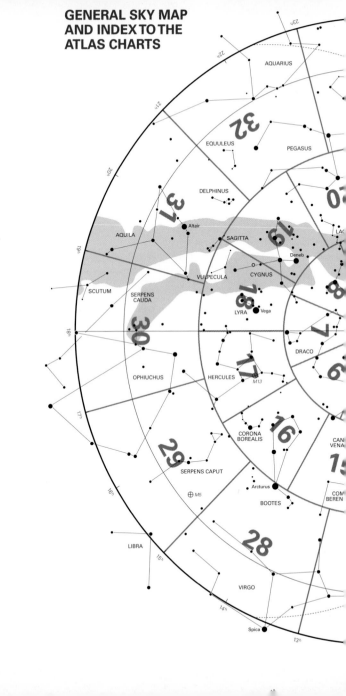

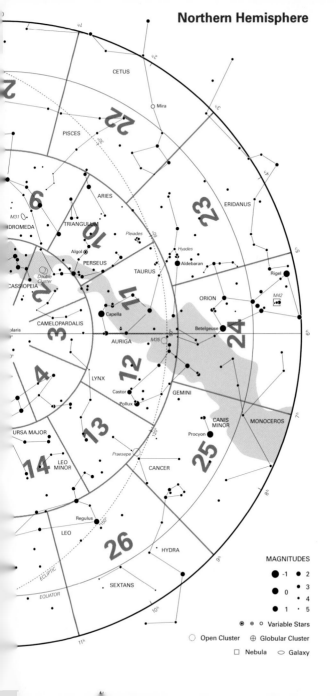

Northern Hemisphere

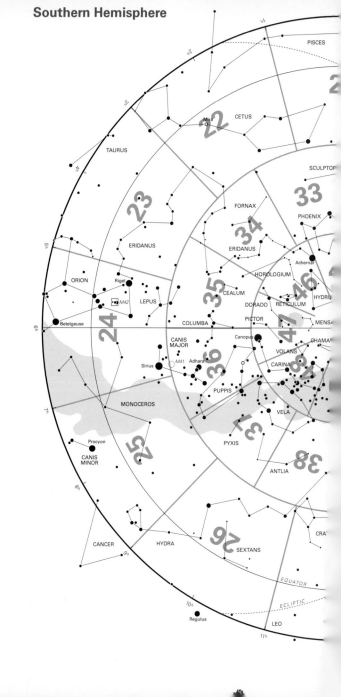

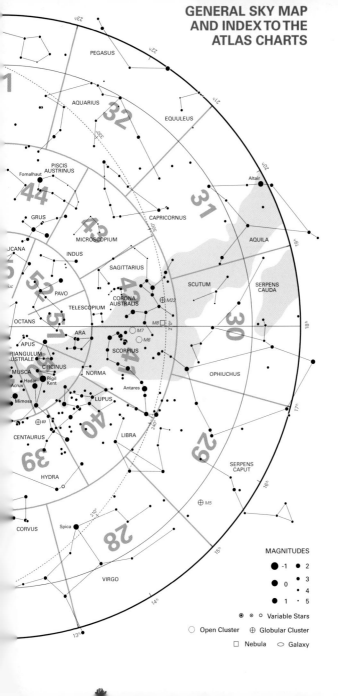

GENERAL SKY MAP AND INDEX TO THE ATLAS CHARTS

PEGASUS

AQUARIUS

EQUULEUS

PISCIS
AUSTRINUS

Fomalhaut

Altair

CAPRICORNUS

GRUS

AQUILA

MICROSCOPIUM

INDUS

SAGITTARIUS

SCUTUM

SERPENS
CAUDA

PAVO

TELESCOPIUM

CORONA
AUSTRALIS

⊕ M22

M8 □

OCTANS

ARA

⟡ M7

M6

APUS

SCORPIUS

OPHIUCHUS

TRIANGULUM
AUSTRALE

CIRCINUS

MUSCA

NORMA

Rigil
Kent

Acrux

Hadar

Antares

Mimosa

LUPUS

⊕ ω

CENTAURUS

LIBRA

HYDRA

SERPENS
CAPUT

⊕ M5

CORVUS

Spica

VIRGO

MAGNITUDES

● -1	● 2
● 0	• 3
	· 4
● 1	· 5

◉ ◎ ○ Variable Stars

⟡ Open Cluster   ⊕ Globular Cluster

□ Nebula   ○ Galaxy

Fig. 7-1 The Bubble Nebula (NGC 7635) in Cassiopeia. (Tom Montemayor, McDonald Observatory, University of Texas)

ATLAS CHART 1. BUBBLE NEBULA, TYCHO'S SUPERNOVA The Milky Way in Cassiopeia, with its rich fields of clouds, gas, dust, and star clusters, is an interesting area to scan with low-power telescopes or binoculars. Several open clusters are easy to find because they lie close to bright stars. NGC 457 is an especially bright open cluster in the same field of view as φ (phi) Cas, which lies southwest of δ (delta) Cas in the W of Cassiopeia. Even a small telescope gives a good view of the stars in this cluster; more stars are visible with telescopes of higher power. Close by is the open cluster NGC 436. Since binoculars or telescopes with small apertures show only a few stars here, NGC 436 is more suitable for viewing with larger telescopes. M103 (NGC 581) is a fan-shaped, 7th-magnitude cluster located northeast of δ (delta) Cas.

About 5° northwest of β (beta) Cas, near the border of Cepheus, is M52 (NGC 7654), a rich, 7th-magnitude open cluster. Close to M52 is the Bubble Nebula, NGC 7635, which has a high total brightness even though it is spread over so much sky that its average surface brightness is low. The bubble was formed by the stellar wind flowing outward from a very hot star (spectral type Of). Southwest of β (beta) Cas, about halfway between ρ (rho) and σ (sigma) Cas, is the open cluster NGC 7789. Its diameter is about the same as the moon's. You can detect this cluster with binoculars, and telescopes resolve many of its 1,000 stars.

Northeast of β (beta) Cas is κ (kappa) Cas, with the two open clusters NGC 133 and NGC 146 nearby. NGC 146 has about 50 stars within a 6 arc min diameter; this cluster should be viewed with high power because of the richness of the Milky Way in this part of the sky.

Tycho's Supernova of 1572, $0^h25^m +63°$, appeared slightly to the northwest of NGC 146, growing so bright that it was visible to the naked eye for about six months. Since the appearance of a new star showed that the sky changed, contrary to Ptolemaic theory, this object was important for the acceptance of Copernicus's heliocentric theory. The remnant of the supernova is now only faintly visible with large telescopes, appearing as thin filaments that form an incomplete ring 8 arc min across. Radio telescopes, however, detect strong signals.

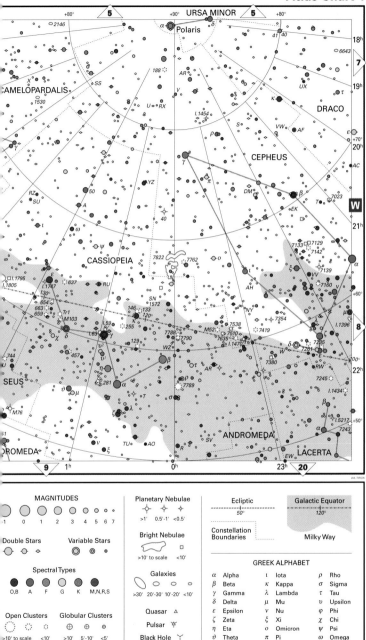

MAGNITUDES								
-1	0	1	2	3	4	5	6	7

Double Stars

Variable Stars

Spectral Types

O,B A F G K M,N,R,S

Open Clusters

>10' to scale   <10'

Globular Clusters

>10'   5'-10'   <5'

Planetary Nebulae

>1'   0.5'-1'   <0.5'

Bright Nebulae

>10' to scale   <10'

Galaxies

>30'   20'-30'   10'-20'   <10'

Quasar △

Pulsar ☼

Black Hole ☌

Ecliptic
- - - - - - - -
50°

Constellation Boundaries

Galactic Equator

120°

Milky Way

GREEK ALPHABET

α	Alpha	ι	Iota	ρ	Rho		
β	Beta	κ	Kappa	σ	Sigma		
γ	Gamma	λ	Lambda	τ	Tau		
δ	Delta	μ	Mu	υ	Upsilon		
ε	Epsilon	ν	Nu	φ	Phi		
ζ	Zeta	ξ	Xi	χ	Chi		
η	Eta	ο	Omicron	ψ	Psi		
ϑ	Theta	π	Pi	ω	Omega		

Fig. 7-2 The double open cluster in Perseus, h and χ (chi) Persei (NGC 869 and 884). (Richard E. Hill)

ATLAS CHART 2. POLARIS, DOUBLE CLUSTER IN PERSEUS Polaris, α (alpha) Ursa Minoris, the North Star, lies about 1° from the true north celestial pole and circles the pole once every 24 hours. Polaris is a double star, with a 9th-magnitude companion at a distance of 18 arc sec. The pair can be resolved by a good small telescope.

In the Milky Way in Perseus, we find the famous "double cluster," a pair of open clusters also known as h and χ (chi) Persei, and marked on the chart as NGC 869 and NGC 884. The double cluster is a favorite of amateurs because it is so easy to observe. Binoculars reveal its many stars. A small telescope used with a low-power, wide-angle eyepiece (field of 1°) shows both clusters. Each cluster has a diameter of about 70 light-years. NGC 869 is the younger of the two, only about 10 million years old, which makes it one of the youngest clusters known; it lies over 7,000 light-years away from us.

About 1½° east is the smaller open cluster NGC 957. At lower right in Perseus is M76 (NGC 650), a planetary nebula whose integrated magnitude is 11.5. Near the chart's right border, at 58°, we find the binary η (eta) Cas. Its components show a beautiful color contrast, sometimes reported as gold and purple or as yellow and red.

As we move from Cassiopeia to Camelopardalis, we go farther from the Milky Way and into less interesting regions of the sky.

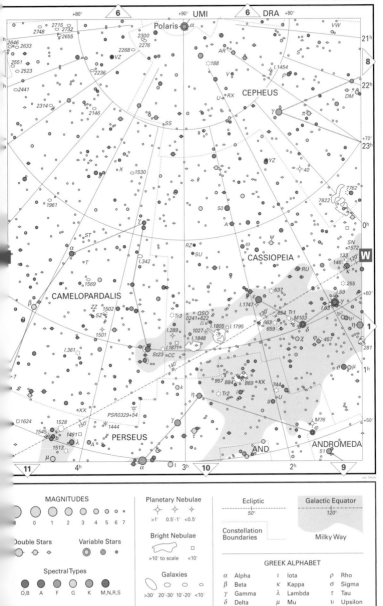

MAGNITUDES								Planetary Nebulae	Ecliptic	Galactic Equator

MAGNITUDES
0 1 2 3 4 5 6 7

Double Stars

Variable Stars

Spectral Types
O,B A F G K M,N,R,S

Open Clusters
>10' to scale <10'

Globular Clusters
>10' 5'-10' <5'

Planetary Nebulae
>1' 0.5'-1' <0.5'

Bright Nebulae
>10' to scale <10'

Galaxies
>30' 20'-30' 10'-20' <10'

Quasar △

Pulsar ✷

Black Hole ⸸

Ecliptic
50°

Constellation Boundaries

Galactic Equator
120°

Milky Way

GREEK ALPHABET

α	Alpha	ι	Iota	ρ	Rho
β	Beta	κ	Kappa	σ	Sigma
γ	Gamma	λ	Lambda	τ	Tau
δ	Delta	μ	Mu	υ	Upsilon
ε	Epsilon	ν	Nu	φ	Phi
ζ	Zeta	ξ	Xi	χ	Chi
η	Eta	ο	Omicron	ψ	Psi
ϑ	Theta	π	Pi	ω	Omega

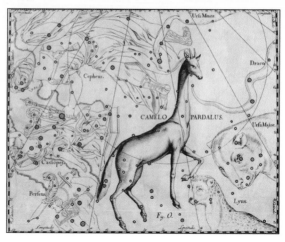

Fig. 7-3 *Camelopardalis, the Giraffe, from the star atlas of Hevelius (1690). The constellation is drawn backward from the way it appears in the sky because Hevelius drew the celestial sphere as it would appear from the outside. (Jay M. Pasachoff)*

ATLAS CHART 3. CAMELOPARDALIS Camelopardalis, the Giraffe, has very few bright stars. In Lynx, a trio of multiple stars lies just right of 7^h and just below +60°. The multiple system farthest to the right is an interesting triple. The double star 19 Lyn (ADS6012), near 7^h23^m +55°, is surrounded by several faint companions.

The galaxy NGC 2403 (7^h34^m +65°40') is one of the nearest spiral galaxies outside of our Local Group of galaxies. Binoculars show it as a large hazy spot, while large telescopes hint at spiral structure. NGC 2403 is 8 million light-years away and 37,000 light-years across. The galaxy is easily seen even with small telescopes at low power, and the view with larger telescopes is striking.

Farther down in Camelopardalis, at 4^h08^m +62°, is the open cluster NGC 1502, readily visible through small telescopes. Larger telescopes show it to be tightly packed with many stars of differing magnitudes. On nights with good seeing, the cluster is visible even through binoculars.

Six degrees above NGC 1502 is the large beautiful spiral galaxy IC 342. Though it can be seen with small telescopes, larger ones are needed to show any spiral structure. The planetary nebula NGC 1501, located in Camelopardalis slightly below NGC 1502, is about 1 arc min in diameter and slightly oval. Because it is dim, at least a medium aperture is needed for a good view.

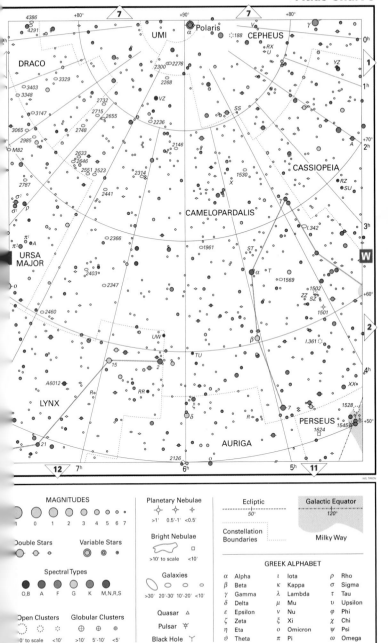

Fig. 7-4 M81 (NGC 3031), a type Sb spiral galaxy in Ursa Major. (CCD image on a 24-inch f/5 telescope by Tim Hunter and James McGaha)

ATLAS CHART 4. M81 AND M82 The galaxies M81 (NGC 3031) and M82 (NGC 3034) appear at the center of this chart, just short of 10^h and below 70°. They are members of the Ursa Major cluster of galaxies. M81 is one of the prettiest spirals in the sky. M82 (fig. 5-44) appears as though gas is exploding from it, but there has been continual controversy over whether we are seeing gas exploding or light reflected from relatively stationary gas. These galaxies are two of the most easily observed galaxies in the sky and are visible through a small telescope or a good pair of binoculars. Nearby galaxies NGC 3077 and NGC 2976 are easily visible with low-power instruments.

QSO 0957+561, at lower left, is the "double quasar," the first known example of a "gravitational lens." The light and radio waves from this quasar are bent by the gravity of an intervening galaxy as they pass it, making us see at least two images separated by only 6 arc sec. The optical objects are 17th magnitude and have a redshift of 1.41 (141 percent), making them some of the most distant ones known.

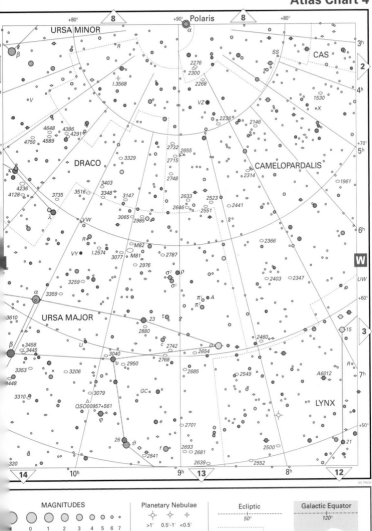

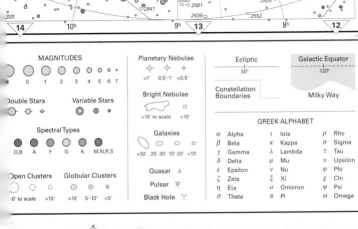

MAGNITUDES							
0	1	2	3	4	5	6	7

Double Stars **Variable Stars**

Spectral Types

O,B A F G K M,N,R,S

Open Clusters **Globular Clusters**

0' to scale <10' >10' 5'-10' <5'

Planetary Nebulae

>1' 0.5'-1' <0.5'

Bright Nebulae

>10' to scale <10'

Galaxies

>30' 20'-30' 10'-20' <10'

Quasar △

Pulsar ☼

Black Hole ⌖

Ecliptic

50'

Constellation Boundaries

Galactic Equator
120'

Milky Way

GREEK ALPHABET

α	Alpha	ι	Iota	ρ	Rho
β	Beta	κ	Kappa	σ	Sigma
γ	Gamma	λ	Lambda	τ	Tau
δ	Delta	μ	Mu	υ	Upsilon
ε	Epsilon	ν	Nu	φ	Phi
ζ	Zeta	ξ	Xi	χ	Chi
η	Eta	o	Omicron	ψ	Psi
ϑ	Theta	π	Pi	ω	Omega

Fig. 7-5 *The Owl Nebula, M97 (NGC 3587) in Ursa Major. (CCD image on a 24-inch f/5 telescope by Tim Hunter and James McGaha)*

ATLAS CHART 5. BIG DIPPER, OWL NEBULA At lower right in Ursa Major—about 2½° southeast of β (beta) UMa—is M97 (NGC 3587), the Owl Nebula (fig. 7-5), one of the largest of the planetary nebulae. Small telescopes show a featureless circular disk, but the image in large telescopes suggests the face of an owl. Within 1° of M97 is the galaxy M108 (NGC 3556), shown in figure 5-49.

Mizar, ζ (zeta) UMa, is the second star from the end of the handle of the Big Dipper (fig. 1-2) and is located near $13^h30^m +55°$. Mizar is one of the most interesting doubles in the sky. Its companion, Alcor (labeled *g* on the chart), lies about 12 arc min away (a distance equal to about one-third of the moon's diameter) and is just barely visible with the naked eye. Both components of Mizar, called Mizar A and Mizar B, are in turn double stars, though their components cannot be resolved by telescopes. The motions of the components show up in spectra. The American Indians referred to Mizar and Alcor as the Horse and Rider.

Magnitudes around the Big Dipper, from handle to bowl and around the bowl, are 1.9 for Alkaid, 2.1 for Mizar and 4.0 for Alcor, 1.8 for Alioth, 3.3 for Megrez, 2.4 for Phecda, 2.4 for Merak, and 1.8 for Dubhe.

M109 (NGC 3992), a bright barred spiral galaxy in Ursa Major, lies near γ (gamma) Ursa Majoris, at 12^h and above +50°. Another easily observed galaxy, NGC 3953, lies nearby. These and other galaxies are part of the Ursa Major cluster of galaxies. Another cluster of galaxies can be observed in Draco.

An interesting variable to observe is RY in Draco, at $13^h +65°$, which varies with a six-month period.

The Hubble Deep Field (p. 212), the area of the sky observed for 10 days with the Hubble Space Telescope to reveal especially faint, far-away ("deep" into space) objects, is no larger than a grain of sand held at the end of your outstretched arm. It is in Ursa Major near $12^h 37^m +62°$.

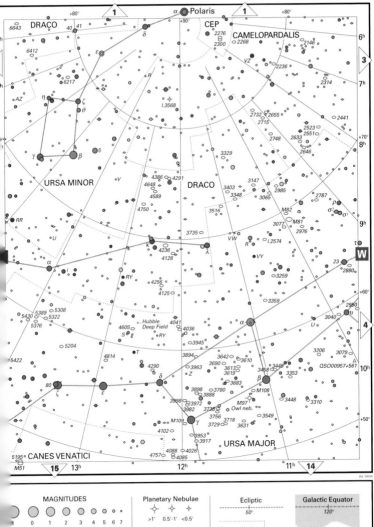

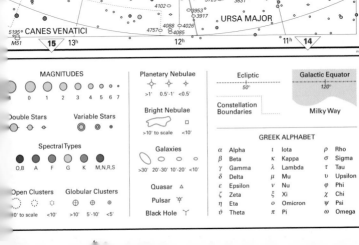

Fig. 7-6 M101 (NGC 5457), a type Sc spiral galaxy in Ursa Major. (Tim Puckett)

ATLAS CHART 6. M101, LITTLE DIPPER M101 (NGC 5457) is an exceptionally beautiful spiral galaxy in Ursa Major, near 14^h +54°. It is seen face-on, and at almost 8th magnitude, it is one of the brightest galaxies in the sky. M101's spiral arms can barely be seen on a very clear night through a medium-sized telescope, but because the galaxy is so spread out, it is hard to find unless conditions are perfect. M102 was one of Messier's few mistakes; it was really a second reference to M101. Many other members of the Ursa Major cluster of galaxies also appear on this chart, near 15^{h}20^m +56°.

The Little Dipper in Ursa Minor lies in the central region of this chart. The bowl contains β (Kochab), γ (Pherkad), η, and ζ, and the handle is ε, δ, and α (Polaris) UMi. Apparent magnitudes, in the same order, are 2.1, 3.1, 5.0, 4.3, 4.2, 4.4, and 2.0. These stars can thus make a good test of the quality of your sky, and of your eyes' sensitivity.

Few other objects of special interest are in this region. One double star worth observing is κ (kappa) Boötes, near 14^{h}13^m +52°.

Thuban, alpha Draconis, was the pole star when the Pyramids were being built, and so was worshiped by the Egyptians.

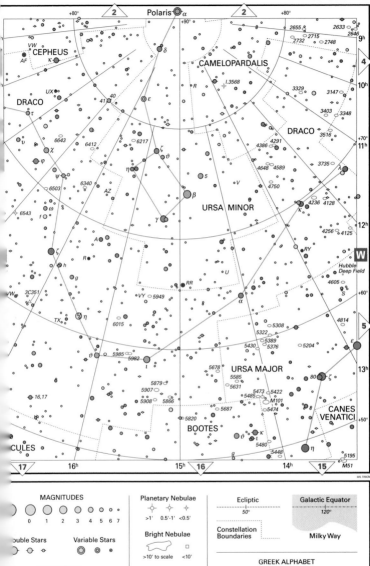

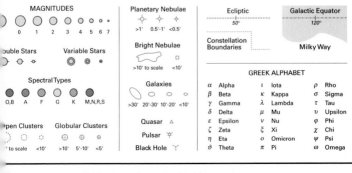

MAGNITUDES							
	0	1	2	3	4	5	6	7

Double Stars   **Variable Stars**

Spectral Types

O,B   A   F   G   K   M,N,R,S

Open Clusters   **Globular Clusters**

' to scale   <10'   >10'   5'-10'   <5'

Planetary Nebulae

>1'   0.5'-1'   <0.5'

Bright Nebulae

>10' to scale   <10'

Galaxies

>30'   20'-30'   10'-20'   <10'

Quasar △

Pulsar ☼

Black Hole Υ

Ecliptic

50°

Constellation Boundaries

Galactic Equator

120°

Milky Way

GREEK ALPHABET					
α	Alpha	ι	Iota	ρ	Rho
β	Beta	κ	Kappa	σ	Sigma
γ	Gamma	λ	Lambda	τ	Tau
δ	Delta	μ	Mu	υ	Upsilon
ε	Epsilon	ν	Nu	φ	Phi
ζ	Zeta	ξ	Xi	χ	Chi
η	Eta	ο	Omicron	ψ	Psi
ϑ	Theta	π	Pi	ω	Omega

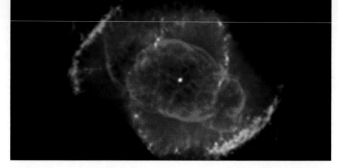

Fig. 7-7 *The Cat's Eye Nebula, NGC 6543. This Hubble Space Telescope view reveals complicated shells of matter that is too blurred to be detectable in ground-based views. (J. P. Harrington and K. J. Borkowski, University of Maryland; and NASA; courtesy of Space Telescope Science Institute)*

ATLAS CHART 7. DRACO The four stars in the head of Draco—ξ, ν, β, and γ Dra (xi, nu, beta, and gamma Dra)—form a conspicuous asterism, the Lozenge, at the bottom of this chart. They are not far from the bright star Vega (see Atlas Chart 18) and are close to the foot of Hercules. The Lozenge and other areas of Draco contain several interesting doubles. Binoculars resolve the star ν (nu) Dra in the Lozenge, and a good small telescope resolves the nearby star μ (mu) Dra, at 17^h +55°.

Several other doubles appear toward the center of the chart, including φ (phi) Dra, near $18^h 20^m$ +71°, and the pair at 18^h +80°, whose members have Flamsteed numbers 40 and 41 Dra. Due east (left) of φ (phi) Dra is ε (epsilon) Dra, a good double to observe at moderate magnifications since its components are only about 3 arc sec apart.

The planetary nebula NGC 6543, the Cat's Eye Nebula, at magnitude 8.8, is one of the brightest in the sky. It is located almost in the center of this chart, about halfway between δ (delta) and ζ (zeta) Dra. NGC 6543 was the first planetary nebula to be observed with a spectroscope; the fact that emission lines were present settled the controversy about whether planetaries were numerous stars or—as they turned out to be—clouds of diffuse gas. A small telescope at moderate magnification shows the disk, but more powerful telescopes are needed to show the internal structure, a bright irregular helix. Since NGC 6543 is a circumpolar object for most northern hemisphere observers, you can view it throughout the year.

The quasar 3C 351 lies at 17 +60°. It is magnitude 15.3 and has a redshift of 0.371 (37.1 percent), making it about 7 billion light-years away.

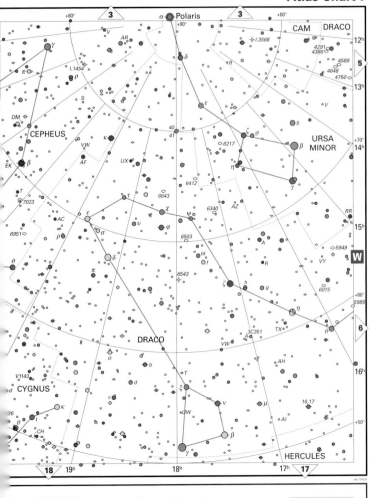

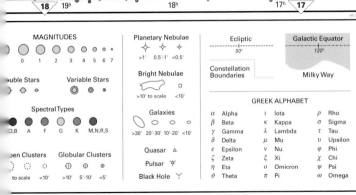

MAGNITUDES									Planetary Nebulae			Ecliptic		Galactic Equator	
0	1	2	3	4	5	6	7		>1'	0.5'-1'	<0.5'			120°	

Double Stars   **Variable Stars**

Spectral Types
O,B   A   F   G   K   M,N,R,S

Open Clusters   **Globular Clusters**
to scale   <10'   >10'   5'-10'   <5'

Bright Nebulae
>10' to scale   <10'

Galaxies
>30'   20'-30'   10'-20'   <10'

Quasar △
Pulsar ☿
Black Hole ⅄

Constellation Boundaries   Milky Way

GREEK ALPHABET

α	Alpha	ι	Iota	ρ	Rho
β	Beta	κ	Kappa	σ	Sigma
γ	Gamma	λ	Lambda	τ	Tau
δ	Delta	μ	Mu	υ	Upsilon
ε	Epsilon	ν	Nu	φ	Phi
ζ	Zeta	ξ	Xi	χ	Chi
η	Eta	ο	Omicron	ψ	Psi
ϑ	Theta	π	Pi	ω	Omega

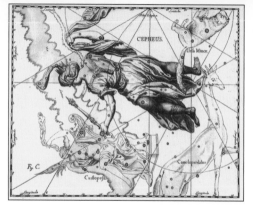

Fig. 7-8 *From the Atlas of Johannes Hevelius (1687). The images are flopped. (Jay M. Pasachoff)*

ATLAS CHART 8. DELTA CEPHEI The Milky Way passes through Cepheus and Cygnus in this chart and presents a very rich star field for observing. The region contains a number of open clusters, planetary nebulae, and gaseous nebulae. Look in Cepheus, near $21^h 50^m$ +58°, for the variable star μ (mu) Cep, one of the reddest stars visible to the naked eye. It is sometimes called the Garnet Star because of its deep red color. Just south of μ Cep is a large faint structure, IC 1396, a region of extended nebulosity. $1^h 15^m$ to the right of μ Cep and slightly north, at about $20^h 30^m$ +60°, are the open cluster NGC 6939 and the spiral galaxy NGC 6946 (fig. 5-52). Above μ Cep, near 22^h +64°, is one of the best double stars in this constellation, ξ (xi) Cep.

Near the galactic equator, at $22^h 30^m$ +58°, is the famous star δ (delta) Cephei, a variable star whose discovery led to the first measurements of the distances to galaxies and thus to our subsequent understanding of the immense scale of the universe (chapter 5). This star is the prototype of the *Cepheid variables* (chapter 6). δ Cephei has a 5.4-day period, during which it varies by about one magnitude in brightness. By observing δ Cephei every night, you can easily detect the variation. Small telescopes also reveal that δ Cephei is a double star.

Toward the top of the chart, near $21^h 30^m$ +78°, is a red variable, S Cep, which is visible through small telescopes. Closer to Polaris is the open cluster NGC 188, one of the oldest open clusters known, at 14 billion years of age. About 70 stars can be seen in this cluster with a medium-sized telescope at low power.

This completes the set of 8 Atlas Charts covering the region from +50° to the north celestial pole. The next 12 Atlas Charts cover the region from +20° to +50°.

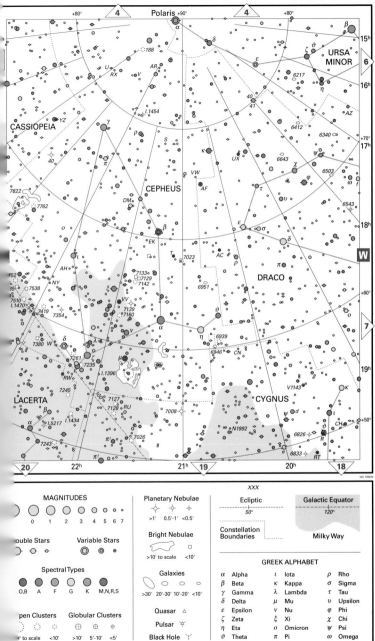

MAGNITUDES							
0	1	2	3	4	5	6	7

Double Stars

Variable Stars

Spectral Types

O,B A F G K M,N,R,S

Open Clusters   **Globular Clusters**

' to scale <10' >10' 5'-10' <5'

Planetary Nebulae

>1' 0.5'-1' <0.5'

Bright Nebulae

>10' to scale <10'

Galaxies

>30' 20'-30' 10'-20' <10'

Quasar △

Pulsar ☆

Black Hole Υ

XXX

Ecliptic
--- 50° ---

Constellation
Boundaries

Galactic Equator
··· 120° ···

Milky Way

GREEK ALPHABET

α	Alpha	ι	Iota	ρ	Rho
β	Beta	κ	Kappa	σ	Sigma
γ	Gamma	λ	Lambda	τ	Tau
δ	Delta	μ	Mu	υ	Upsilon
ε	Epsilon	ν	Nu	φ	Phi
ζ	Zeta	ξ	Xi	χ	Chi
η	Eta	ο	Omicron	ψ	Psi
ϑ	Theta	π	Pi	ω	Omega

Fig. 7-9 M31 (NGC 224), a type Sb spiral galaxy in Andromeda. M32 is its superimposed elliptical companion, and NGC 205 is a slightly more distant elliptical companion. (© Jerry Lodriguss; 5.1-inch f/6 refractor, hypered ISO 400 film, composite of two 45-minute exposures)

ATLAS CHART 9. M31 (ANDROMEDA GALAXY) M31 (NGC 224), the Great Galaxy in Andromeda (fig. 7-9), lies in the center of the chart, near 0^h45^m +41°. At a distance of 2.2 million light-years from earth, M31 is the largest neighboring galaxy. It is visible to the naked eye as a hazy glow. It appears in telescopes as a yellowish oval glow, brighter toward the center; the dust lane below its center may be seen or photographed. If you slowly sweep the field with medium power you will see more detail in outer regions of this spiral galaxy. M31 has two faint elliptical companions, M32 (NGC 221) and NGC 205, sometimes known as M110. M32 usually appears as a bright roundish haze, while NGC 205 looks a bit larger though not as bright.

At left in Triangulum is the face-on spiral galaxy M33 (NGC 598), the brightest spiral in the northern sky (figs. 2-17 and 5-39) except for M31. M33 covers an area about the same size as the moon but can be difficult to find.

3C 48, one of the first quasars discovered, lies at about 1^h40^m +34°, above M33. It has a magnitude of 16.2 and a redshift of 0.367 (36.7 percent), making it about 7 billion light-years away.

M76 (NGC 650), a prominent planetary nebula, lies near the boundary of Perseus and Andromeda. M76 is difficult to find, though it is easiest to observe in early fall.

Some noteworthy double stars in this area of the sky are: ι (iota) Tri (yellow and blue) at left center; 1 Ari (ADS1 457, white and green), near β (beta) Ari at 1^h50^m +22°; and γ (gamma) Ari (both components bluish white). γ (gamma) And at 2^h +43° is one of the most beautiful doubles in the sky, with orange and green.

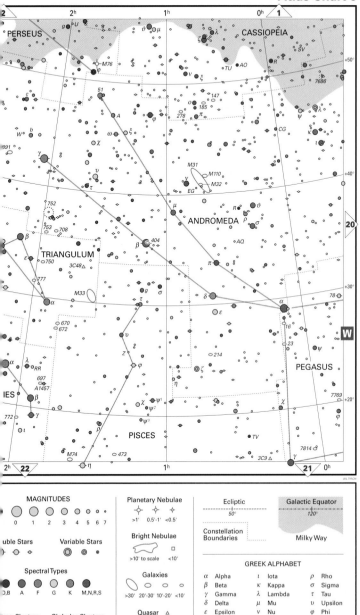

PERSEUS

CASSIOPEIA

ANDROMEDA

TRIANGULUM

PISCES

PEGASUS

M31
M110
M32
EG

M33

M76

M74

MAGNITUDES							
0	1	2	3	4	5	6	7

Double Stars

Variable Stars

Spectral Types

O,B   A   F   G   K   M,N,R,S

Open Clusters

Globular Clusters

to scale   <10'   >10'   5'-10'   <5'

Planetary Nebulae		
>1'	0.5'-1'	<0.5'

Bright Nebulae

>10' to scale   <10'

Galaxies

>30' 20'-30' 10'-20' <10'

Quasar △

Pulsar ☿

Black Hole

Ecliptic
- - - - - -
50°

Constellation
Boundaries

Galactic Equator
· · · · · ·
120°

Milky Way

GREEK ALPHABET

α	Alpha	ι	Iota	ρ	Rho
β	Beta	κ	Kappa	σ	Sigma
γ	Gamma	λ	Lambda	τ	Tau
δ	Delta	μ	Mu	υ	Upsilon
ε	Epsilon	ν	Nu	φ	Phi
ζ	Zeta	ξ	Xi	χ	Chi
η	Eta	o	Omicron	ψ	Psi
ϑ	Theta	π	Pi	ω	Omega

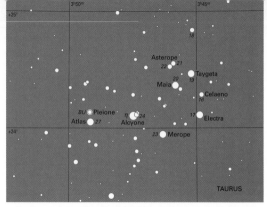

Fig. 7-10 A chart of the Pleiades. (Wil Tirion)

ATLAS CHART 10. THE PLEIADES, ALGOL Algol, β (beta) Per, is sometime called the Demon Star (chapter 6). It is the most famous of the eclipsing binaries and the easiest of these variable stars to observe. The eclipses occur approximately every 69 hours, dropping the total brightness of the system from about magnitude 2.2 to minimum of magnitude 3.4 within a five-hour period; the system returns to its normal brightness after about 20 minutes. The eclipses are fun to observe with the naked eye (see fig. 6-1 and appendix 8).

M34 (NGC 1039) is an open cluster in Perseus, about halfway between the two bright stars Algol (β Per) and γ (gamma) And. It appears to the naked eye as a hazy object. About 80 stars are loosely grouped in M34, and many of these are blue-white doubles. M34 is beautiful in binoculars and small telescopes with wide fields.

M45, the Pleiades, one of the most spectacular and most obvious open clusters in the sky, is in Taurus. Six or seven of its stars, arranged roughly in the shape of a little dipper, are visible to the naked eye. Binoculars clearly show numerous faint stars (see fig. 7-10 above), and even a small telescope shows more than 100 stars. Wisps of nebulosity, reflection nebulae, cover the group and are especially bright near the more luminous stars (see fig. 5-18) The nebulous background shows clearly on long-exposure photographs, but you will probably not be able to see it visually.

North of the Pleiades, in Perseus, are NGC 1499 — the California Nebula (fig. 5-33) — a large faint nebula even more difficult to find than M76; and IC 348, a bright diffuse nebula. NGC 1220, a small open cluster that appears in binoculars as a hazy patch, is at the top center of the chart, near γ (gamma) Per. NGC 1528, a large open cluster visible to the naked eye on clear nights is at the top left of the chart.

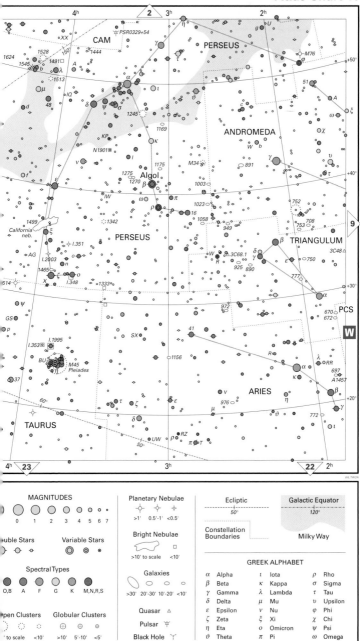

MAGNITUDES							
0	1	2	3	4	5	6	7

Double Stars

Variable Stars

Spectral Types

O,B A F G K M,N,R,S

Open Clusters

Globular Clusters

to scale <10' >10' 5'-10' <5'

Planetary Nebulae

>1' 0.5'-1' <0.5'

Bright Nebulae

>10' to scale <10'

Galaxies

>30' 20'-30' 10'-20' <10'

Quasar △

Pulsar ☆

Black Hole

Ecliptic

50'

Constellation Boundaries

Galactic Equator

120°

Milky Way

GREEK ALPHABET					
α	Alpha	ι	Iota	ρ	Rho
β	Beta	κ	Kappa	σ	Sigma
γ	Gamma	λ	Lambda	τ	Tau
δ	Delta	μ	Mu	υ	Upsilon
ε	Epsilon	ν	Nu	φ	Phi
ζ	Zeta	ξ	Xi	χ	Chi
η	Eta	ο	Omicron	ψ	Psi
ϑ	Theta	π	Pi	ω	Omega

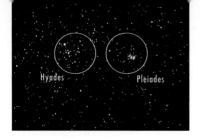

Fig. 7-11 *The Hyades and the Pleiades. Compare them with their Atlas Chart mapping at lower right of the facing page.* (Richard E. Hill)

ATLAS CHART 11. CRAB NEBULA, HYADES, ALDEBARAN With binoculars or a rich-field telescope, sweep along the Milky Way through Perseus, Auriga, and Gemini. Many rich star fields, including numerous nebulae and open clusters, will reward your search. M36, M37 and M38 are all open clusters in Auriga. In the same region of Auriga, several gaseous nebulae of interest can be seen. IC 405 is sometimes called the Flaming Star; IC 410 is a cluster with surrounding nebulosity.

The most famous object on the chart is M1 (NGC 1952), the Crab Nebula in Taurus. The Crab is the remnant of a supernova, a star whose explosion the Chinese recorded in A.D. 1054. The Crab is still rapidly expanding, with gas traveling outward at about 1,000 km per second. Small telescopes disappointingly show only an oval nebulosity; increasing the magnification reveals more shape. Only with a large aperture might you detect some of the filamentary structure that appears in photographs (fig. 5-12).

A spectacular open cluster, the Hyades, outlines the bull's face in Taurus. It appears at the bottom edge of the chart, near 4^h30^m +15°. The Hyades is an older and looser cluster than the Pleiades and contains fewer stars. Use low-power or good binoculars to see this extended V-shaped group; it is so large that it is not circled on the chart. The stars in this cluster are approximately 150 light years away and formed about a billion years ago. Their white color contrasts with the red of the star Aldebaran, α (alpha) Tau. Aldebaran is less than half as far from us as the Hyades stars.

Northeast of the Hyades, near 4^h45^m +18°, is the open cluster NGC 1647; it includes about 50 stars spread across an area about the size of the moon's diameter.

In Auriga, at about 5^h10^m +45°, is Capella, α (alpha) Aur, the sixth brightest star in the sky. Near Capella, ε (epsilon) Aur is an eclipsing variable with a period of 27 years, one of the longest periods known. Below ε Aur is ζ (zeta) Aur, another eclipsing variable with a fairly long period of $2\frac{2}{3}$ years.

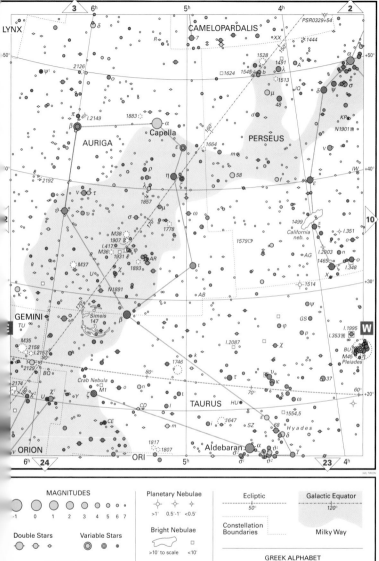

MAGNITUDES

-1 · 0 · 1 · 2 · 3 · 4 · 5 · 6 · 7

Double Stars Variable Stars

Spectral Types

O,B · A · F · G · K · M,N,R,S

Open Clusters Globular Clusters

>10' to scale · <10'   >10' · 5'-10' · <5'

Planetary Nebulae

>1' · 0.5'-1' · <0.5'

Bright Nebulae

>10' to scale · <10'

Galaxies

>30' · 20'-30' · 10'-20' · <10'

Quasar △

Pulsar ☿

Black Hole ⅄

Ecliptic
_ _ _ _
50°

Constellation
Boundaries

Galactic Equator
······
120°

Milky Way

GREEK ALPHABET

α	Alpha	ι	Iota	ρ	Rho
β	Beta	κ	Kappa	σ	Sigma
γ	Gamma	λ	Lambda	τ	Tau
δ	Delta	μ	Mu	υ	Upsilon
ε	Epsilon	ν	Nu	φ	Phi
ζ	Zeta	ξ	Xi	χ	Chi
η	Eta	ο	Omicron	ψ	Psi
ϑ	Theta	π	Pi	ω	Omega

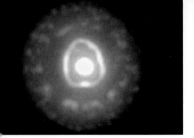

Fig. 7-12 The Eskimo Nebula
(NGC 2392), a planetary nebula.
The image has North up; look at it
upside down. (Edward A. Grafton)

ATLAS CHART 12. CASTOR AND POLLUX Gemini is marked by the heavenly twins, Castor and Pollux. The fainter of the two, Castor, α (alpha) Gem, is one of the most beautiful double stars easily resolved in small telescopes. The two stars, of magnitudes 1.9 and 2.9, slowly revolve around each other with a period of 510 years; a third, fainter star called Castor C is also present. Castor is actually a sextuple system, since each of its three components has been found by spectroscopy to be double. Castor C is not only a spectroscopic double but also an eclipsing binary and is known as YY Gem. Pollux, β (beta) Gem, is brighter than Castor by about ½ magnitude and is more yellowish.

Southwest of Pollux is δ (delta) Gem, a bright double star. Nearby is NGC 2392, the Eskimo or Clown Face Nebula (fig. 7-12). The distinctive greenish "face" of this planetary nebula is visible only in large telescopes. ζ (zeta) Gem, to the right of the Eskimo Nebula, near 7^h +20°, is one of the brightest Cepheid variables. It changes in brightness by more than ½ magnitude during its 10-day period.

Toward the other side of the chart, near 6^h9^m +22°, is η (eta) Gem, a variable red giant. Slightly above η Gem is M35 (NGC 2168), a beautiful open cluster, and its companion NGC 2158. M35 is about 40 arc min in diameter and contains about 200 stars. It is sometimes visible to the naked eye as a hazy, faint patch; any telescope or even binoculars will reveal a fine open cluster. It can be seen best with an eyepiece with at least 1° field. High power tends to reduce the beauty of M35 but is fine for observing its companion about ½° to the southwest, NGC 2158. NCG 2158 is a very rich open cluster; in its center, about 40 stars are crowded into an area only 4 arc sec across, making it look almost like a globular cluster. Although M35 and NGC 2158 appear close to each other, they are completely unrelated physically: M35 is only 2,200 light-years away, while NGC 2158 is 16,000 light-years away, near the edge of our galaxy.

NGC 2419 in Lynx, about 7° north of Castor, is so far from the center of our galaxy that it has been called an "intergalactic tramp." On a clear dark night, you can see it in a small telescope.

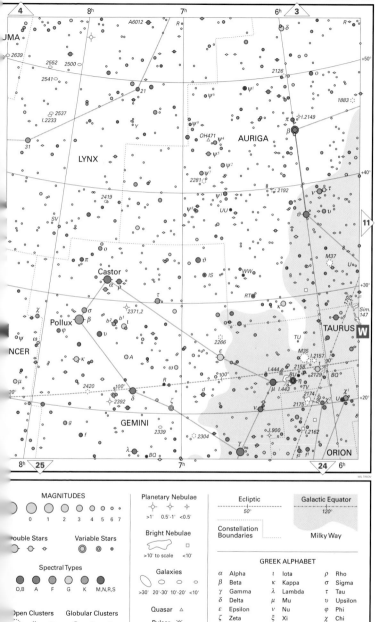

MAGNITUDES							
0	1	2	3	4	5	6	7

Double Stars   **Variable Stars**

Spectral Types

O,B   A   F   G   K   M,N,R,S

Open Clusters   **Globular Clusters**

0' to scale   <10'   >10'   5'-10'   <5'

Planetary Nebulae

>1'   0.5'-1'   <0.5'

Bright Nebulae

>10' to scale   <10'

Galaxies

>30'   20'-30'   10'-20'   <10'

Quasar △

Pulsar ☆

Black Hole

Ecliptic
- - - - - - -
50°

Constellation Boundaries

Galactic Equator
120°

Milky Way

GREEK ALPHABET

α	Alpha	ι	Iota	ρ	Rho
β	Beta	κ	Kappa	σ	Sigma
γ	Gamma	λ	Lambda	τ	Tau
δ	Delta	μ	Mu	υ	Upsilon
ε	Epsilon	ν	Nu	φ	Phi
ζ	Zeta	ξ	Xi	χ	Chi
η	Eta	ο	Omicron	ψ	Psi
ϑ	Theta	π	Pi	ω	Omega

Fig. 7-13 M44 (NGC 2632), also known as Praesepe, the Beehive Cluster. This guided 25-minute exposure was taken with a 200 mm f/4 lens. (Chris Cook)

ATLAS CHART 13. PRAESEPE The four stars, γ, η, θ, and δ (gamma, eta, theta, and delta) Cancri (Cnc) at lower right outline the irregular shape that forms the body of Cancer, the Crab. Within the crab's body lies the open cluster M44 (NGC 2632), also called Praesepe or the Beehive Cluster (fig. 7-13). M44 covers more than 80 arc min—an area almost three times the diameter of the moon—and is one of the most spectacular clusters in the sky. Although M44 is visible to the naked eye only as a hazy patch, a small telescope or binoculars will resolve it into its component stars, many of which are multiples. It is about 520 light-years from us. Praesepe means "the manger"; γ (gamma) and δ (delta) Cnc, which are nearby, are known as "the asses," from a story in Greek mythology.

To the west of M44 is ζ (zeta) Cnc (at about 8^h55^m +18°), an unusual triple star with three yellow components. This system can be viewed in medium-sized or large telescopes. ι^1 (iota¹) Cnc (at about 8^h47^m +29°) is an interesting orange-green double, outstanding in even a small telescope because of the beautiful color contrast between its components. Located slightly to the northeast, ι^2 (iota²) Cnc is a triple.

R Leo Minoris is a long-period variable, ranging in brightness from magnitude 6.3 to magnitude 13.2 with a period of 372 days. Its deep red color, especially when near maximum, makes this star an interesting telescopic object.

A noteworthy galaxy in Lynx is NGC 2683 at 8^h55^m +33°, a galaxy seen almost edge-on. Near the boundary between Lynx (Lyn) and Leo Minor, at 9^h20^m + 35°, are several other galaxies, easy for beginners to find by sweeping near α (alpha) Lyn. The brightest of these galaxies is NGC 2859, a barred spiral in Leo Minor. NGC 2903 is an isolated spiral in Leo.

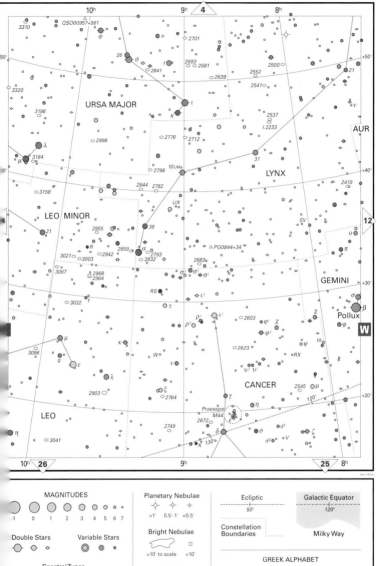

URSA MAJOR

LEO MINOR

LYNX

AUR

GEMINI

Pollux

W

CANCER

LEO

Praesepe
M44

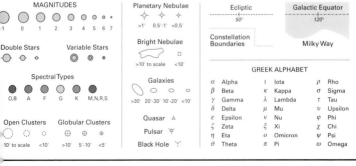

MAGNITUDES								
-1	0	1	2	3	4	5	6	7

Double Stars

Variable Stars

Spectral Types

O,B A F G K M,N,R,S

Open Clusters

10' to scale   <10'

Globular Clusters

⊕ ⊕ ⊕
>10' 5'-10' <5'

Planetary Nebulae

✧ ✧ ✧
>1' 0.5'-1' <0.5'

Bright Nebulae

>10' to scale   <10'

Galaxies

>30' 20'-30' 10'-20' <10'

Quasar △

Pulsar ⚲

Black Hole ⚯

Ecliptic

50'

Constellation Boundaries

Galactic Equator

120°

Milky Way

GREEK ALPHABET

α	Alpha	ι	Iota	ρ	Rho
β	Beta	κ	Kappa	σ	Sigma
γ	Gamma	λ	Lambda	τ	Tau
δ	Delta	μ	Mu	υ	Upsilon
ε	Epsilon	ν	Nu	φ	Phi
ζ	Zeta	ξ	Xi	χ	Chi
η	Eta	o	Omicron	ψ	Psi
ϑ	Theta	π	Pi	ω	Omega

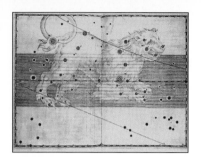

Fig. 7-14 *Leo, the Lion, from the star atlas of Bayer (1603). The backward question-mark shape of its brightest stars are at lower right of the facing page.*
(Jay M. Pasachoff)

ATLAS CHART 14. LEO Leo Minor and Leo contain a number of interesting doubles. One of the most striking is γ (gamma) Leo, one of the stars in the sickle that makes up Leo, actually a double with magnitudes 2.2 and 3.5. This golden yellow pair is easily resolved and provides a good view even in small telescopes. ζ (zeta) Leo, located in the sickle above γ Leo, is a bright pair of stars far enough apart to be detected with binoculars.

Nearby, at about 11^h +25°, is another double star, 54 Leo, that can be resolved with very low power into a pair with magnitudes 4.5 and 6.3. The star Denebola—β (beta) Leo—at the tip of the lion's tail and just below the bottom of the chart at left, is a beautiful blue-orange pair. However, it is merely an optical double—the stars are actually far apart in space. The pulsar CP 1133 nearby was one of the original four pulsars discovered by Jocelyn Bell (now Jocelyn Bell Burnell) and Antony Hewish in 1968.

To the lower left of the chart's center is the double star ξ (xi) Ursae Majoris; it is only 26 light-years away and is a good object to observe with small telescopes. The 60-year elliptical orbit of the two visual components, magnitudes 4.3 and 4.8, is tilted 57° to our line of sight. The pair reached their maximum separation of 2.9 arc min as seen from earth in 1980. Each of these two components is in turn a double. The stars in one of the pairs are separated by only 0.5 A.U. (Astronomical Unit)—half the distance between the earth and the sun—and orbit each other in 1.8 years. The stars in the other pair are separated by only the distance from the earth to the moon and orbit each other every four days.

Since this chart includes an area close to the north galactic pole (which is on the next chart), we are looking as far away as possible from the Milky Way and can thus see many galaxies (fig. 7-14). Scanning the nearby region of Leo Minor reveals a hard-to-see group of galaxies. NGC 3158, at 10^h20^m +38°, is the brightest of this group and should be visible in a medium-sized telescope.

We will discuss the cluster of galaxies at upper left with the next chart. M97 and M108 are discussed with Atlas Chart 5.

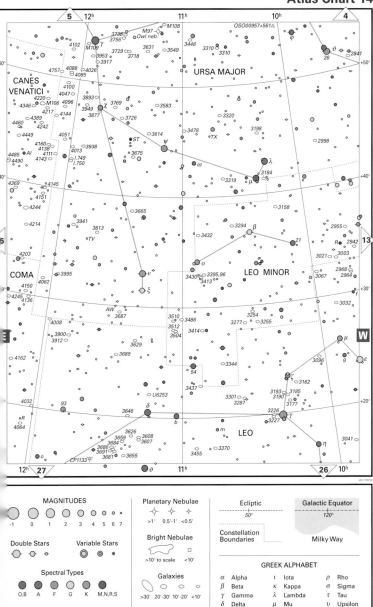

MAGNITUDES
-1 0 1 2 3 4 5 6 7

Double Stars

Variable Stars

Spectral Types
O,B   A   F   G   K   M,N,R,S

Open Clusters
>10' to scale   <10'

Globular Clusters
>10'   5'-10'   <5'

Planetary Nebulae
>1'   0.5'-1'   <0.5'

Bright Nebulae
>10' to scale   <10'

Galaxies
>30'   20'-30'   10'-20'   <10'

Quasar △

Pulsar ⚡

Black Hole Y

Ecliptic
50°

Constellation Boundaries

Galactic Equator
120°

Milky Way

GREEK ALPHABET					
α	Alpha	ι	Iota	ρ	Rho
β	Beta	κ	Kappa	σ	Sigma
γ	Gamma	λ	Lambda	τ	Tau
δ	Delta	μ	Mu	υ	Upsilon
ε	Epsilon	ν	Nu	φ	Phi
ζ	Zeta	ξ	Xi	χ	Chi
η	Eta	ο	Omicron	ψ	Psi
ϑ	Theta	π	Pi	ω	Omega

Fig. 7-15 *The Whirlpool Galaxy, M51 (NGC 5194), a type Sb spiral, with the irregular companion galaxy NGC 5195 at the end of one of its spiral arms. (CCD image on a 24-inch f/5 telescope by Tim Hunter and James McGaha)*

ATLAS CHART 15. WHIRLPOOL GALAXY, BLACK-EYE GALAXY Characteristic of this part of the sky is the large number of galaxies, most of them millions of light-years distant. Relatively close to us is one of the brightest galaxies, M51 (NGC 5194)—also known as the Whirlpool Galaxy (fig. 5-40)—at upper left-center in Canes Venatici. Located a few degrees south of the handle of the Big Dipper, M51 is a face-on spiral about 15 million light-years distant, with a companion galaxy, NGC 5195. Other galaxies nearby to the south and west include M63 (NGC 5055), M94 (NGC 4736), and M106 (NGC 4258), all spirals of type Sb that can be observed with small telescopes. Far beyond are the clusters of galaxies in Ursa Major and Canes Venatici.

At 12^h52^m +27° in Coma Berenices, we come to the region around the north galactic pole. Since we are looking away from the Milky Way at this point, there is minimum interference from gas and dust in our galaxy, and many galaxies are visible here. M64 (NGC 4826), at lower center in Coma Berenices, is a spiral known as the Black-eye Galaxy because the dark dust lane across its center is so prominent. The galaxy is relatively bright and can be detected in binoculars. Also located here are the spiral galaxy M100 (NGC 4321)—shown in Fig. 5-50—and the elliptical galaxy M85 (NGC 4382). Ten degrees to the north are NGC 4559 and NGC 4565, an edge-on spiral tilted only 4° from our line of sight. These galaxies are all in the Virgo Cluster of galaxies (see Atlas Charts 27 and 28). Much farther away is the Coma Cluster of galaxies, in which only NGC 4889 and NGC 4874, near 13^h + 28°, are bright enough to be seen by most amateur observers.

In this area of the sky we also find several globular clusters. At magnitude 6.4, the globular cluster M3 (NGC 5272)—located at 13^h40^m +28° in Canes Venatici—is one of the brightest in the sky (fig. 5-23). Two other globular clusters, M53 (NGC 5024) and NGC 5053, appear lower in the sky in Coma Berenices. M53, near α (alpha) Com, is twice as far away as M3 and therefore twice as large, since both star clusters have roughly the same apparent size in the sky.

Mel 111 (at 12^h20^m +27° in Coma Berenices) is a large diffuse open cluster of 5th- to 10th-magnitude stars.

This part of the sky is best placed for viewing in the spring.

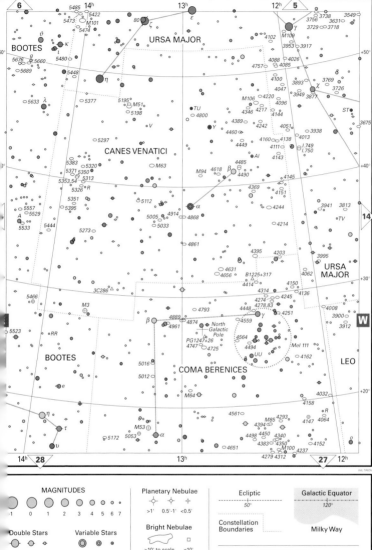

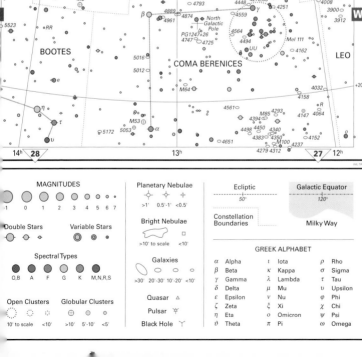

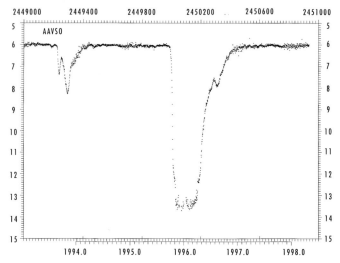

Fig. 7-16 Light curve for R Coronae Borealis. (AAVSO)

ATLAS CHART 16. ARCTURUS At lower right is the third brightest star in the sky, Arcturus, α (alpha) Boötes. It is 37 light-years away and about 25 times the diameter of the sun. ξ (xi) Boo (at 14^h51^m +19°) is a double star with widely spaced components that orbit each other with a period of 152 years. Farther to the northeast, μ (mu) Boo (at 15^h25^m +37°) is a widely spaced white-orange pair with fainter companions.

To the south, in Corona Borealis (the Northern Crown), η (eta) CrB is a close double, with components orbiting each other in about 42 years. ζ (zeta) CrB is a wide pair of blue stars, separated from each other by more than 6 arc sec. σ (sigma) CrB is also a wide pair. γ (gamma) CrB is a very close pair, orbiting in about 100 years. o (omicron) CrB has several faint companions and a period of 215 years.

R Coronae Borealis (R CrB) is one of the most unusual stars in the sky. It stays at 6th magnitude for a period ranging from months to years and then precipitously drops to 11th magnitude or even fainter for a period that lasts only weeks (see light curve above). At the same time, its spectrum of absorption lines is obscured and a spectrum of emission lines appears. R CrB is a carbon-rich star, and apparently it sometimes throws off clouds of carbon soot that obscure its photosphere, making spectral lines from its higher and relatively hotter chromosphere appear as emission lines. The clouds of soot have been detected in the infrared. A finding chart for R CrB appears at the end of chapter 6.

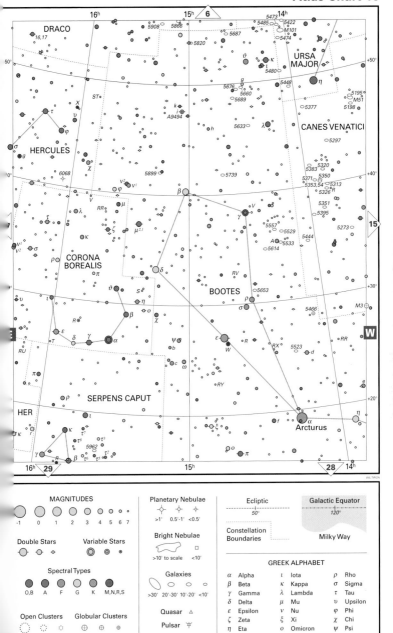

MAGNITUDES								
-1	0	1	2	3	4	5	6	7

Double Stars

Variable Stars

Spectral Types
O,B A F G K M,N,R,S

Open Clusters
>10' to scale <10'

Globular Clusters
>10' 5'-10' <5'

Planetary Nebulae
>1' 0.5'-1' <0.5'

Bright Nebulae
>10' to scale <10'

Galaxies
>30' 20'-30' 10'-20' <10'

Quasar △

Pulsar ☼

Black Hole Υ

Ecliptic
50'

Constellation Boundaries

Galactic Equator
120°

Milky Way

GREEK ALPHABET

α	Alpha	ι	Iota	ρ	Rho		
β	Beta	κ	Kappa	σ	Sigma		
γ	Gamma	λ	Lambda	τ	Tau		
δ	Delta	μ	Mu	υ	Upsilon		
ε	Epsilon	ν	Nu	φ	Phi		
ζ	Zeta	ξ	Xi	χ	Chi		
η	Eta	ο	Omicron	ψ	Psi		
ϑ	Theta	π	Pi	ω	Omega		

Fig. 7-17 The globular cluster M13 in Hercules looks like a hazy moth-ball in the sky near the center of this wide-field view. (George East)

ATLAS CHART 17. M13 At the center of the chart in Hercules we find the Keystone, an asterism formed by the stars η, ζ, ε, and π (eta, zeta, epsilon, and pi) Her. About one-third of the way from η (eta) to ζ (zeta) Her is the spectacular globular cluster M13 (NGC 6205), shown in figure 2-6. M13 lies about 25,000 light-years away. It is faintly visible to the naked eye as a fuzzy mothball of about 6th magnitude. You can resolve the outer edges of M13 into its member stars with a medium-sized telescope.

About ½° northeast of M13 is the faint spiral galaxy NGC 6207, sometimes visible in the same field at low power.

At 17^{h}15^m +42°, about 7° north of π (pi) Her, is the globular cluster M92 (NGC 6341), about ½ magnitude fainter than M13 but still a good object for small telescopes. A third globular cluster, NGC 6229 (at 16^{h}45^m +47°), lies about 10° above M13; it is much smaller and fainter (9th magnitude) than the other two clusters.

By sweeping with low power near β (beta) Her at the bottom of the chart, you can find NGC 6210, a bluish green planetary nebula. NGC 6210 has a fairly bright inner ring with a faint outer ring that measures 20 by 43 arc sec. The nebula can be detected with a small telescope, although a larger telescope is needed to distinguish the rings. The planetary nebula NGC 6058 (at 16^h +41°) is more difficult to locate.

δ (delta) Her is an optical double for which many different colors have been reported. ρ (rho) Her is a pair of blue-white stars that differ in brightness by 1 magnitude. Only 30 light-years away from us is ζ (zeta) Her, a pair with magnitudes 2.9 and 5.5. All lie on the constellation outline drawn.

The quasar 3C 345 is 16th magnitude and has a redshift of 0.6 (60 percent), placing it about 8 billion light-years away.

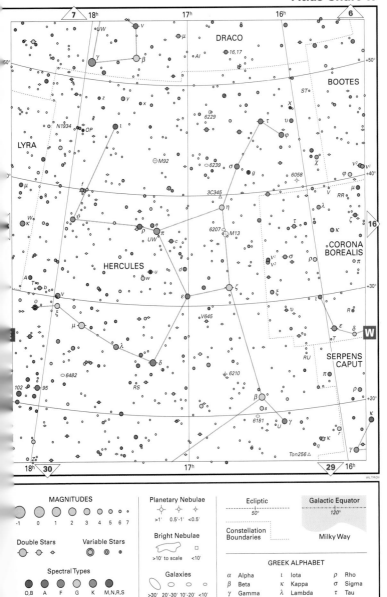

MAGNITUDES

○ ○ ○ ○ ○ ○ ○ ○
-1 0 1 2 3 4 5 6 7

Double Stars

Variable Stars

Spectral Types

O,B   A   F   G   K   M,N,R,S

Open Clusters   **Globular Clusters**

>10' to scale   <10'   >10' 5'-10' <5'

Planetary Nebulae

✧   ✧   ✧
>1'   0.5'-1'   <0.5'

Bright Nebulae

>10' to scale   <10'

Galaxies

>30' 20'-30' 10'-20' <10'

Quasar   △
Pulsar   ☆
Black Hole   ⅄

Ecliptic
- - - - - - -
50°

Constellation
Boundaries
· · · · · · ·

Galactic Equator

120°

Milky Way

GREEK ALPHABET

α	Alpha	ι	Iota	ρ	Rho
β	Beta	κ	Kappa	σ	Sigma
γ	Gamma	λ	Lambda	τ	Tau
δ	Delta	μ	Mu	υ	Upsilon
ε	Epsilon	ν	Nu	φ	Phi
ζ	Zeta	ξ	Xi	χ	Chi
η	Eta	ο	Omicron	ψ	Psi
ϑ	Theta	π	Pi	ω	Omega

ATLAS CHART 18. RING NEBULA, DUMBBELL NEBULA The brightest star on this chart is Vega, α (alpha) Lyrae. Vega is a pure white star of magnitude 0 and is dazzlingly beautiful in a telescope. With ε (epsilon) and ζ (zeta) Lyr, Vega forms a small equilateral triangle. A person with acute eyesight can barely detect the two components of ε (epsilon) Lyr, each of which is about 5th magnitude. Even a small telescope separates the pair and splits each component into a second pair, making four white stars in all. It is "the double double."

β (beta) Lyr, a spectroscopic double and an eclipsing binary, varies between magnitudes 3.3 and 4.3 every 12.9 days. The variations are easy to follow by comparing the magnitude of this double with that of γ (gamma) Lyr, magnitude 3.2 (see finding chart for Algol, end of chapter 6). Other nearby stars in Lyra to compare are η (eta) at magnitude 4.4, θ (theta) at 4.4, ζ¹ (zeta¹) at 4.4, κ (kappa) at 4.3, λ (lambda) at 4.9, and ι (iota) at 5.3.

Albireo—β (beta) Cygni—at the nose of the Swan, is one of the prettiest doubles. Binoculars split it into yellow and green components. The components of δ (delta) Cygni, whose orbital period is 830 years, differ by 3 magnitudes.

RR Lyrae (at $19^h25^m +43°$) is the prototype of a class of variable stars with regular periods of less than one day (chapter 6).

β, γ, δ, and ζ Lyr form a parallelogram. About halfway between β and γ Lyr, on a short side of this parallelogram, is M57 (NGC 6720), the Ring Nebula. This most famous of the planetary nebulae has a shell of gas expanding from a central star. The shell represents the outer quarter of the star's mass and bares the core of the star, which appears bluish because it is so hot. This is the end that our sun will reach one day, in about five billion years. A small telescope shows the nebula clearly (figs. 5-3 and 7-19), although a larger one is necessary to reveal structural details. The best view has been taken from space (fig. 5-4). The colors show up only in long-exposure photographs.

To the south and east of M57, near $19^h20^m +30°$, is the globular cluster M56 (NGC 6779). This small highly concentrated cluster is easily visible with small telescopes. Farther south and east in Vulpecula, near $20^h +23°$, is M27 (NGC 6853), the second-largest planetary nebula, about three light-years in diameter. Because of its shape, M27 is commonly called the Dumbbell Nebula (figs. 5-5 and 7-18). Though its total brightness is great, M27 is spread over such a large area that the average brightness of its surface remains fairly low. Low power shows its shape and greenish color; higher power begins to show some structure. To the west are the pulsars PSR 1937 + 215 and CP 1919, which can be detected only in the radio spectrum. PSR 1937 + 215 gives off more than 600 pulses of radio waves per second.

Fig. 7-18 (above) *An amateur photo. M27, the Dumbbell Nebula (Atlas Chart 18), a planetary nebula. (Richard E. Hill) (below) An image with one of the four huge 8.2 m telescopes of the Very Large Telescope (VLT) in Chile (European Southern Observatory)*

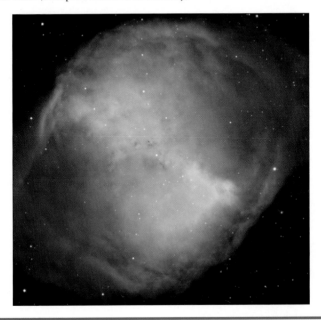

Fig. 7-19 M57, the Ring Nebula (Atlas Chart 18), a planetary nebula. It looks like a hazy white smoke ring to the eye when looking through a telescope. (Richard E. Hill)

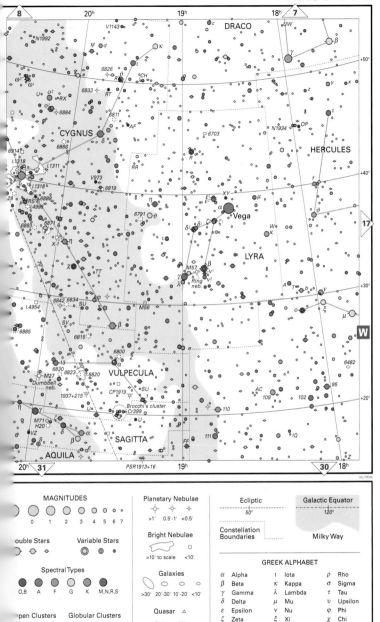

MAGNITUDES

◯ ◯ ◯ ◯ ○ ○ ○ ∘ ·
0   1   2   3   4   5   6   7

Double Stars   **Variable Stars**

Spectral Types

O,B   A   F   G   K   M,N,R,S

Open Clusters   **Globular Clusters**

' to scale   <10'   >10'   5'-10'   <5'

Planetary Nebulae
-◇-   ◇   ◇
>1'   0.5'-1'   <0.5'

Bright Nebulae
>10' to scale   □ <10'

Galaxies
◯   ◯   ○   ∘
>30'   20'-30'   10'-20'   <10'

Quasar △

Pulsar ☆

Black Hole Υ

Ecliptic		Galactic Equator
------ 50' ------		----- 120° -----
Constellation Boundaries	....	Milky Way

GREEK ALPHABET

α	Alpha	ι	Iota	ρ	Rho
β	Beta	κ	Kappa	σ	Sigma
γ	Gamma	λ	Lambda	τ	Tau
δ	Delta	μ	Mu	υ	Upsilon
ε	Epsilon	ν	Nu	φ	Phi
ζ	Zeta	ξ	Xi	χ	Chi
η	Eta	ο	Omicron	ψ	Psi
ϑ	Theta	π	Pi	ω	Omega

In the center of the chart near the bright star Deneb, α (alpha) Cyg, is the North America Nebula (NGC 7000), named because of its shape. Although this nebula and its neighbor, the Pelican Nebula (IC 5067-5070), are extremely hard to observe telescopically, long-exposure photography and electronic imaging make them visible. The Pelican's bill is facing the North America Nebula (fig. 7-20). Another region of nebulosity lies around γ (gamma) Cyg, southwest of Deneb, near 20^{h}20^m +40°.

On the 20^h meridian is Cygnus X-1, invisible in ordinary light but the brightest x-ray source in Cygnus. The x-rays probably come from gas swirling around a black hole; the presence of the black hole is inferred from its gravitational effect on the 9th-magnitude star HDE 226868 that is observed at that location.

Southeast of this region, near 21^h +30°, is another area of diffuse nebulae. NGC 6960 appears close to the double star 52 Cygni, whose orange and blue components are separated by 6½ arc sec. NGC 6960, 6992, 6995, and 6979 are jointly known as the Veil Nebula, which makes up most of the Cygnus Loop, a faint circular nebula about 2½° across. The Cygnus Loop is the remnant of a supernova that exploded more than 100,000 years ago. The Veil Nebula is visible through a wide range of optical instruments, from binoculars to large telescopes, but its structure and beautiful colors show only in long-exposure photographs (figs. 5-13, 5-14, and 7-21).

Within the triangle formed by α, γ, and ε (alpha, gamma, and epsilon) Cyg is the Northern Coalsack, an opaque patch of nebulosity that obscures the Milky Way behind it.

Near the upper right of the chart, at 19^{h}40^m +50°, is the planetary nebula NGC 6826. This nebula is often called the Blinking Nebula because its central star, which is relatively bright, appears to blink on and off if you look rapidly toward and away from it.

M39 (NGC 7092) is an open cluster in northern Cygnus (near 21^{h}30^m +48°), with about 30 stars scattered across its seven light-year diameter. This cluster is a fine object to view with binoculars or low-power, wide-field telescopes.

Near 21^{h}40^m +43° is the variable star SS Cygni, which rises sharply to its maximum about once every 50 days, on a somewhat irregular schedule (fig. 6-5). It is very interesting to follow as it rapidly increases in brightness from 12th to 8th magnitude. You can see its brightness change markedly in two or three hours during certain parts of its cycle. SS Cyg is too faint to be shown on this chart, but its position is marked on a finding chart in chapter 6.

To the southwest, near 21^{h}20^m +39°, is 61 Cyg, the first star whose distance from the sun was measured (from its parallax). 61 Cyg is an orange pair of stars only 11.4 light-years away from us.

Fig. 7-20 (above) *The North America Nebula (NGC 7000) with the Pelican Nebula (IC 5067-5070) facing it. (Alan Dyer)* (below) *Cygnus, the Swan, from Bayer's Atlas (1603). (Jay M. Pasachoff)*

Fig. 7-21 NGC 6992 (top part) and 6995 (bottom part, unlabeled in the Atlas Chart), the eastern part of the Veil Nebula, which makes up the Cygnus Loop, the remnant of an ancient supernova explosion. (Alan Dyer)

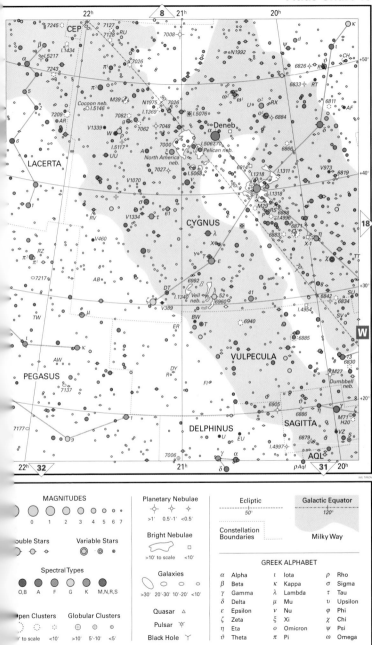

MAGNITUDES
0 1 2 3 4 5 6 7

Double Stars

Variable Stars

Spectral Types

O,B   A   F   G   K   M,N,R,S

Open Clusters   **Globular Clusters**

' to scale   <10'   >10'   5'-10'   <5'

Planetary Nebulae
>1'   0.5'-1'   <0.5'

Bright Nebulae

>10' to scale   <10'

Galaxies

>30'   20'-30'   10'-20'   <10'

Quasar   △

Pulsar   ☆

Black Hole   ⅄

Ecliptic
- - - - -
50'

Galactic Equator
~~~~~~~~
120'

Constellation
Boundaries

Milky Way

GREEK ALPHABET

α	Alpha	ι	Iota	ρ	Rho
β	Beta	κ	Kappa	σ	Sigma
γ	Gamma	λ	Lambda	τ	Tau
δ	Delta	μ	Mu	υ	Upsilon
ε	Epsilon	ν	Nu	φ	Phi
ζ	Zeta	ξ	Xi	χ	Chi
η	Eta	ο	Omicron	ψ	Psi
ϑ	Theta	π	Pi	ω	Omega

Fig. 7-22 NGC 7331, a type Sb spiral galaxy in Pegasus. The dust lane surrounding its disk shows clearly. Three other spirals are visible beyond it. (Tim Puckett)

ATLAS CHART 20. PEGASUS As we move away from the Milky Way into Lacerta, Andromeda, and Pegasus, the star fields thin out. The bright star β (beta) Peg—below the center of the chart, near 23^h $+27°$—marks the northwest corner of the Great Square of Pegasus. β Peg, also called Scheat, is a red-giant star that varies with an irregular period from magnitude 2.3 to 2.7. One corner of the square is not in Pegasus; α (alpha) Andromedae (And) marks the northeast corner of the Square. About halfway between these two stars is the double star 78 Peg, a close pair with magnitudes 5 and 8. The third star in the Square, α (alpha) Peg, is drawn just outside the chart boundary at the bottom center; the fourth star does not show on this chart.

About 5° to the northwest of β Peg is η (eta) Peg. Follow the line from β to η Peg about 7° farther to a pair of double stars, $π^1$ (pi^1) and $π^2$. $π^2$ is an optical double—two yellow stars at different distances from earth that are apparently separated in the sky by only 15 minutes of arc. The pair can be seen well in binoculars.

Just to the right of the center of the chart, about 5° north and slightly west of η Peg, is the spiral galaxy NGC 7331 (fig. 7-22). This galaxy can be found by sweeping the area with low power.

The planetary nebula NGC 7662 lies near 23^h26^m $+42.5°$ in Andromeda, about 3° southwest of ι (iota) And. One of the best planetaries for observing, NGC 7662 appears round, slightly ring-like, and bluish green. About 3 arc min south of NGC 7662 is NGC 7640, a barred spiral galaxy that is fairly easy to locate.

Sweeping the northern area of Lacerta with a low-power telescope reveals several deep-sky objects. Binoculars give a good view of the open cluster NGC 7243, west of α (alpha) Lac, near 22^h15^m $+50°$. Large telescopes show about 50 of the cluster's stars. About 4° southwest is another open cluster, NGC 7209, smaller and fainter than NGC 7243. Many of its stars are 9th to 10th magnitude; they provide a beautiful view in a medium-sized telescope. East of β (beta) Lac is a third open cluster, NGC 7296.

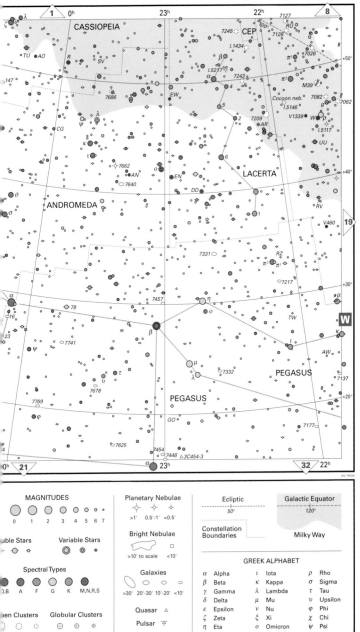

MAGNITUDES							
0	1	2	3	4	5	6	7

Double Stars **Variable Stars**

Spectral Types

O,B   A   F   G   K   M,N,R,S

Open Clusters   **Globular Clusters**

to scale   <10'   >10'   5'-10'   <5'

Planetary Nebulae

>1'   0.5'-1'   <0.5'

Bright Nebulae

>10' to scale   <10'

Galaxies

>30'   20'-30'   10'-20'   <10'

Quasar △

Pulsar ☿

Black Hole ⚹

Ecliptic

--------- 50°

Constellation Boundaries

Galactic Equator

········· 120°

Milky Way

GREEK ALPHABET					
α	Alpha	ι	Iota	ρ	Rho
β	Beta	κ	Kappa	σ	Sigma
γ	Gamma	λ	Lambda	τ	Tau
δ	Delta	μ	Mu	υ	Upsilon
ε	Epsilon	ν	Nu	φ	Phi
ζ	Zeta	ξ	Xi	χ	Chi
η	Eta	ο	Omicron	ψ	Psi
ϑ	Theta	π	Pi	ω	Omega

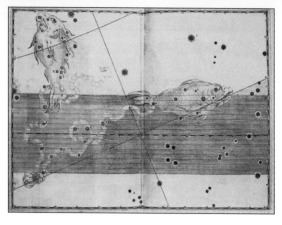

Fig. 7-23 Pisces, the Fish, from the star atlas of Bayer (1603).

ATLAS CHART 21. PISCES The next group of star charts shows the regio
of the sky surrounding the celestial equator. Near the top of th
chart are γ (gamma) and α (alpha) Peg, the southernmost stars o
the Great Square of Pegasus (Atlas Chart 20).

In Pisces, in the middle of the right-hand side of the chart, th
stars ι, θ, γ, κ, and λ (iota, theta, gamma, kappa, and lambda) Ps
form a distinctive asterism shaped like a small irregular pentagor
This asterism, called the Circlet, marks the western fish in Pisce
and lies directly south of the Great Square of Pegasus. The Ci
clet covers an area about 7° by 5° and also includes fainter stars.

Due west of δ (delta) Psc is UU Psc, a wide pair with comp
nents of 6th and 8th magnitudes.

Below the Circlet, at −10°, is the triple system ψ¹, ψ², and ψ
(psi¹, psi², and psi³) in Aquarius. About 4½° south of ψ³ Aqr is th
double star 94 Aqr, a pair of white and yellow-white stars wit
magnitudes 5.3 and 7.3. Even a small telescope at low power re
solves this wide pair.

To the southeast, R Aqr lies south of ω¹ (omega¹) and α
(omega²) Aqr. R Aqr is a long-period (Mira-type) variable th
changes in brightness from about 6th to 12th magnitude an
back to 6th magnitude again during its 387-day period. The dou
ble star 107 Aqr can be found 4° south and slightly east of R Aq
The white and yellow-white components of 6th and 7th magn
tude are separated by about 6 arc sec.

In Cetus we find φ¹, φ², and φ³ (phi¹, phi², and phi³) about 7
north of β (beta) Cet. φ¹ and φ² form an equilateral triangle wit
the planetary nebula NGC 246. The planetary looks like a di
oval through small telescopes but looks like a ring with a 12th
magnitude central star through large telescopes.

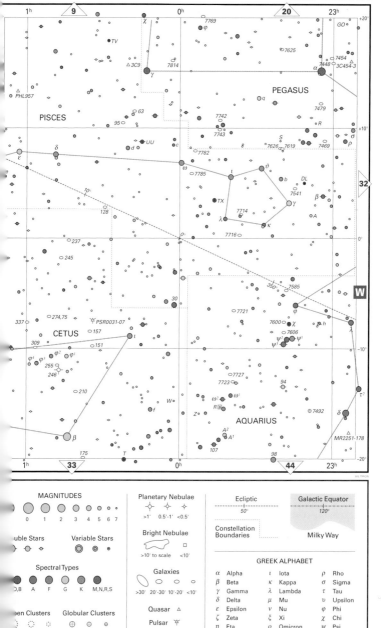

ATLAS CHART 22. CETUS, M77 Just above the center of the chart, the "cords" of Pisces—the cords holding the two fish together—converge to form an acute angle at α (alpha) Psc, also called El Rischa, the Knot. With components of magnitudes 4.2 and 5.1 separated by about 2 arc sec, α Psc can be observed with a medium-sized telescope.

The cords point to the lower left toward Mira, o (omicron) Ceti, the variable star about halfway down the neck of Cetus, the Whale. The first variable discovered, in 1596, Mira ("the Wonderful") is the prototype of long-period variables (chapter 6). At its maximum magnitude of about 2.0, it is the brightest star in this constellation, shining with a deep red color. During its 332-day cycle, Mira slowly fades to 10th magnitude, far below naked-eye visibility, then gradually brightens. (See figs. 6-6 and 6-14.)

About 6°—three fingers' width at arm's length—northeast of Mira is δ (delta) Cet, and about ½° southeast of this star is M77 (NGC 1068), one of the brighter spiral galaxies. M77's bright nucleus can be seen with a small telescope, but observations of its face-on spiral structure require a large telescope.

α (alpha) Cet, at left center, is an optical double. Binoculars can resolve the two components, separated by 16 arc min.

ζ (zeta) Cet lies near 1^{h}50^m −10°, toward the bottom of the neck of Cetus. About 5° northwest of ζ Cet is NGC 615, a spiral galaxy.

This field contains two of the stars closest to us. The flare star UV Ceti near the bottom of the chart is the fifth-nearest star system. It is too faint to see easily when at its normal 13th magnitude, but sometimes it brightens within about five minutes to almost 6th magnitude, a change of nearly 600 times in brightness. To its upper left is τ (tau) Ceti, the twentieth-closest star to earth. τ Cet, magnitude 3.5, is the nearest solar-type star (that is, a star with the same type of spectrum, temperature, and other properties as our sun). It is thus one of the prime objects in the search for life elsewhere in the universe.

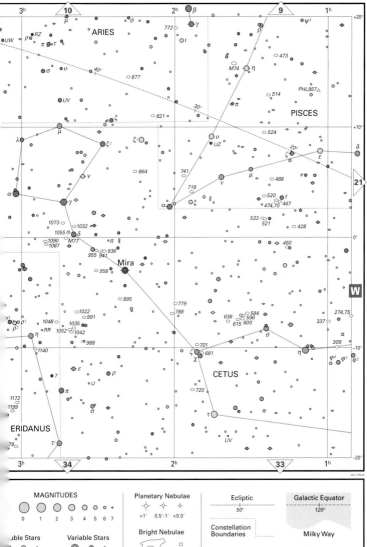

Fig. 7-25 Taurus, the Bull, from the star atlas of Hevelius (1687). The constellation is drawn backward from the way it appears in the sky because Hevelius drew the celestial sphere as it would appear from the outside. (Jay M. Pasachoff)

ATLAS CHART 23. ALDEBARAN, HAYDES As we move toward the western edge of the Milky Way in Taurus and Eridanus, the star fields become richer. On winter evenings, sweep southward with binoculars from the Hyades (at top; see also Chart 11) through the shield of Orion.

Just north of the Hyades is T Tauri (at $4^h22^m +19°32'$, but unmarked because it is too faint to be included on our chart), one of the most interesting stars in the sky. T Tauri varies in brightness because it is extremely young and has not settled down to a steady existence. It fluctuates between 8th and 13th magnitudes. T Tauri stars, the youngest stars, are being studied by optical, infrared, and radio techniques. T Tauri is near a diffuse nebula, NGC 1554-55, known as Hind's Variable Nebula. Over the last century, the nebula faded from view, then brightened considerably.

Slightly northwest of μ (mu) Eri, at $4^h46^m -3°$, is NGC 1637, a galaxy visible with medium-sized telescopes. The planetary nebula NGC 1535 lies 12° southwest, near γ (gamma) Eri, at lower center; it consists of a bright inner ring and a faint outer ring with a 12.2-magnitude central star.

At $4^h15^m -8°$ is the triple system o² (omicron²) Eri, also known as 40 Eridani. It is 16.5 light-years from us. It contains a 9th-magnitude white dwarf, 40 Eri B, the easiest white dwarf to study with a small telescope. Astronomers have measured the mass of the white dwarf and verified the theoretical prediction that its mass is about that of the sun. The system has also provided a test of Einstein's general theory of relativity, since the white dwarf's spectral lines are redshifted by gravity, as Einstein predicted. 11th-magnitude 40 Eri C, which forms a visual binary with the white dwarf, is a small red star, one of the least massive stars known. The pair is separated from the bright (4th-magnitude) 40 Eri A by 84 arc sec.

About 6° northwest of o¹ (omicron¹), just to the right of the chart's center, lies w Eri. This star is also known as 32 Eri (double star number ADS 2850). Its greenish and yellowish components are 5th and 6th magnitude.

The cluster of galaxies at the bottom edge of the chart includes NGC 1300, a fine example of a barred spiral galaxy (fit. 5-56).

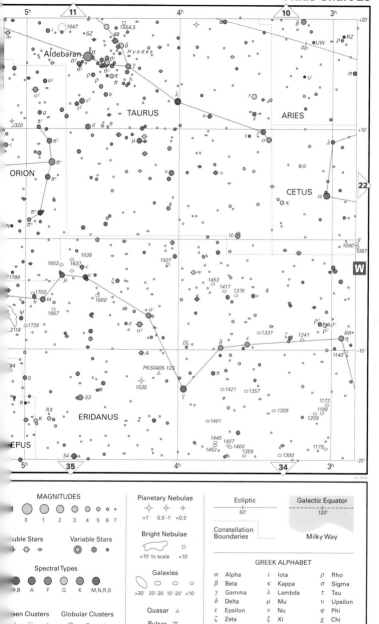

MAGNITUDES								
0	1	2	3	4	5	6	7	

Double Stars Variable Stars

Spectral Types

O,B A F G K M,N,R,S

Open Clusters Globular Clusters

to scale <10' >10' 5'-10' <5'

Planetary Nebulae

>1' 0.5'-1' <0.5'

Bright Nebulae

>10' to scale <10'

Galaxies

>30' 20'-30' 10'-20' <10'

Quasar △

Pulsar ☿

Black Hole ⌘

Ecliptic
- - - - - -
50°

Constellation
Boundaries

Galactic Equator
120°

Milky Way

GREEK ALPHABET

α	Alpha	ι	Iota	ρ	Rho
β	Beta	κ	Kappa	σ	Sigma
γ	Gamma	λ	Lambda	τ	Tau
δ	Delta	μ	Mu	υ	Upsilon
ε	Epsilon	ν	Nu	φ	Phi
ζ	Zeta	ξ	Xi	χ	Chi
η	Eta	ο	Omicron	ψ	Psi
ϑ	Theta	π	Pi	ω	Omega

The entire region of Orion and Monoceros is interesting to sweep with a low-power telescope or binoculars. The region of Orion is one of the most fascinating areas in the sky because it contains both the youngest known stars and many beautiful nebulae.

Orion's belt, three stars in a line, is one of the easiest asterisms to locate (fig. 2-8). In Orion's shoulder is Betelgeuse, α (alpha) Ori, a supergiant star whose reddish color is apparent to the naked eye. The tenth-brightest star in the sky, Betelgeuse varies in brightness by about 1 magnitude during its 5.8-year period. Rigel, the bright bluish star β (beta) Ori, marks Orion's heel. Rigel, a blue-white double, is the seventh-brightest star in the sky. Because it is so bright, the secondary is hard to find.

From Orion's belt extends his sword (fig. 2-10). M42 (NGC 1976), the Great Nebula in Orion, is a magnificent luminous cloud surrounding θ^1 (theta^1) Ori in the sword. It is often simply called the Orion Nebula (fig. 2-11). Even the smallest optical aid reveals the wispy structure of this nebula. Small telescopes readily resolve θ^1 into four components, which are called the Trapezium. The four stars (magnitudes 5.1, 6.7, 6.7, and 7.9) are very hot and provide the energy to illuminate the nebula. Larger telescopes also show two additional components of 11th magnitude and reveal more of the nebula's structure. The colors, though, show only with time exposures. Even a short time exposure with a tripod-mounted 35 mm camera, using an ordinary 50 mm or 35 mm lens, shows the nebula as red and allows the principal stars in Orion to be picked out. Photographs taken through equatorially driven telescopes look very different depending on the exposure time; short exposures show only the bright inner core of the nebula, while longer ones overexpose the inner parts but show outer contours.

The Orion Nebula lies about 1,500 light-years from the sun. The nebula is a blister on the side of the Orion Molecular Cloud that is closest to us. In the Orion Molecular Cloud, dust shields molecules from being broken apart by the ultraviolet light from the hot stars. Astronomers study the molecules with radio telescopes; objects that are probably stars in formation are studied in the infrared.

NGC 1973-75-77 refers to the nebulosity and open cluster around the northern end of Orion's sword. NGC 1981 is an open

Fig. 7-26 (right) The region of the Orion Nebula, M42. (© Anglo-Australian Observatory/Royal Observatory, Edinburgh, photograph from UK Schmidt plates by David Malin)

cluster slightly farther north, with about 12 widely scattered members. NGC 1980 is a faint nebula around ι (iota) Ori, south of the Orion Nebula.

Extending downward from the star at the far left of Orion's belt —ζ (zeta) Ori—is IC 434, which contains the famous Horsehead Nebula (fig. 7-28). The horsehead shape is formed by dark, absorbing gas and dust that blocks our view of a bright reddish emission nebula beyond. Northeast of the Horsehead, the nebula NGC 2023 is partially covered by the same dust cloud. Many beautiful photographs of the Orion Nebula and the Horsehead Nebula have been taken by amateur astronomers.

M78 (NGC 2068) is a wispy cloud north of Orion's belt, found by sweeping the area with low power. Long exposures show Barnard's Loop, which rings the northeast side of the Orion complex.

With low power, sweep the triangle formed by β (beta) Ori, η (eta) Ori (about 5° up toward Orion's belt), and β (beta) Eridani (Eri). The triangle contains many faint irregular patches of nebulosity. IC 2118, the Witch Head Nebula, lies about 1½° west of Rigel. γ (gamma) Ori has faint nebulosity around it.

Continue sweeping northeast from the Orion and Horsehead Nebulae into Monoceros. About 2° east of ε (epsilon) Mon is NGC 2244, a loose open cluster visible to the naked eye and resolvable into its member stars with good binoculars. NGC 2244 has about 24 member stars in an area 40 arc min across. Around NGC 2244 is NGC 2237, the Rosette Nebula. It is relatively large —1° across—and hard to locate. With low power it often shows just as a halo or glow around the open cluster; photographed in color, it looks spectacular (fig. 5-34).

Nearby, farther northeast, NGC 2264 is a nebulous cluster around S Mon, a variable double star. The cluster contains about 20 stars in an area close to the size of the full moon. The Cone Nebula (fig. 5-35), like the Rosette, is difficult to find, but can be located with a medium-sized telescope.

To the south is NGC 2261, Hubble's Variable Nebula. This nebula is so interesting that it was the first object photographed with the huge 5 m telescope when it opened at Palomar Observatory. The nebula reflects light from the irregular variable R Mon, which varies from 10th to 13th magnitude. The nebula has the shape of a comet's tail. The shape sometimes changes in a matter of weeks. It is hard to observe with small telescopes.

Moving farther south in the sky, near 6ʰ50ᵐ 0° is the open cluster NGC 2301, which consists of several overlapping groups of stars. This cluster has about 50 stars, including many bright ones.

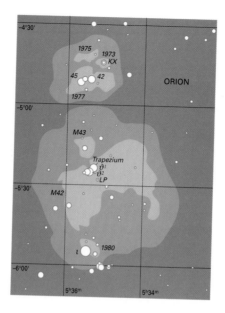

Fig. 7-27 A chart of the Orion Nebula region. (Wil Tirion)

Toward the bottom of the chart in Canis Major is Sirius, α (alpha) CMa, the Dog Star. Millennia ago, its dates of first visibility in the evening sky gave its name to the "dog days" of summer. At magnitude −1.46, Sirius is the brightest star in the sky.

In Orion's sword, ι (iota) and θ² (theta²) are interesting doubles, as is η (eta) to the northwest. ζ (zeta) Ori, the easternmost star in Orion's belt, is a triple system with blue-white components of magnitudes 2, 4, and 10. σ (sigma) Ori, south of ζ, is a multiple with at least 5 components. The two brightest components, of magnitudes 4.0 and 6.0, are separated by only ¼ arc sec and so are difficult to resolve, but a small telescope can separate the other components. Two other components are 8th- and 10th-magnitude stars; they are about 12 arc sec away, on opposite sides of the brighter pair. Another companion (7th magnitude) is 40 arc sec away.

The components of the double star ε (epsilon) Mon, 7° southeast of Betelgeuse, show as blue and gold. The two components are separated by about 13 arc sec and are accompanied by a third companion. Low power should be used to find ε Mon because the surrounding star field is so rich.

About 12° south is β (beta) Mon, an easily resolved triple that is visible to the naked eye. Its three blue components are of magnitudes 4.7, 5.2, and 6.1.

Fig. 7-28 Orion, showing the Orion Nebula (M42) and the Horsehead Nebula, IC434. (Ronald Royer)

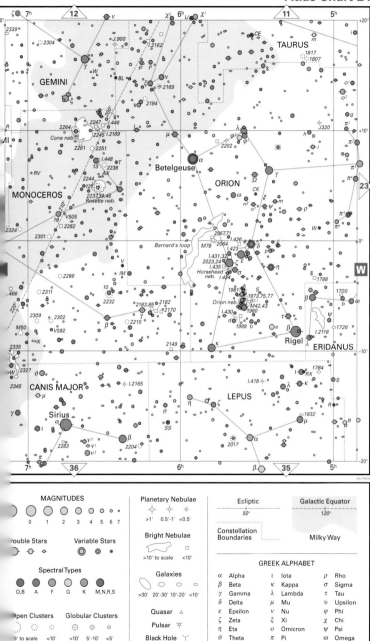

MAGNITUDES								
0	1	2	3	4	5	6	7	

Double Stars | **Variable Stars**

Spectral Types

O,B | A | F | G | K | M,N,R,S

Open Clusters | **Globular Clusters**

0' to scale | <10' | >10' | 5'-10' | <5'

Planetary Nebulae

>1' | 0.5'-1' | <0.5'

Bright Nebulae

>10' to scale | <10'

Galaxies

>30' | 20'-30' | 10'-20' | <10'

Quasar △
Pulsar ☆
Black Hole Υ

Ecliptic
50°

Constellation
Boundaries

Galactic Equator
120°

Milky Way

	GREEK ALPHABET			
α	Alpha	ι	Iota	ρ Rho
β	Beta	κ	Kappa	σ Sigma
γ	Gamma	λ	Lambda	τ Tau
δ	Delta	μ	Mu	υ Upsilon
ε	Epsilon	ν	Nu	φ Phi
ζ	Zeta	ξ	Xi	χ Chi
η	Eta	ο	Omicron	ψ Psi
ϑ	Theta	π	Pi	ω Omega

Fig. 7-29 *Monoceros plus the head of Canis Major, from the atlas of Hevelius (1687). The constellation is drawn backward from the way it appears in the sky because Hevelius drew the celestial sphere as it would appear from the outside. (Jay M. Pasachoff)*

ATLAS CHART 25. PROCYON A number of open clusters appear in this region, which is worth sweeping with low power. The most spectacular is the Beehive Cluster (M44), called Praesepe (fig. 7-13), which is described opposite Chart 13. Southeast of ζ (zeta) Mon, near $8^h13^m -6°$ in Hydra, the rich open cluster M48 (NGC 2548) is bright enough to be visible to the naked eye but difficult to locate. On the right of the chart, near $7^h03^m -8°$ in Monoceros, M50 (NGC 2682) is a rich open cluster.

A beautiful open cluster in Puppis is M46 (NGC 2437), somewhat fainter than M67 (which is west of α (alpha) Cancri) but containing more stars. On the northern edge of M46 is the planetary nebula NGC 2438, an irregular patchy ring. Nearby is M47 (NGC 2422), an open cluster that is faintly visible to the naked eye because of its many 6th-magnitude stars. Both M46 and M47 are only about 20 million years old. NGC 2539 is another rich open cluster in Puppis. Low power should be used to view the open cluster NGC 2506 in Monoceros, slightly northwest of NGC 2539.

Near $7^h40^m -18°$ in Puppis is the planetary nebula NGC 2440, best viewed with high power.

At the lower right of the chart (at −12°), extending from Canis Major into Monoceros, is IC 2177, a nebulous patch more than 3° across. This nebula was called the Eagle Nebula in older references, though that name is now usually applied to M16 (NGC 6611) instead (Chart 30). Toward the upper tip of IC 2177 are two open clusters, NGC 2335 and NGC 2343.

Procyon, α (alpha) Canis Minoris, is the eighth-brightest star in the sky (magnitude 0.38) and one of the closest stars to us (about 11 light-years away). Procyon has a faint (13th-magnitude) white-dwarf companion, the second-closest white dwarf to us.

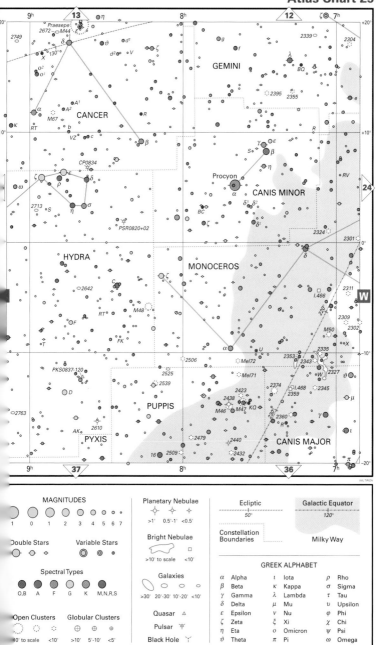

MAGNITUDES	Planetary Nebulae	Ecliptic	Galactic Equator

MAGNITUDES
1 0 1 2 3 4 5 6 7

Double Stars

Variable Stars

Spectral Types
O,B   A   F   G   K   M,N,R,S

Open Clusters
10' to scale   <10'

Globular Clusters
>10'   5'-10'   <5'

Planetary Nebulae
>1'   0.5'-1'   <0.5'

Bright Nebulae
>10' to scale   <10'

Galaxies
>30'   20'-30'   10'-20'   <10'

Quasar △

Pulsar ☿

Black Hole

Ecliptic
50'

Constellation Boundaries

Galactic Equator
120'

Milky Way

GREEK ALPHABET

α	Alpha	ι	Iota	ρ	Rho
β	Beta	κ	Kappa	σ	Sigma
γ	Gamma	λ	Lambda	τ	Tau
δ	Delta	μ	Mu	υ	Upsilon
ε	Epsilon	ν	Nu	φ	Phi
ζ	Zeta	ξ	Xi	χ	Chi
η	Eta	o	Omicron	ψ	Psi
ϑ	Theta	π	Pi	ω	Omega

As we move away from the obscuring dust clouds of the Milky Way, the more distant galaxies become visible. In Leo we see an outer edge of the huge cluster of galaxies—the Virgo Cluster—that extends through Virgo, Coma Berenices, and Corvus. Three of the brighter galaxies in Leo are M95 (NGC 3351), M96 (NGC 3368), and M105 (NGC 3379), located about 30 million light-years from the Milky Way Galaxy. M95 (magnitude 9.7) is a barred spiral and M96 (magnitude 9.2) is a regular spiral; both are relatively large. M105 is a small elliptical galaxy of magnitude 9.3. In a medium-power telescope, about 9° east of Regulus, these galaxies appear together.

Regulus, α (alpha) Leo, lies directly on the ecliptic. Regulus is at the base of the "sickle" that many people see as part of the constellation's shape; it marks the lion's heart. The twenty-first-brightest star in the sky at magnitude 1.35, Regulus is visible most of the year, and can be seen in the eastern sky in spring. Its faint companion is sometimes detectable with binoculars.

γ (gamma) Leo, the brightest star in the sickle, is a binary located about 8° northeast of Regulus. One of the best doubles in the sky, γ Leo consists of two yellow components of magnitudes 2.2 and 3.5 that orbit with a period of 620 years. The components are fairly close, so they are difficult to resolve with a small telescope. The gap between the pair is currently widening, though; these stars will be about 5 arc sec apart by the year 2100.

This chart includes Wolf 359, the third-closest star to the sun, after the α (alpha) Centauri system and Barnard's star. At magnitude 13.5, Wolf 359 is too faint to be marked on this chart, but it is located near VY Leo, at $10^h56^m +7°$. It is one of the least luminous stars known, but has been discovered because its rapid proper motion—motion across the sky (p. 25)—makes its closeness (only 7.8 light-years from earth) apparent.

The radiant of the Leonid meteor shower (table 20), which reaches its maximum frequency on November 17 each year, lies about 2° northwest of γ Leo. This shower peaks every 33 years, as the earth passes through the part of a defunct comet's orbit where the remaining particles from the comet are bunched. The last spectacular display of Leonids occurred in 1999.

In the Lion's foreleg, R Leo is one of the brightest of the long-period variables; it ranges from magnitude 11.3 to 4.4 over a period of 312 days. Located about 5° west of Regulus, R Leo forms a triangle with two westward stars of magnitudes 9.0 and 9.6.

Although the constellation Hydra is not particularly rich in stars, sweeping the region with low-power binoculars does show some objects of interest. NGC 3242 is a planetary nebula located about 1.8° south of μ (mu) Hydra at the lower left of the chart. In a small telescope the planetary appears as a pale blue disk.

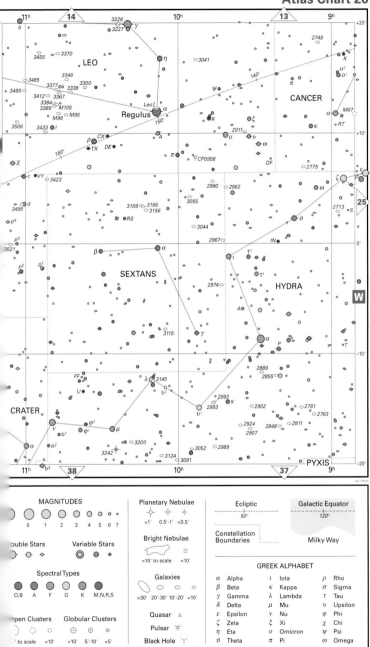

LEO

CANCER

Regulus

SEXTANS

HYDRA

W

25

CRATER

PYXIS

MAGNITUDES							
0	1	2	3	4	5	6	7

Double Stars

Variable Stars

Spectral Types

O,B · A · F · G · K · M,N,R,S

Open Clusters · **Globular Clusters**

to scale · <10' · >10' · 5'-10' · <5'

Planetary Nebulae

>1' · 0.5'-1' · <0.5'

Bright Nebulae

>10' to scale · <10'

Galaxies

>30' · 20'-30' · 10'-20' · <10'

Quasar △

Pulsar ☆

Black Hole ⅄

Ecliptic
50°

Constellation
Boundaries

Galactic Equator
120°

Milky Way

GREEK ALPHABET					
α	Alpha	ι	Iota	ρ	Rho
β	Beta	κ	Kappa	σ	Sigma
γ	Gamma	λ	Lambda	τ	Tau
δ	Delta	μ	Mu	υ	Upsilon
ε	Epsilon	ν	Nu	φ	Phi
ζ	Zeta	ξ	Xi	χ	Chi
η	Eta	ο	Omicron	ψ	Psi
ϑ	Theta	π	Pi	ω	Omega

Here we find the greatest concentration of galaxies in the Virgo Cluster of galaxies that extends north into Canes Venatici and south into Corvus. The Virgo Cluster is an average of about 65 million light-years away. Our Milky Way Galaxy is part of the Local Group which, in turn, is an outlying part of the Virgo Cluster. Many of the galaxies in the main part of the Virgo Cluster are shown on an expanded chart (27A) of the area outlined at upper-left.

Just above the outlined area is M85 (NGC 4382), an elliptical galaxy in Coma Berenices, at 12^h25^m +18°. One of the brightest galaxies in the Virgo Cluster, M85 shows in a small telescope as a bright oval patch.

In the Leo Cluster of galaxies at upper right, M65 (NGC 3623) and M66 (NGC 3627) are two bright spiral galaxies about 20 arc min apart, located about halfway between θ (theta) and ι (iota) Leo. Both galaxies appear similar to the Andromeda Galaxy. At magnitude 9.0, M66 is about 0.3 magnitude brighter than M65. M65 and M66 can be seen in the same field of view with low-power telescopes. On a clear night, you may be able to glimpse both galaxies with a good pair of binoculars. Two nearby spirals, NGC 3628 and NGC 3593, may be visible in the same field.

On the border of Virgo and Corvus, near 12^h40^m −12°, is M104 (NGC 4594), another of the brightest members of the cluster. M104 is called the Sombrero Galaxy (fig. 5-51) because of its distinctive large center and the dark lane of dust that runs along the equatorial plane of this spiral galaxy, seen only 6° from edge-on. These features are hard to see in small telescopes, though the galaxy can be readily detected as a hazy oval patch. In larger telescopes, the dust lane can be detected under good observing conditions.

In Corvus, about 4° southwest of γ (gamma) Corvi, near 12^h −18°, are a pair of peculiar galaxies, NGC 4038 and NGC 4039, known as the Antennae. Each of these elliptical galaxies appears to have a long curving tail; computer models show that the tails may have formed by gravity as a result of a collision of the two galaxies. The Antennae are a source of radio waves and lie about 90 million light-years away.

One of the finest double stars in the sky and a favorite of amateurs is γ (gamma) Vir (at 12^h42^m −1°). Its two almost equal components (magnitudes 3.5) are separated by less than 3 arc seconds and closing rapidly.

Located about 4° south of σ (sigma) Leo, near the ecliptic (at about 11^h27^m +3°), is τ (tau) Leo, a pair of yellow and blue stars of 5th and 8th magnitudes, separated by 90 arc sec. δ (delta) Crv, whose components are magnitudes 3.0 and 9.2, is also easily resolved.

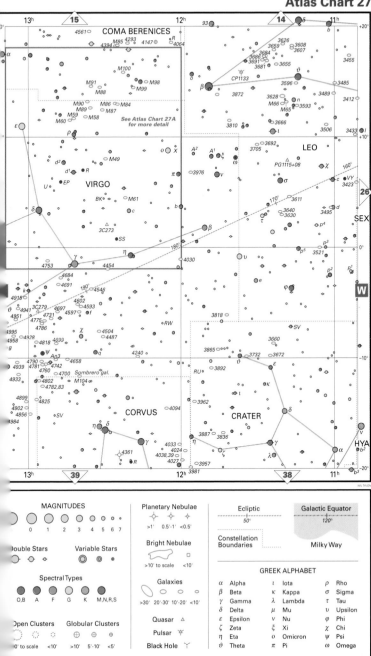

MAGNITUDES

0 1 2 3 4 5 6 7

Double Stars

Variable Stars

Spectral Types

O,B A F G K M,N,R,S

Open Clusters

10' to scale <10'

Globular Clusters

>10' 5'-10' <5'

Planetary Nebulae

>1' 0.5'-1' <0.5'

Bright Nebulae

>10' to scale <10'

Galaxies

>30' 20'-30' 10'-20' <10'

Quasar △

Pulsar ☿

Black Hole ⌣

Ecliptic
50°

Constellation
Boundaries

Galactic Equator
120°

Milky Way

GREEK ALPHABET

α	Alpha	ι	Iota	ρ	Rho
β	Beta	κ	Kappa	σ	Sigma
γ	Gamma	λ	Lambda	τ	Tau
δ	Delta	μ	Mu	υ	Upsilon
ε	Epsilon	ν	Nu	φ	Phi
ζ	Zeta	ξ	Xi	χ	Chi
η	Eta	ο	Omicron	ψ	Psi
ϑ	Theta	π	Pi	ω	Omega

This chart shows the heart of the Virgo Cluster. About 75 percent of the brightest galaxies in this cluster are spirals.

Since the Virgo Cluster is the closest rich cluster of galaxies to us, it is the easiest to observe. In small telescopes most of the galaxies appear as faint patches of light. But on a clear dark night in the summer, more than 100 galaxies can be viewed with a medium-sized telescope. It is best to locate a galaxy with low power first and then switch to higher magnifications.

Some outstanding galaxies in Coma Berenices include: M98 (NGC 4192) at 12^h19^m +15°, a bright, elongated spiral seen nearly edge-on; M99 (NGC 4254)—about 1.5° southeast of M98, near 12^h19^m +14°—a bright, round spiral seen face-on; and M88 (NGC 4501)—about 4° east of M99, near 12^h32^m +14°—a spiral that can be located by searching for the two faint stars near its edge. M100 (NGC 4321), near 12^h23^m +16°, is the largest spiral in the Virgo Cluster.

In Virgo, noteworthy galaxies include: M49 (NGC 4472), M58 (NGC 4579), M59 (NGC 4621), M60 (NGC 4649), M61 (NGC 4303), M84 (NGC 4374), M86 (NGC 4406), M89 (NGC 4552), and M90 (NGC 4569). M49, one of the brightest ellipticals, is fairly easy to find near 12^h30^m +8°. M61, about 6° southwest of M49, near 12^h22^m +4°30', is a beautiful face-on spiral. M84 and M86 are nearly identical ellipticals separated by less than 20 arc minutes. (fig. 5-62). M90 is an edge-on spiral with a bright nucleus.

One of the brightest and most interesting galaxies in the Virgo Cluster is M87 (NGC 4486), a huge peculiar elliptical galaxy at 12^h31^m +13'. The best photographs taken with large telescopes show the more than 1,000 globular clusters that surround this galaxy. Short-exposure photographs taken with large telescopes have also revealed an unusual jet of gas (fig. 5-46). M87 has been identified as the radio source Virgo A, and many astronomers think that a giant black hole containing millions of times more mass than the sun lies at its center.

Near 12^h30^m +2°, about 3½° northeast of η (eta) Vir, is 3C 273 (fig. 5-67), the brightest and one of the first known quasars. The easiest quasar to observe, 3C 273 ranges in magnitude from about 12.2 to 13.0. Depending on its magnitude and on observing conditions, 3C 273 can sometimes be glimpsed in medium to large telescopes with the aid of an accurate finding chart. It appears almost starlike, not nebulous, and is thus "quasi-stellar." From the enormous redshift in its spectrum, astronomers have calculated that this bluish object is receding at the tremendous speed of 50,000 km per second, 16 percent of the speed of light. Hubble's Law tells us that 3C 273 is out among the distant galaxies.

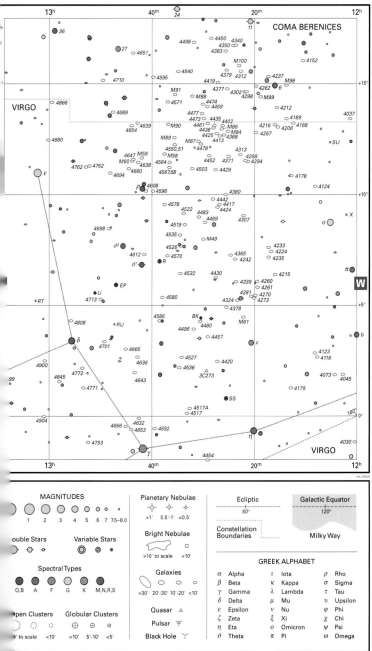

Fig. 7-30 *The constellation Virgo, the Virgin, from the star atlas of Hevelius (1687). The constellation is drawn backward from the way it appears in the sky because Hevelius drew the celestial sphere as it would appear from the outside. (Jay M. Pasachoff)*

ATLAS CHART 28. ARCTURUS, SPICA Arcturus, α (alpha) Boötes, at magnitude 0.0 the third-brightest single star in the sky, and 1st-magnitude Spica, α (alpha) Virginis, are the most prominent objects in this region. Arcturus is a cool red giant. Spica, the sixteenth-brightest star, is extremely hot and therefore bluish. It is an eclipsing spectroscopic binary that drops about one magnitude from its usual magnitude of 1.0 during a four-day period. These objects are easily found in the sky from the curve in the handle of the Big Dipper. Following the curve backward from the bowl, you "arc to Arcturus." Then, from Arcturus, you can continue straight on: "spike to Spica." These two stars are so bright that they stand out.

In the Virgo Cluster of galaxies, NGC 5364 is a spiral galaxy of 10th magnitude. NGC 5746 (at $14^h45^m +2°$) is an edge-on spiral of 11th magnitude, best seen with small telescopes; its companion, NGC 5740, a little over $\frac{1}{4}°$ south, is a spiral of magnitude 11.9. To the left of the chart, the galaxies NGC 5806 (spiral), NGC 5813 (elliptical), and NGC 5831 (elliptical) are visible in medium-sized telescopes. The dim galaxies NGC 5838, NGC 5854, and NGC 5864 are slightly farther east. The nearby spiral NGC 5846 is brighter. Another spiral galaxy located slightly to the southeast, NGC 5850, is dimmer but about twice as large as NGC 5846.

Of the double stars shown on this chart, ζ (zeta) Boo, at upper left, is a beautiful pair of white stars with magnitudes 4.5 and 4.6, which orbit each other with a period of 123 years. They can be resolved with a medium-sized telescope. π (pi) Boo, above ζ Boo, is another double, more easily resolved.

S Vir, about halfway between Spica and ζ (zeta) Vir, is a long-period variable ranging from 6th to 13th magnitude, with an average period of 377 days. As it nears maximum brightness, its red color makes it an especially interesting telescopic object.

This chart contains 3C 279, one of the brightest quasars. It has reached an absolute magnitude as high as −31, which is 10,000 times brighter than the Andromeda Galaxy. 3C 279 is on the ecliptic at extreme right.

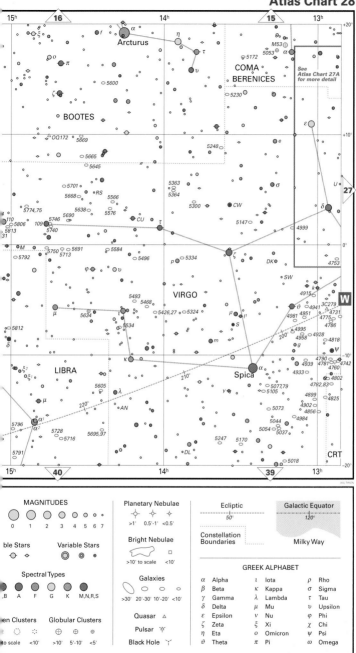

COMA BERENICES

See Atlas Chart 27A for more detail

BOOTES

VIRGO

LIBRA

Arcturus

Spica

M53

27

W

39

40

CRT

MAGNITUDES							
0	1	2	3	4	5	6	7

Double Stars

Variable Stars

Spectral Types

O,B A F G K M,N,R,S

Open Clusters   **Globular Clusters**

to scale <10'   >10' 5'-10' <5'

Planetary Nebulae

>1' 0.5'-1' <0.5'

Bright Nebulae

>10' to scale <10'

Galaxies

>30' 20'-30' 10'-20' <10'

Quasar △

Pulsar

Black Hole

Ecliptic

50°

Constellation Boundaries

Galactic Equator

120°

Milky Way

GREEK ALPHABET					
α	Alpha	ι	Iota	ρ	Rho
β	Beta	κ	Kappa	σ	Sigma
γ	Gamma	λ	Lambda	τ	Tau
δ	Delta	μ	Mu	υ	Upsilon
ε	Epsilon	ν	Nu	φ	Phi
ζ	Zeta	ξ	Xi	χ	Chi
η	Eta	ο	Omicron	ψ	Psi
ϑ	Theta	π	Pi	ω	Omega

ATLAS CHART 29. GLOBULAR CLUSTERS M5, M10, M12 In Serpens, the globular cluster M5 (NGC 5904) is one of the finest in the sky, surpassed only by 47 Tuc, ω (omega) Cen, and possibly M13. During the summer months it is easily located in the sparse area near $15^h18^m +2°$, slightly northwest of the naked-eye star 5 Ser. Under very good conditions, M5 (fig. 7-31) is also visible to the naked eye. With binoculars, it appears as a hazy star, while small telescopes show it as a bright, circular glow. Larger telescopes begin to resolve its stars. Nearby, 5 Ser is a double with 5th- and 10th-magnitude components that are separated by 11 arc sec.

M10 and M12, two very similar globular clusters in Ophiuchus, are also outstanding. Both brighter than 7th magnitude, they are visible to the naked eye. Located about 5° northeast of ζ (zeta) Oph, near $16^h57^m -4°$, M10 (NGC 6254) is a rich cluster with a diameter of about 12 arc min, and medium-sized telescopes begin to resolve its stars. M12 (NGC 6218) lies about 3° northwest of M10 and is a bit larger. Since M12 has a relatively loose structure, its stars are more readily resolved.

M107 (NGC 6171), a third globular cluster in Ophiuchus, lies about 3° south of ζ (zeta) Oph. At magnitude 9 and about 8 arc min across, M107 is much fainter and smaller than M10 or M12.

See Chart 28 for the galaxies in Virgo shown here west of M5.

During the summer months scan the region of Serpens Caput (the Serpent's Head); the tail, Serpens Cauda, is on Chart 30. Use binoculars or a low-power telescope near β (beta) Ser. δ (delta) Ser is a fine double with components of magnitudes 4.1 and 5.2, easily resolved by a small telescope. About 1° east of β (beta) Ser near $15^h46^m +15°$, R Ser is a long-period variable, ranging from magnitude 5.6 to 14.0 over an average period of 356 days. Its color near maximum, when it can sometimes be seen with the naked eye, is crimson.

S Her, a long-period variable located near $16^h50^m +15°$ in Hercules, ranges from magnitude 5.9 to 13.6 in 307 days.

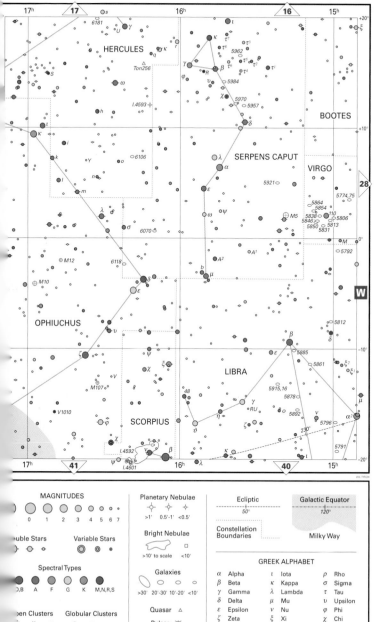

HERCULES

BOOTES

SERPENS CAPUT

VIRGO

OPHIUCHUS

LIBRA

SCORPIUS

W

MAGNITUDES							
0	1	2	3	4	5	6	7

Double Stars

Variable Stars

Spectral Types

O,B A F G K M,N,R,S

Open Clusters   Globular Clusters

to scale <10'   >10' 5'-10' <5'

Planetary Nebulae
>1' 0.5'-1' <0.5'

Bright Nebulae
>10' to scale <10'

Galaxies
>30' 20'-30' 10'-20' <10'

Quasar △
Pulsar ☼
Black Hole Υ

Ecliptic
50°

Constellation
Boundaries

Galactic Equator
120°

Milky Way

GREEK ALPHABET					
α	Alpha	ι	Iota	ρ	Rho
β	Beta	κ	Kappa	σ	Sigma
γ	Gamma	λ	Lambda	τ	Tau
δ	Delta	μ	Mu	υ	Upsilon
ε	Epsilon	ν	Nu	φ	Phi
ζ	Zeta	ξ	Xi	χ	Chi
η	Eta	ο	Omicron	ψ	Psi
ϑ	Theta	π	Pi	ω	Omega

Fig. 7-32 M16. (Tim Puckett)
Compare the Hubble Space Telescope image (fig. 5-30).

ATLAS CHART 30. OMEGA NEBULA The star fields begin to increase in brilliance toward the Milky Way in Aquila, Scutum, and Sagittarius,
with good views in binoculars or low-power telescopes.

The open clusters NGC 6633 in Ophiuchus (at $18^h28^m +7°$)
and IC 4756 in Serpens (at $18^h40^m +6°$) are easily resolved at relatively low power. About 5° southwest of ζ (zeta) Aquilae at upper
left, you can see the open cluster NGC 6709 with medium power.
Another rich field lies about 1° northeast of β (beta) Oph, in the
open cluster IC 4665 (at $17^h47^m +6°$).

The rich open cluster M11 (NGC 6705), the Wild Duck, is located near $18^h50^m -6°$ on the northern edge of the Scutum star
cloud, a bright region of the Milky Way. M11 is magnificent in
both small and large telescopes (fig. 5-24); it can sometimes be
seen with the naked eye. A medium-sized telescope resolves its
edges, while larger telescopes resolve hundreds of its stars. About
3½° south of M11 is the open cluster M26 (NGC 6694).

In Sagittarius, M23 (NGC 6494) is a large open cluster that
appears prominent against a dark nebula. Its bright stars are outstanding even when viewed through binoculars. The open clusters M18 (NGC 6613), M24 (NGC 6603), and M25 (IC 4725)
also lie among the dark nebulae in Sagittarius. M24 is actually a
part of the Milky Way and includes the open cluster NGC 6603.
Nearby are the reflection nebulae NGC 6589 and NGC 6590.

An interesting planetary, NGC 6572, is in Ophiuchus. Use
high power to observe the globular clusters M9 (NGC 6333),
NGC 6356, and NGC 6342.

In Serpens, near $18^h20^m -14°$, is M16 (NGC 6611), the Eagle
Nebula, an easily located nebulous cluster (fig. 5-29). See also
the fabulous high-resolution view of M16's center (fig. 5-30), a
photograph that has become very well known as a symbol of the
Hubble Space Telescope. To the southeast in Sagittarius near
$18^h20^m -16°$ is the Omega Nebula, M17 (NGC 6618), also
called the Horseshoe or Swan Nebula (fig. 5-31). The star background in this region is particularly rich. In Scutum, the nebula
IC 1287, 2° south of α (alpha) Sct, requires slightly higher power.

Barnard's Star (just above the Chart's center, at $18^h +5°$), magnitude 9.5, is the second-closest star to us, counting the whole α
Centauri system as the first. It has the greatest apparent angular
motion across the sky (proper motion) of any star (fig. 2-3).

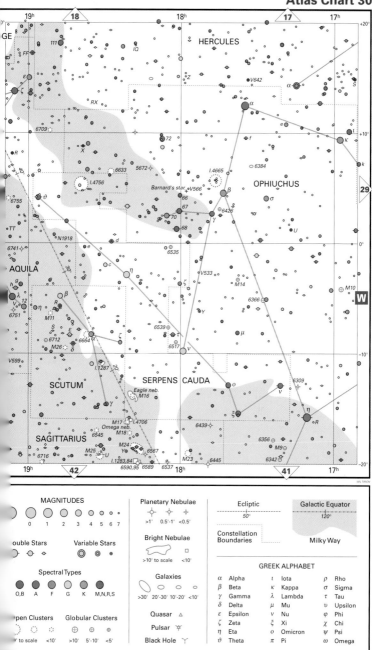

MAGNITUDES								
0	1	2	3	4	5	6	7	

Double Stars   **Variable Stars**

Spectral Types

O,B   A   F   G   K   M,N,R,S

Open Clusters   **Globular Clusters**

to scale   <10'   >10'   5'-10'   <5'

Planetary Nebulae

>1'   0.5'-1'   <0.5'

Bright Nebulae

>10' to scale   <10'

Galaxies

>30'   20'-30'   10'-20'   <10'

Quasar △

Pulsar ☿

Black Hole ✶

Ecliptic

- - - - - 50°

Constellation Boundaries

Galactic Equator

120°

Milky Way

GREEK ALPHABET

α	Alpha	ι	Iota	ρ	Rho
β	Beta	κ	Kappa	σ	Sigma
γ	Gamma	λ	Lambda	τ	Tau
δ	Delta	μ	Mu	υ	Upsilon
ε	Epsilon	ν	Nu	φ	Phi
ζ	Zeta	ξ	Xi	χ	Chi
η	Eta	ο	Omicron	ψ	Psi
ϑ	Theta	π	Pi	ω	Omega

ATLAS CHART 31. ALTAIR From Cygnus into Aquila, the Milky Way appears to be divided lengthwise by enormous dust clouds. These clouds form a band called the Great Rift, which lies along the central plane of our galaxy in a lane like the equatorial lanes visible in photographs of edge-on galaxies. The dark nebulae and rich star fields provide beautiful views during the summer months.

Near the top of the chart in Sagitta (the Arrow), about halfway between γ (gamma) and δ (delta) Sge, is the rich globular cluster M71 (NGC 6838), which is magnitude 8.3.

Above the center of the chart is α (alpha) Aquilae, Altair, the twelfth-brightest star in the sky. About 2° north of Altair is γ (gamma) Aql. Slightly northwest of γ Aql is B143, a dark nebula about 30 arc min in diameter that is visible in medium-sized telescopes. About 5° to the west, the planetary nebula NGC 6803 shows a small disk, as does the fainter planetary, NGC 6804, nearby. The planetary nebula NGC 6891, 7° northeast of Altair, appears as a bright disk surrounded by a fainter ring. Eight degrees west-southwest of Altair is the planetary nebula NGC 6781.

At lower center, close to the Milky Way in Sagittarius, is NGC 6822, an irregular dwarf galaxy and one of the few members of the Local Group that is readily visible. Use a rich-field telescope to view it. An intervening dust cloud reduces its brightness.

Slightly northwest of NGC 6822 and visible in the same field is NGC 6818, the Little Gem, one of the closest planetary nebulae. Use a large telescope to observe its 13th-magnitude central star.

Double stars in this region include: π (pi) Aql, a 6th-magnitude double about 3° north of Altair; γ (gamma) Del (at $20^h47^m +16°$), a wide double located at the northeast corner of the lozenge-shaped asterism formed by α, β, δ, and γ (alpha, beta, delta, and gamma) Del; and 15 Aql, a wide double lying northeast of the Scutum star cloud, about 1 arc min north and slightly west of λ (lambda) Aql (at $19^h05^m -05°$).

Among the variable stars on the chart are η (eta) Aql, a Cepheid variable that ranges from magnitude 3.5 to 4.4 during a period of slightly more than seven days; σ (sigma) Aql, an eclipsing variable that drops about 0.2 magnitudes within a period of 47 hours; R Aql, a reddish long-period variable often visible to the naked eye at maximum brightness; and U Sagittae, an Algol-type eclipsing binary. The brightness of U Sge ranges from 6.6 to 9.2 with a period of three days, nine hours.

One of the most unusual objects in our galaxy, SS433 (at $19^h15^m +05°$), is too faint (14th magnitude) for small telescopes. Its spectrum shows that it is emitting jets of gas in opposite directions at 25 percent of the speed of light, a speed otherwise unknown in our galaxy. Astronomers have concluded that it is a binary system containing a neutron star.

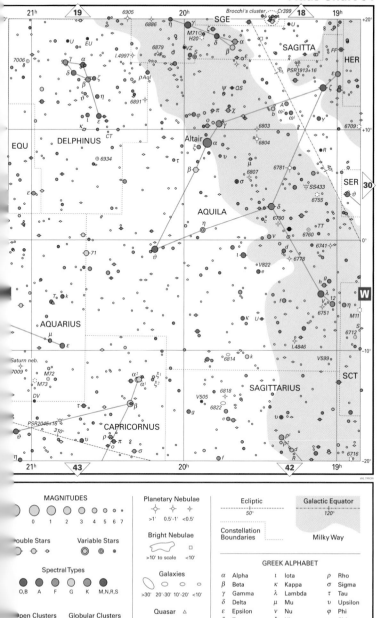

MAGNITUDES								
0	1	2	3	4	5	6	7	

Double Stars

Variable Stars

Spectral Types

O,B A F G K M,N,R,S

Open Clusters   **Globular Clusters**

>10' to scale   <10'   >10' 5'-10' <5'

Planetary Nebulae

>1'   0.5'-1'   <0.5'

Bright Nebulae

>10' to scale   <10'

Galaxies

>30' 20'-30' 10'-20' <10'

Quasar △

Pulsar ☆

Black Hole Y

Ecliptic

50'

Constellation Boundaries

Galactic Equator

120'

Milky Way

GREEK ALPHABET

α	Alpha	ι	Iota	ρ	Rho	
β	Beta	κ	Kappa	σ	Sigma	
γ	Gamma	λ	Lambda	τ	Tau	
δ	Delta	μ	Mu	υ	Upsilon	
ε	Epsilon	ν	Nu	φ	Phi	
ζ	Zeta	ξ	Xi	χ	Chi	
η	Eta	ο	Omicron	ψ	Psi	
ϑ	Theta	π	Pi	ω	Omega	

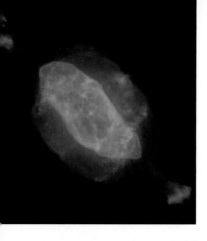

Fig. 7-33 NGC 7009, the Saturn Nebula, a planetary nebula in Aquarius. This image is in pseudo-color, in which emission from cooler gas is shown in red, somewhat hotter gas in green, and the hottest gas in blue. It was taken with the Hubble Space Telescope and has much higher resolution than can be achieved from the ground. The nebula is 1,400 light-years away. (Bruce Balick, University of Washington) See also fig. 5-6, p. 151.

ATLAS CHART 32. SATURN NEBULA NGC 7009, one of the brightest planetary nebulae, is located near 21^h −11°, about 1° west of ν (nu) Aquarii (Aqr). It consists of a bright inner ring surrounded by an outer disk with faint extensions. It is sometimes called the Saturn Nebula (fig. 7-33).

Slightly southwest of NGC 7009 is M72 (NGC 6981), a small globular cluster difficult to resolve even with large telescopes and high power. The open cluster M73 (NGC 6994) is nearby.

Two globular clusters on this chart, M15 (NGC 7078) and M2 (NGC 7089), are among the brightest in the sky. Both are clearly visible to the naked eye, and even the smallest optical aid shows them as hazy patches that are condensed toward the centers. M15 (at 21^{h}30^m +12°) is located about 4° northwest of ε (epsilon) Pegasi. An unusually rich and compact cluster, M15 has been identified as an x-ray source. M2 is located just below the celestial equator, about 5° north of β (beta) Aqr. Since M2 has such a highly condensed center, even large telescopes resolve only its outer edges.

The globular cluster NGC 7006 in Delphinus is very remote; it lies about 185,000 light-years from the solar system, about the same distance as the Magellanic Clouds. The cluster is very difficult to resolve; in a medium-sized telescope, it appears as a hazy spot about 1 arc min in diameter.

East of α (alpha) Aqr, the four stars γ, ζ, η, and π (gamma, zeta, eta, and pi Aqr) form a small Y-shaped asterism called the Water Jar, a characteristic feature of this constellation. The star ζ Aqr at the center of the Y is one of the finest doubles in the sky, with components of magnitudes 4.3 and 4.5.

This ends the set of equatorial charts; we now turn to charts of intermediate southern declinations.

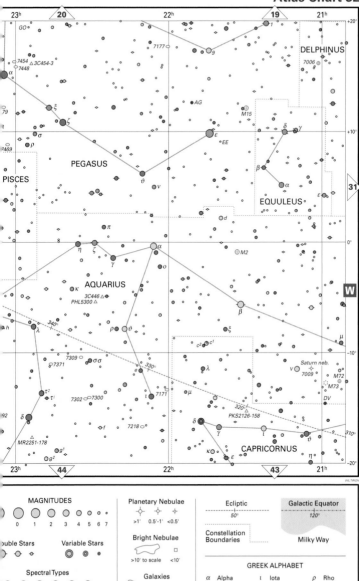

MAGNITUDES										Planetary Nebulae				Ecliptic		Galactic Equator	

	Planetary Nebulae	Ecliptic	Galactic Equator
MAGNITUDES	✧ ✧ ✧	— — — —	
0 1 2 3 4 5 6 7	>1' 0.5'-1' <0.5'	50°	120°
Double Stars Variable Stars	Bright Nebulae	Constellation Boundaries	Milky Way
	⬭ □		
	>10' to scale <10'		
Spectral Types	Galaxies	GREEK ALPHABET	

GREEK ALPHABET

α	Alpha	ι	Iota	ρ	Rho
β	Beta	κ	Kappa	σ	Sigma
γ	Gamma	λ	Lambda	τ	Tau
δ	Delta	μ	Mu	υ	Upsilon
ε	Epsilon	ν	Nu	φ	Phi
ζ	Zeta	ξ	Xi	χ	Chi
η	Eta	ο	Omicron	ψ	Psi
ϑ	Theta	π	Pi	ω	Omega

Spectral Types: O,B A F G K M,N,R,S

Galaxies: >30' 20'-30' 10'-20' <10'

Quasar △
Pulsar ☿
Black Hole ⬝

Open Clusters   Globular Clusters
>' to scale <10'   >10' 5'-10' <5'

WIL TIRION

Fig. 7-34 NGC 253, a spiral galaxy seen at a low angle. The dark dust lanes show clearly. (© Anglo-Australian Observatory, photograph by David Malin)

ATLAS CHART 33. SOUTH GALACTIC POLE The stellar background in this region is thin, and even galaxies are rare. The galaxy NGC 253, near the south galactic pole in Sculptor, is a large and bright spiral (fig. 7-34), often considered the finest in the sky except for the Andromeda Galaxy. Because NGC 253 is at such a southern declination, observers in midnorthern latitudes need a medium-sized telescope and good viewing conditions to see it well.

The spiral galaxy NGC 247, about 4° north of NGC 253, is nearly as large as NGC 253 but about two magnitudes fainter. NGC 247 may be found most easily by looking with low power about 3° south of β (beta) Ceti and then waiting for the galaxy to drift into view. NGC 247 and NGC 253 are very close to us (for galaxies) — only about 12 million light-years away.

About 4° northwest of α (alpha) Phe, near the border between Sculptor and Phoenix, NGC 55 is another prominent galaxy in the southern sky, though difficult for northern observers. It is a large nearly edge-on spiral that shows in small telescopes as an oval patch of light. All three galaxies belong to the Sculptor group, most of which are large loose spirals.

The globular cluster NGC 288 near the south galactic pole is relatively uninteresting.

SX Phe, about 7° west of α Phe, is one of the best-known dwarf Cepheid variables. During its 79-minute period, it varies from magnitude 6.8 to about 7.5.

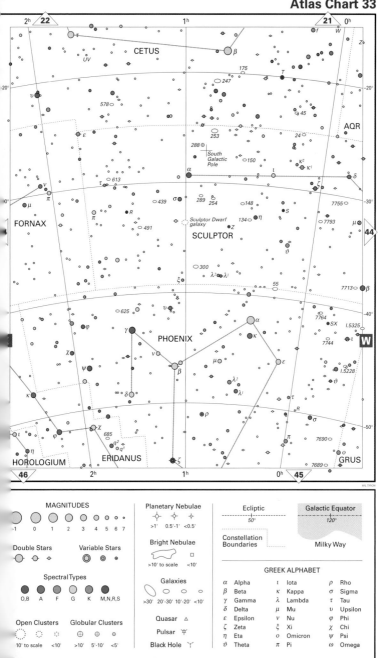

MAGNITUDES
-1 0 1 2 3 4 5 6 7

Double Stars   **Variable Stars**

Spectral Types
O,B A F G K M,N,R,S

Open Clusters   **Globular Clusters**
10' to scale <10'   >10' 5'-10' <5'

Planetary Nebulae
>1' 0.5'-1' <0.5'

Bright Nebulae
>10' to scale   <10'

Galaxies
>30' 20'-30' 10'-20' <10'

Quasar △
Pulsar ☿
Black Hole ⅄

Ecliptic
50'

Constellation Boundaries

Galactic Equator
120'

Milky Way

GREEK ALPHABET

α	Alpha	ι	Iota	ρ	Rho
β	Beta	κ	Kappa	σ	Sigma
γ	Gamma	λ	Lambda	τ	Tau
δ	Delta	μ	Mu	υ	Upsilon
ε	Epsilon	ν	Nu	φ	Phi
ζ	Zeta	ξ	Xi	χ	Chi
η	Eta	ο	Omicron	ψ	Psi
ϑ	Theta	π	Pi	ω	Omega

Fig. 7-35 The central part of the Fornax Cluster of galaxies, 55 million light-years away. The barred spiral galaxy NGC 1365 is in one corner; most of the galaxies are elliptical (© Anglo-Australian Observatory/Royal Observatory, Edinburgh, photograph from UK Schmidt plates by David Malin)

ATLAS CHART 34. NGC 1300 Left of the center of the chart, near the border between Fornax and Eridanus, is the Fornax Cluster of galaxies. The cluster consists of a number of faint galaxies as well as 18 bright ones. Nine of these bright galaxies can be seen in the same 1° field of view.

NGC 1316, an elliptical or So galaxy at a distance of about 55 million light-years, is the brightest member of the group. It has been identified as the radio source Fornax A and may be undergoing some type of nuclear outburst. Its companion galaxy, NGC 1317, is a spiral with a bright nucleus.

NGC 1365, the second-brightest galaxy of the Fornax Cluster, is an excellent example of a barred spiral galaxy. Its central bar extends about 45,000 light-years and appears to be about 3 arc minutes long.

Near 3^h20^m −20° is NGC 1300, the prototype of a barred spiral galaxy with moderately wide arms (called SBb).

Just to the right of the chart's center, southwest of β (beta) Fornacis, is the globular cluster NGC 1049. It is part of the Fornax System, a faint dwarf galaxy that is a member of our Local Group of galaxies. The system contains five groups of stars that somewhat resemble the globular clusters in our galaxy. The clusters in Fornax, though, differ in density, brightness, and size. For example, NGC 1049 (at 2^h40^m −34°) is 50 times larger than the largest globular cluster in our galaxy. It is 630,000 light-years away.

θ (theta) Eri (at 2^h58^m −40°) is a wide double that can be resolved with binoculars. Its components are bright, of magnitudes 3.4 and 4.5. The double star f Eri (at 3^h48^m −38°) has widely spaced components of magnitudes 4.8 and 5.3.

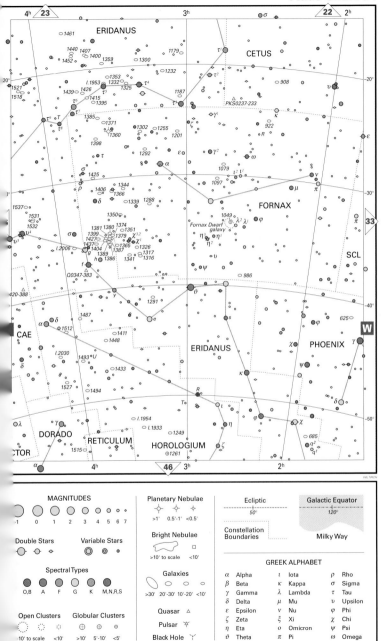

MAGNITUDES

-1 0 1 2 3 4 5 6 7

Double Stars

Variable Stars

Spectral Types

O,B   A   F   G   K   M,N,R,S

Open Clusters

>10' to scale   <10'

Globular Clusters

>10'   5'-10'   <5'

Planetary Nebulae

>1'   0.5'-1'   <0.5'

Bright Nebulae

>10' to scale   <10'

Galaxies

>30'   20'-30' 10'-20' <10'

Quasar △

Pulsar ☿

Black Hole ⅄

Ecliptic
50°

Constellation Boundaries

Galactic Equator
120°

Milky Way

GREEK ALPHABET

α	Alpha	ι	Iota	ρ	Rho
β	Beta	κ	Kappa	σ	Sigma
γ	Gamma	λ	Lambda	τ	Tau
δ	Delta	μ	Mu	υ	Upsilon
ε	Epsilon	ν	Nu	φ	Phi
ζ	Zeta	ξ	Xi	χ	Chi
η	Eta	ο	Omicron	ψ	Psi
ϑ	Theta	π	Pi	ω	Omega

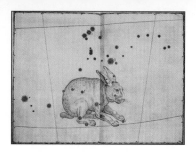

Fig. 7-36 Lepus, the Hare, from the atlas of Bayer (1603). (Jay M. Pasachoff)

ATLAS CHART 35. M79, LEPUS, COLUMBA, CAELUM, ERIDANUS In the constellation Lepus, the Hare, we find M79 (NGC 1904), one of the few globular clusters visible during the winter. Under good conditions it can be detected even with binoculars. M79 is not very impressive with small or medium-sized telescopes; only larger instruments begin to resolve its edges into their component stars. Located near 5^h24^m $-25°$, M79 forms an isosceles triangle with β (beta) Lep and ε (epsilon) Lep. M79 is an 8th-magnitude cluster about 3 arc min in diameter and lies about 50,000 light-years away.

In the constellation Columba, the Dove, about 1° east of α (alpha) Col, is the 12th-magnitude spiral galaxy NGC 2090. Because of its extreme southern declination, NGC 2090 is difficult for northern-hemisphere observers to find; it can be glimpsed only when the sky above the southern horizon is exceptionally clear. If you can find NGC 2090, try to find the brighter spiral NGC 1792, which lies even farther south. NGC 1792 lies on the border of Columba and Caelum, about 3° south and slightly east of the 4.6-magnitude star γ (gamma) Caeli.

Farther south of NGC 1792, near 5^h14^m $-40°$ in Columba, the globular cluster NGC 1851 is barely visible to the eye but is an interesting object through a telescope.

Among the double stars visible in this region, γ (gamma) Lep is an easily resolved double that offers a nice color contrast even in small telescopes. Its components are magnitudes 3.7 and 6.3, separated by 95 arc sec.

About ½° to the southwest of M79 is ADS 3954, a double star with components of magnitudes 5.4 and 6.6, separated by 3.2 arc sec. This double can be resolved with small telescopes.

μ (mu) Columbae is a well-known "runaway star" that seems to be moving at high speed—for a star in our galaxy—away from the Orion Molecular Cloud, the star nursery that lies behind the Orion Nebula. Two similar hot stars are known. All three lie at the same distance from the Orion Molecular Cloud, where infrared and radio studies have found stars in formation (see p. 167). The hot stars may have been ejected from that region during the last few million years.

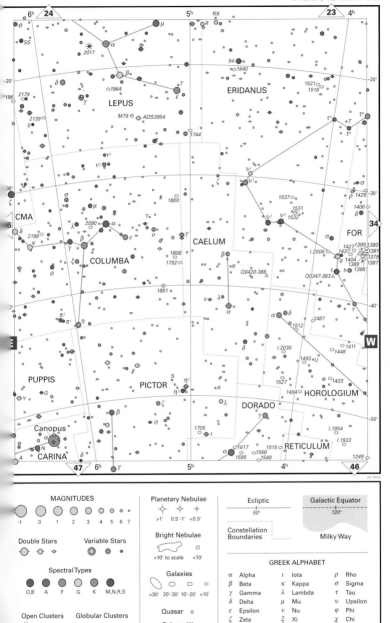

MAGNITUDES	Planetary Nebulae	Ecliptic	Galactic Equator
	>1′ 0.5′-1′ <0.5′	*50°*	*120°*
-1 0 1 2 3 4 5 6 7	Bright Nebulae	Constellation Boundaries	Milky Way
Double Stars Variable Stars	>10′ to scale <10′		

Spectral Types

O,B   A   F   G   K   M,N,R,S

Galaxies

>30′   20′-30′   10′-20′   <10′

Open Clusters   Globular Clusters

>10′ to scale   <10′   >10′   5′-10′   <5′

Quasar △

Pulsar ☿

Black Hole ˚

GREEK ALPHABET

α	Alpha	ι	Iota	ρ	Rho
β	Beta	κ	Kappa	σ	Sigma
γ	Gamma	λ	Lambda	τ	Tau
δ	Delta	μ	Mu	υ	Upsilon
ε	Epsilon	ν	Nu	φ	Phi
ζ	Zeta	ξ	Xi	χ	Chi
η	Eta	ο	Omicron	ψ	Psi
ϑ	Theta	π	Pi	ω	Omega

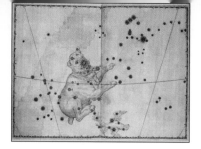

ATLAS CHART 36. SIRIUS, M93, M41, CANIS MAJOR, PUPPIS In the winter Milky Way, many beautiful star fields are visible low in the southern sky. In particular, Canis Major and Puppis are worth sweeping with a telescope or binoculars.

The open cluster M41 (NGC 2287) is one of the finest in the sky, located near $6^h47^m -21°$, about 4° south of Sirius in Canis Major. At magnitude 4.5 it is sometimes visible with the naked eye. M41 is a good object for viewing with binoculars and is easy to resolve with medium power. It is a fairly rich cluster, with about 25 bright stars and many fainter ones in a field about the size of the full moon.

A second open cluster suitable for viewing with a low-power telescope is M93 (NGC 2447), about 2° northwest of 3rd-magnitude ξ (xi) Pup, near the galactic equator at $7^h45^m -24°$. M93 lies in a dense star field at a distance of about 3,300 light-years. It is a very rich cluster, with about 300 component stars. About 2° on the other side of x Pup is the diffuse nebula NGC 2467.

A third outstanding open cluster in this region is NGC 2477, located 15° south of M93 in Puppis, near $7^h52^m -39°$. It is rich and compact, with about 300 stars in a field 20 arc min across.

ε (epsilon) Canis Majoris lies above the center of the chart, near $7^h -29°$. Its 7th-magnitude companion is about 7 arc sec away.

Northeast, near τ (tau) CMa, is the variable star UW CMa, near $7^h19^m -25°$. About 5° to the east, the double star n Pup has two widely separated components of almost equal magnitude.

k Pup, near $7^h39^m -27°$, is a bright double easily resolved with small telescopes. Its bluish components are about magnitudes 4.5 and 4.6, separated by 9.8 arc sec. Near $7^h29^m -43°$, σ (sigma) Pup is a double with orange and white components of magnitudes 3.3 and 9.4, separated by 22.4 arc sec.

About 3° southwest of σ Pup, near $7^h14^m -45°$, L² Pup is one of the brightest red variables. Since its minimum brightness is about 6th magnitude, L² Pup is visible to the naked eye during most of its cycle. Another variable star visible to the naked eye is V Pup, about 3° southwest of γ (gamma) Velorum. V Pup is an eclipsing variable, ranging in magnitude from 4.7 to 5.2.

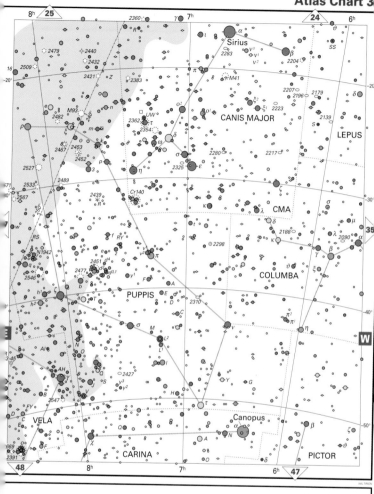

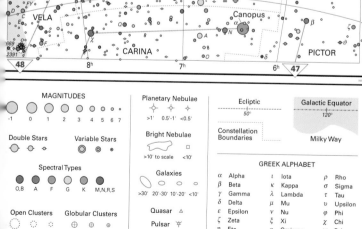

MAGNITUDES								
-1	0	1	2	3	4	5	6	7

Double Stars

Variable Stars

Spectral Types

O,B A F G K M,N,R,S

Open Clusters

>10' to scale   <10'

Globular Clusters

>10'   5'-10'   <5'

Planetary Nebulae

>1'   0.5'-1'   <0.5'

Bright Nebulae

>10' to scale   <10'

Galaxies

>30'   20'-30'   10'-20'   <10'

Quasar △

Pulsar ☼

Black Hole ⅄

Ecliptic
50°

Constellation
Boundaries

Galactic Equator
120°

Milky Way

GREEK ALPHABET			
α	Alpha	ι	Iota	ρ	Rho
β	Beta	κ	Kappa	σ	Sigma
γ	Gamma	λ	Lambda	τ	Tau
δ	Delta	μ	Mu	υ	Upsilon
ε	Epsilon	ν	Nu	φ	Phi
ζ	Zeta	ξ	Xi	χ	Chi
η	Eta	ο	Omicron	ψ	Psi
ϑ	Theta	π	Pi	ω	Omega

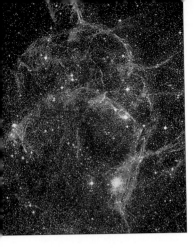

ATLAS CHART 37. GUM NEBULA In this region of the southern sky, we are looking toward the Milky Way along the galactic equator. Unlike other regions of the Milky Way, though, this region includes few interesting objects.

The Gum Nebula is too large to mark on the chart; it is a huge emission nebula that extends through the constellations Puppis and Vela over an area at least $35°$ in diameter. It is named after its discoverer, Colin Gum. The object that provides the energy for the central part of the nebula is probably PSR 0833-45 (the pulsar that also provides the energy for the x-ray source Vela X), near 8^h30^m $-45°$. The pulsar emits both radio and optical pulses with a period of 0.09 seconds, and lies about 1,500 light-years distant.

ζ (zeta) Pup, left of 8^h at $-40°$, is a star of spectral type O, one of the hottest stars in the sky. About $5°$ north-northeast of ζ Pup, near 8^h12^m $-35°$, is the site where Nova Puppis 1942 erupted. One of the brightest novae in modern times, it rapidly reached a maximum of about 0.3 magnitude.

About $4°$ east-northeast of α (alpha) Pyxidis is T Pyx, near 9^h05^m $-32°$. T Pyx is one of the few recurrent novae. It is normally about 14th magnitude but may rise to magnitude 6.3 or 7 at maximum. More eruptions have been observed in this source than in any other. Its five maxima were recorded in 1890, 1902, 1920, 1944, and 1966. Given this average interval of 19 years, we might have expected another eruption in about 1985, but as of 1999, none had occurred. Check T Pyx regularly.

NGC 2997, a spiral galaxy in Antlia, is shown as fig. 7-39.

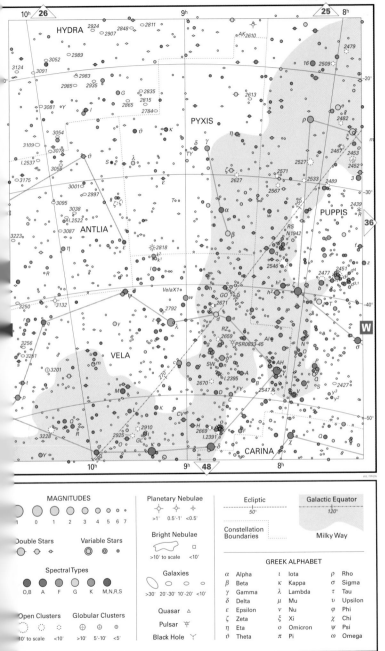

HYDRA

PYXIS

PUPPIS

ANTLIA

VELA

CARINA

36

W

10ʰ   9ʰ **48**   8ʰ

MAGNITUDES		Planetary Nebulae	Ecliptic	Galactic Equator
1 0 1 2 3 4 5 6 7		⊕ ⊕ ⊕ >1' 0.5'-1' <0.5'	⎯ ⎯ ⎯ 50°	⎯⎯⎯ 120°
Double Stars	**Variable Stars**	**Bright Nebulae**	**Constellation Boundaries**	
		⬭ ▫ >10' to scale <10'		**Milky Way**
Spectral Types		**Galaxies**		
O,B A F G K M,N,R,S		⬭ ⬭ ◦ ◦ >30' 20'-30' 10'-20' <10'		
		Quasar △		
Open Clusters	**Globular Clusters**	**Pulsar** ☿		
10' to scale <10'	⊕ ⊕ ⊕ >10' 5'-10' <5'	**Black Hole** ⅄		

GREEK ALPHABET

α	Alpha	ι	Iota	ρ	Rho
β	Beta	κ	Kappa	σ	Sigma
γ	Gamma	λ	Lambda	τ	Tau
δ	Delta	μ	Mu	υ	Upsilon
ε	Epsilon	ν	Nu	φ	Phi
ζ	Zeta	ξ	Xi	χ	Chi
η	Eta	ο	Omicron	ψ	Psi
ϑ	Theta	π	Pi	ω	Omega

Fig. 7-39 NGC 2997, a spiral galaxy in Antlia near 9^h40^m −31°, shows on the right edge of this chart and in left center of the preceding chart. It is inclined at about 45° to our line of sight. We see the bluish arms, colored by its hot stars; the reddish nebulae in the spiral arms; and the yellowish nucleus. (© Anglo-Australian Observatory, photograph by David Malin)

ATLAS CHART 38. NGC 3132 The number of stars in this area diminishes with increasing distance from the Milky Way. Antlia, the Water Pump, is one of the least interesting regions of the sky, both to the telescope and the naked eye. Hydra and the northern part of Centaurus also contain few noteworthy objects.

One notable object is NGC 3132, one of the few planetary nebulae visible in the spring, located right on the Vela/Antlia border, near 10^h07^m −40°. This planetary is about 8th magnitude in total brightness and has a bright central star of about 10th magnitude. It is nearly the same size as the Ring Nebula in Lyra. Though brighter than the Ring Nebula, NGC 3132 is harder to find and not as well known because of its southern declination and the lack of nearby stars to aid in finding it. One method for locating NGC 3132 is to wait for Regulus in Leo to reach your meridian and then point your telescope about 52° south.

NGC 3132 shows more detail than the Ring Nebula. On photographs, its elliptical disk looks as though several oval rings have been superimposed and tilted at different angles. Because of this appearance, it is sometimes called the Eight-Burst Nebula.

N Hydrae (ADS 8202), at 11^h32^m −29°, is a wide pair of closely matched yellow stars of magnitudes 5.8 and 5.9. Another interesting double is β (beta) Hya (at 11^h53^m −34°). Its components of magnitudes 4.7 and 5.5 can be resolved in a medium-sized telescope.

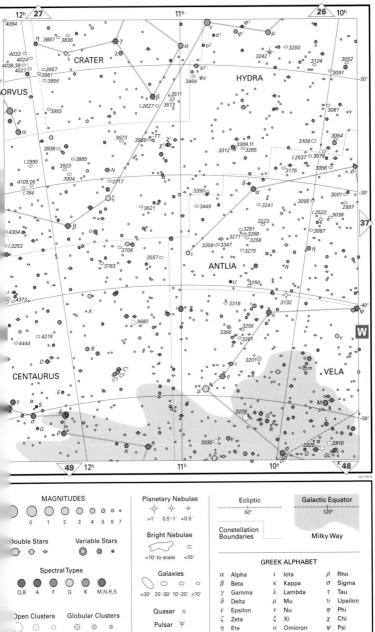

MAGNITUDES

| 0 | 1 | 2 | 3 | 4 | 5 | 6 | 7 |

Double Stars   Variable Stars

Spectral Types

O,B   A   F   G   K   M,N,R,S

Open Clusters   Globular Clusters

10' to scale   <10'   >10'   5'-10'   <5'

Planetary Nebulae

>1'   0.5'-1'   <0.5'

Bright Nebulae

>10' to scale   <10'

Galaxies

>30'   20'-30'   10'-20'   <10'

Quasar   △

Pulsar   ⚡

Black Hole

Ecliptic
50°

Galactic Equator
120°

Constellation
Boundaries   Milky Way

GREEK ALPHABET

α	Alpha	ι	Iota	ρ	Rho
β	Beta	κ	Kappa	σ	Sigma
γ	Gamma	λ	Lambda	τ	Tau
δ	Delta	μ	Mu	υ	Upsilon
ε	Epsilon	ν	Nu	φ	Phi
ζ	Zeta	ξ	Xi	χ	Chi
η	Eta	ο	Omicron	ψ	Psi
ϑ	Theta	π	Pi	ω	Omega

Here we find one of the most spectacular objects within range of amateur telescopes: NGC 5139, ω (omega) Centauri, one of the brightest and largest globular clusters (figs. 2-24 and 7-40). Its apparent diameter is about the size of the full moon. Because it is so far south of the celestial equator, ω Cen is visible in the U.S. only from the southernmost states. Observers there can sometimes detect it low on the horizon on clear spring and summer evenings, about 36° almost due south of Spica, near 13^h27^m −47°. Binoculars may help you find it. To the naked eye, ω Cen looks like a fuzzy 4th-magnitude star. Through small telescopes or binoculars, it looks like a hazy patch. The cluster contains more than one million stars. To locate it from a southern site, find ε Cen 7° northwest of β Cen, and then note that ε, ζ, and ω form an isosceles triangle.

An outstanding galaxy in the summer sky is the spiral M83 (NGC 5236), near 13^h37^m −30° on the Hydra/Centaurus border, about 18° south of Spica and nearly halfway between γ (gamma) Hya and θ (theta) Cen. A spiral seen face-on (fig. 5-45), it is one of the 10 largest galaxies and one of the 25 brightest. M83 is hard for northern observers to see because it lies very low in the southern sky, in a field rich with star clouds and dust.

About 3° south of M83 is NGC 5253, another bright galaxy. A medium-sized telescope shows it as an oval with a bright center. Two of the brightest extragalactic supernovae ever recorded erupted in this galaxy. The first, in 1895, reached magnitude 7.2; another, in 1972, reached magnitude 7.9.

Perhaps the most famous galaxy in this field is NGC 5128, located at 13^h25^m −43°, about 4½° north of ω Cen. NGC 5128 is a giant peculiar galaxy, one of the largest and most luminous galaxies known. It looks like a glowing sphere of 7th magnitude crossed by a thick dust lane (figs. 5-60 and 61) and shows well in telescopes and sometimes in binoculars. NGC 5128 has been identified as the powerful radio source Centaurus A. Its radio waves and x-rays may result from matter falling into a giant black hole.

The globular cluster M68 (NGC 4590) is relatively bright even though it is small and highly concentrated. It is located at 12^h40^m −27° near ADS 8612, a 5th-magnitude double star slightly to the southwest—the only nearby star visible to the naked eye. M68 appears as a fuzzy glowing ball of 8th magnitude, about 3 arc min across. To locate it, sweep with binoculars from δ (delta) Crv through β (beta) Crv and continue about 4° farther south. NGC 5367 is a nebula with an embedded open cluster.

R Hya is a notable long-period variable about 3° east of γ Hya, at 13^h30^m −23°. The third long-period variable discovered, R Hya is one of the easiest to observe. It ranges from magnitude 3 to 11 with a period of 390 days and is noted for its deep red color.

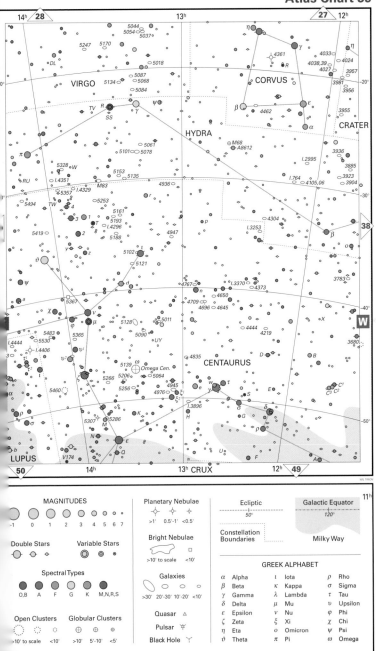

MAGNITUDES
-1 0 1 2 3 4 5 6 7

Double Stars   **Variable Stars**

Spectral Types
O,B   A   F   G   K   M,N,R,S

Open Clusters   **Globular Clusters**
>10' to scale   <10'   >10'   5'-10'   <5'

Planetary Nebulae
>1'   0.5'-1'   <0.5'

Bright Nebulae
>10' to scale   <10'

Galaxies
>30'   20'-30'   10'-20'   <10'

Quasar △

Pulsar ☿

Black Hole

Ecliptic
50°

Constellation Boundaries

Galactic Equator
120°

Milky Way

GREEK ALPHABET

α	Alpha	ι	Iota	ρ	Rho
β	Beta	κ	Kappa	σ	Sigma
γ	Gamma	λ	Lambda	τ	Tau
δ	Delta	μ	Mu	υ	Upsilon
ε	Epsilon	ν	Nu	φ	Phi
ζ	Zeta	ξ	Xi	χ	Chi
η	Eta	ο	Omicron	ψ	Psi
ϑ	Theta	π	Pi	ω	Omega

Fig. 7-40 ω (omega) Centauri, NGC 5139, one of the largest, brightest globular clusters (Chart 39). It is 17,000 light-years from us. (Daniel Good)

ATLAS CHART 40. CENTAURUS, LUPUS, LIBRA Few interesting objects lie in this area of the southern sky, which is dominated by sparse regions in Centaurus, Lupus, and Libra. We discussed the globular cluster ω Centauri with Atlas Chart 39.

The globular cluster NGC 5694 is located in the upper right-hand part of the chart, at the eastern tip of the tail of Hydra, almost halfway between σ (sigma) Lup and π (pi) Hya, near 14^h40^m $-27°$. NGC 5694 is difficult to resolve because of its small apparent diameter. Its total magnitude is 10.2, though none of its 10 brightest stars is brighter than about 16th magnitude. Lying about 100,000 light-years away, on the far side of our galaxy, NGC 5694 is one of the most distant globular clusters.

Toward the bottom of the chart, near 14^h08^m $-48°$, is the open cluster NGC 5460. Its total magnitude is 5.6, and none of its members is brighter than 8th magnitude. Though sparse, this cluster covers an area about the size of the full moon. NGC 5460 is best observed with a rich-field telescope or binoculars.

Though it is very hard to observe visually, an extended region of nebulosity can be detected in Scorpius, between δ (delta) and π (pi) Sco. δ Sco is an extremely hot star, spectral type B0. It lies about 540 light-years away and is one of the brightest members of the Scorpio-Centaurus Association. The Scorpio-Centaurus Association is a scattered group of stars that extends over 90° of the southern sky and includes many of the brightest stars in the constellations Scorpius, Lupus, Centaurus, and Crux. This group of stars is part of a larger group of about 100 stars often called the Local Star Cloud, located in one of the spiral arms in our galaxy. Typically, the stars in the group are spectral type B, about 20 million years old. The reddish Antares, α (alpha) Sco, about 7° southeast of δ Sco (Atlas Chart 41), is the brightest member of this group.

Many double stars lie in the constellation Lupus. One of these is π (pi) Lupi (near 15^h05^m $-47°$), with components of magnitudes 4.1 and 6.0. ε (epsilon) Lup, east of λ (lambda) Lup, near 15^h23^m $-45°$, can be resolved with a medium-sized telescope. κ (kappa) Lup, southeast of π Lup, near 15^h12^m $-49°$, is a very wide pair, of magnitudes 4.6 and 4.7. ξ (xi) Lup, northeast in Lupus, near 15^h57^m $-34°$, is a wide double, with stars of magnitudes 5.3 and 5.8.

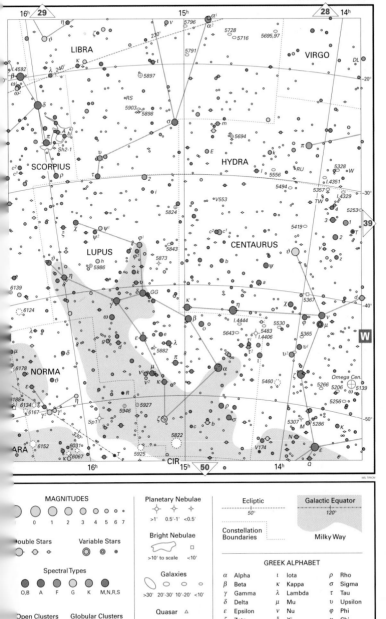

MAGNITUDES							
0	1	2	3	4	5	6	7

Double Stars   **Variable Stars**

Spectral Types

O,B   A   F   G   K   M,N,R,S

Open Clusters   **Globular Clusters**

0' to scale   <10'   >10'   5'-10'   <5'

Planetary Nebulae

>1'   0.5'-1'   <0.5'

Bright Nebulae

>10' to scale   <10'

Galaxies

>30'   20'-30'   10'-20'   <10'

Quasar △

Pulsar ☆

Black Hole

Ecliptic
50°

Constellation
Boundaries

Galactic Equator
120°

Milky Way

GREEK ALPHABET

α	Alpha	ι	Iota	ρ	Rho
β	Beta	κ	Kappa	σ	Sigma
γ	Gamma	λ	Lambda	τ	Tau
δ	Delta	μ	Mu	υ	Upsilon
ε	Epsilon	ν	Nu	φ	Phi
ζ	Zeta	ξ	Xi	χ	Chi
η	Eta	ο	Omicron	ψ	Psi
ϑ	Theta	π	Pi	ω	Omega

Fig. 7-41 The region around ρ Ophiuchus reveals its colors with a long exposure. Globular cluster M4 is at lower right center. The star Antares is in the nebula at lower left. (© Anglo-Australian Observatory/Royal Observatory, Edinburgh, photograph from UK Schmidt plates by David Malin)

ATLAS CHART 41. ANTARES, SOUTHERN MILKY WAY One of the most exciting regions of the entire sky, this area offers outstanding sights for summer observers. Brilliant star fields lie among the dark clouds and winding dust lanes of the Milky Way, and many open and globular clusters are present. First sweep the area with binoculars or a telescope at low power and then switch to higher magnifications for individual objects.

A good starting point is the most prominent object in this region—Antares, α (alpha) Scorpii, near 16^h29^m −26°. At magnitude 0.96, the reddish Antares (its name means "compared to Mars," the reddish planet) is the fifteenth-brightest star. It becomes visible in midspring or early summer in the southeastern sky and has an emerald green companion. IC 4606, a reddish nebula about five light-years in diameter, surrounds Antares. IC 4604, a reflection nebula around ρ (rho) Ophiuchi, about 3° north-northwest of Antares, is too faint for visual observations but shows up as a rich bluish purple cloud in long-exposure photographs through large telescopes. In between, IC 4603 is a reflection nebula around a fainter star. (fig. 7-41)

Less than 2° due west of Antares is M4 (NGC 6121), one of the largest and, at 7,000 light-years away, nearest globular clusters. At 6th magnitude, M4 is visible with binoculars and can sometimes be detected with the naked eye. M4 is one of the easiest globular clusters to locate. Because of its very loose structure, small telescopes can resolve its edges, and medium-sized telescopes under high power can completely resolve its stars.

Less than 1° northeast of M4 in the same field is NGC 6144, a smaller and fainter globular cluster. To observe this cluster, use high power.

A little over 4° northwest of Antares, sweep about halfway between Antares and β (beta) Sco to find M80 (NGC 6093), a globular cluster that looks smaller and brighter than M4. It lies on the western border of a dark cloud. Because M80 is very compact, it is much harder to resolve than M4.

About 8° east of Antares is M19 (NGC 6273), a compact globular cluster. To find M19, point at Antares, then sweep east and slightly north. Alternatively, point your telescope at Antares and wait about 33 minutes for M19 to drift into view. It is one of the most oval globular clusters, with an apparent diameter of 5 arc min and a magnitude of 7.2. Its outer edges can be easily resolved.

About 7° southeast of Antares and about 5° south of M19 is another pretty globular, M62 (NGC 6266). M62 is about the same size and slightly brighter than M19, though it looks round in a small telescope. However, M62 is one of the most asymmetrical clusters, as large telescopes show. Its nucleus lies southeast of the center of the cluster, making it look somewhat like a comet.

Slightly east and about 1½° north of M19 is NGC 6284, a small bright globular cluster that looks like a 9th-magnitude star in a medium telescope. The 8th-magnitude globular cluster NGC 6293 can be found about 1¾° east-southeast of M19.

East of Antares and north of M19, the southern part of Ophiuchus is one of the most striking regions of the Milky Way, with a vast expanse of dark nebulae lit by bright stars. Here we are looking through the spiral arms of our galaxy toward the galactic center. A good star to help you orient yourself in this area is θ (theta) Ophiuchi, a 3rd-magnitude star near $17^h22^m -25°$. About 2° east and below θ Oph, the huge Pipe Nebula extends about 7°. It is readily visible to the naked eye and is one of the largest dark clouds in the Milky Way. Absorption from the Pipe Nebula obscures the stars in an area several degrees across.

For northern-hemisphere observers, the finest portions of the Milky Way lie in Sagittarius. Moonless nights in the summer are ideal for sweeping the region with binoculars or viewing with medium-sized telescopes. The brightest part is the Great Sagittarius Star Cloud, just north of γ (gamma) Sgr. Farther west, star clouds are obscured by the Great Rift, which marks the equatorial plane of our galaxy. In this region, we are looking directly toward the center of our galaxy (the *galactic center*), about 30,000 light-years from earth. The galactic center is hidden from our optical view by interstellar dust and dark nebulae. It lies about 1½° southwest of the 4th-magnitude Cepheid variable X Sgr, a foreground object. The galactic center can be detected as part of the radio source Sgr A. Infrared, x-rays, and gamma rays from the galactic center also penetrate the absorbing dust, providing ways to study this part of our galaxy, which may contain a giant black hole.

About 2½° north and slightly west of γ (gamma) Sgr is one of the few dark nebulae that can be seen in amateur telescopes.

Farther southwest along the galactic equator in Scorpius, M6 (NGC 6405) and M7 (NGC 6475) are two large bright open clusters, both beautiful in small telescopes. Looking somewhat like the Pleiades, M6 (at 17^h40^m $-32°$) has a total magnitude brighter than 4 and contains many stars between 8th and 12th magnitude. Even a small telescope resolves M6 into its member stars.

Easily located about 3° southeast of M6, near 17^h50^m $-35°$, M7 is a brilliant cluster of 3rd magnitude, sometimes visible to the naked eye on a dark night. To the naked eye, M7 looks like a hazy glow with a rich star cloud in the background. It is one of the few open clusters that look impressive in binoculars. Medium-sized telescopes with higher power completely resolve the cluster. M7 is much larger than M6, with many stars from 6th to 8th magnitude. On the northwest edge of M7 is the globular cluster NGC 6453, visible in the same field of view under low power.

Still farther southwest—near the galactic equator at 16^h55^m $-41°$, about ¼° north of ζ (zeta) Sco—is NGC 6231, another fine open cluster (though it is unfavorably located for northern-hemisphere observers). Under good conditions, it can be seen with the naked eye by observers who are far enough south; it looks like a miniature version of the Pleiades because of its central group of seven or so brilliant white stars. NGC 6231 lies about 6,000 light-years away; if it were as close to us as the Pleiades, it would look almost the same size but would be about 50 times brighter.

NGC 6231 appears to be the center of a large group of bright stars of spectral types O and B. About 1° north and slightly east of NGC 6231 lies H12, the richest part of this group, with about 200 stars. A small telescope shows the stars of H12 as a trail extending northeast from NGC 6231, ending in a bright nebulous region. This extended group of stars marks one of the spiral arms of our galaxy, which lies more than 5,000 light-years from us—closer to the galactic center than the arm that contains our sun. IC 4628 is a faint nebula about 1¼° north of NGC 6231. A much larger loop of nebulosity about 300 light-years in diameter circles this entire association of stars in a giant irregular ring about 4° wide. The nebula is one of the giant regions of glowing hydrogen that outline the spiral arms in our own galaxy and other spirals.

Other open clusters in this region include: NGC 6242, a compact cluster about 1° north of H12; NGC 6281, about 2¼° farther to the northeast, a compact and elongated cluster; NGC 6124, a rich cluster about 6° west of H12 near 16^h25^m $-41°$; and NGC 6193, a large cluster located about 7° south-southwest of ζ (zeta) Sco.

The remnant of the supernova that Kepler saw as a nova in 1604 is at upper left, 17^h30^m $-22°$. It is faint even in photographs.

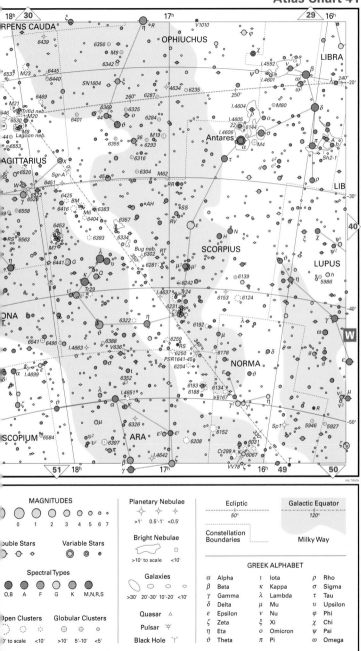

MAGNITUDES

0   1   2   3   4   5   6   7

Double Stars

Variable Stars

Spectral Types

O,B   A   F   G   K   M,N,R,S

Open Clusters

Globular Clusters

>' to scale   <10'   >10'   5'-10'   <5'

Planetary Nebulae

>1'   0.5'-1'   <0.5'

Bright Nebulae

>10' to scale   <10'

Galaxies

>30'   20'-30'   10'-20'   <10'

Quasar   △

Pulsar   ☆

Black Hole   Y

Ecliptic

50°

Constellation Boundaries

Galactic Equator

120°

Milky Way

GREEK ALPHABET

α	Alpha	ι	Iota	ρ	Rho
β	Beta	κ	Kappa	σ	Sigma
γ	Gamma	λ	Lambda	τ	Tau
δ	Delta	μ	Mu	υ	Upsilon
ε	Epsilon	ν	Nu	φ	Phi
ζ	Zeta	ξ	Xi	χ	Chi
η	Eta	ο	Omicron	ψ	Psi
ϑ	Theta	π	Pi	ω	Omega

ATLAS CHART 42. LAGOON NEBULA, TRIFID NEBULA In this region we are still looking toward the center of our galaxy from our vantage point more than halfway out on an outer spiral arm. Moonless nights in July are best for viewing the bright star clouds, open and globular clusters, and curious nebulae (fig. 7-42) in this area of the Milky Way, visible low above the southern horizon. Sagittarius is located by its "Teapot" of stars: σ, τ, ζ, and φ make its handle; to the southwest δ, γ, and ε make the spout; and λ marks the top.

γ (gamma) Sgr marks the top of the Archer's arrow in the figure of Sagittarius, on the right-hand side of the chart near 18^h06^m −30°. North and west of γ Sgr, just off the mainstream of the Milky Way near 18^h02^m −23°, are two outstanding emission nebulae—the Lagoon Nebula and the Trifid Nebula. Both are about 2,500 light-years away and, like M42 (the Orion Nebula), get their energy from ultraviolet radiation from young, hot stars in their midst.

M8 (NGC 6523), the Lagoon Nebula, is the more spectacular of the two, visible to the naked eye as a cometlike glow. It is an irregular nebula about the size of the full moon, crossed by a broad absorbing band of dust that gives the nebula its name (fig. 7-42). Smaller dust clouds can also be seen. Because of the nebula's great extent, it is best viewed with a wide-field eyepiece. The Lagoon Nebula contains dust and gas clouds where new stars form and a region of glowing hydrogen resulting from the presence of the hot stars that can be seen in its midst. The eastern half of the nebula contains the prominent star cluster NGC 6530, a scattered open cluster about 10 arc min in diameter. It is one of the youngest clusters known, not more than a few million years old, and includes several T Tauri stars (variable stars with irregular fluctuations, indicating that they have not yet settled down to a regular existence).

About 1½° north-northwest of the Lagoon Nebula is M20 (NGC 6514), the Trifid Nebula. The lower part of the Trifid is a diffuse emission nebula with gas ionized by a hot central star. It is less than 100,000 years old. In photographs we see the reddish radiation from the cloud of glowing hydrogen, which is divided into three parts by dust lanes (figs. 5-28, 5-32, and 7-42). Where the three dark lanes overlap, we find a central triple star. These lanes are fairly easy to detect with a medium-sized telescope at medium power. The bluish upper part of M20, unconnected to the lower part, is a reflection nebula.

M21 is a small rich open cluster of magnitude 5.9, located in the same low-power field as M20 (the Trifid). About 1° southeast of M8 (the Lagoon) is NGC 6544, a small very remote globular cluster that is difficult to find. One degree farther southeast is

Fig. 7-42 M8 (NGC 6523), the Lagoon Nebula, and M20 (NGC 6514), the Trifid Nebula, in Sagittarius. (Alan Dyer)

NGC 6553, another globular cluster; it is very difficult to resolve in part because thick obscuring dust dims it by about 6 magnitudes. About 1½° east-northeast of M8 are several irregular nebulae, possibly connected to the Lagoon Nebula by a faint gaseous haze.

λ (lambda) Sgr, a few degrees to the east of M8 near 18^h28^m −25°, also lies in a rich field. M22 (NGC 6656) is an impressive globular cluster, 2° northeast of λ Sgr, near 18^h36^m −24°. At 5th magnitude, M22 is the third-brightest and most easily resolved globular cluster in the sky. Its only equal in the northern sky is M13, the Hercules Cluster (see Atlas Chart 17). M22's brightest stars are 11th magnitude, and it has a total population of at least half a million stars. It is about 10,400 light-years away. Since it is less than 1° from the ecliptic, we can sometimes see a planet in the same field. About 1° west-northwest of M22 is NGC 6642, a small faint globular cluster that looks like a hazy spot.

Slightly northwest of λ (lambda) Sgr, at 18^h24^m −25° in the same low-power field, is M28 (NGC 6626), a bright globular cluster that looks like a star in a rich star background. It is one of the most compact globulars, not very striking in a medium-sized telescope because it is hard to resolve. Slightly closer on the east side of λ Sgr is NGC 6638, a small remote globular cluster that is difficult to resolve.

About 5° southwest of λ Sgr is δ (delta) Sgr, with the faint globular cluster NGC 6624 located slightly to the southeast. About 3° southeast of δ and about 2½° northeast of ε (epsilon) Sgr, near 18^h30^m −32°, is M69 (NGC 6637), a small 8th-magnitude globular cluster resolved only with large telescopes. About 2° east of M69 and almost halfway between ζ (zeta) and ε Sgr is the globular cluster M70 (NGC 6681), similar in size and brightness to M69. A third, smaller globular cluster is NGC 6652, about 1° southeast of M69. All three globular clusters look like bright stars in a uniformly rich star field and lie on the far side of our galaxy.

Northeast of M70 and about 2° southwest of ζ Sgr, near 18^h55^m −31°, is M54 (NGC 6715), a small, bright globular cluster of 8th magnitude. Because M54 is highly concentrated, though, only large telescopes begin to resolve it.

About 7¼° south of ζ Sgr is a field of bright and dark nebulae and obscuring dust in Corona Australis. NGC 6726-27 is the brightest part of the field. Slightly southeast is the reflection nebula NGC 6729, whose variations in brightness follow the variations of its illuminating star, R CrA.

Farther east in Sagittarius the richness of the star fields diminishes rapidly. Sweeping through this region does reveal one noteworthy globular cluster, M55 (NGC 6809), near 19^h40^m −31°, about 7° east from ζ Sgr and slightly to the south. Readily visible to the naked eye, M55 looks like a hazy star; in a small telescope it looks like a circular glow about 10 arc min in diameter. Since M55 appears so low in the sky to northern observers, it is difficult to resolve when viewed from northern latitudes.

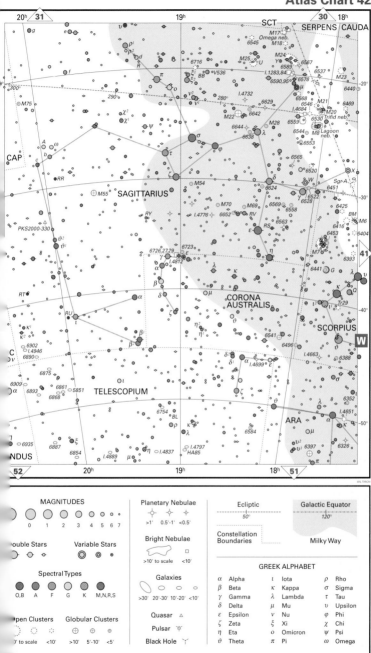

MAGNITUDES

0 1 2 3 4 5 6 7

Double Stars

Variable Stars

Spectral Types

O,B A F G K M,N,R,S

Open Clusters

>10' to scale   <10'

Globular Clusters

>10'   5'-10'   <5'

Planetary Nebulae

>1'   0.5'-1'   <0.5'

Bright Nebulae

>10' to scale   <10'

Galaxies

>30'   20'-30'   10'-20'   <10'

Quasar △

Pulsar ☿

Black Hole Y

Ecliptic
50°

Constellation Boundaries

Galactic Equator
120°

Milky Way

GREEK ALPHABET

α	Alpha	ι	Iota	ρ	Rho
β	Beta	κ	Kappa	σ	Sigma
γ	Gamma	λ	Lambda	τ	Tau
δ	Delta	μ	Mu	υ	Upsilon
ε	Epsilon	ν	Nu	φ	Phi
ζ	Zeta	ξ	Xi	χ	Chi
η	Eta	o	Omicron	ψ	Psi
ϑ	Theta	π	Pi	ω	Omega

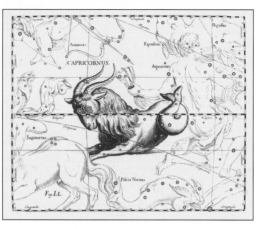

Fig. 7-43 Capricornus, the Goat from the star atlas of Hevelius (1687). The constellation is drawn backward from the way it appears in the sky because Hevelius drew the celestial sphere as it would appear from the outside. (Jay M. Pasachoff)

ATLAS CHART 43. CAPRICORNUS In the rather barren area near the borders of Capricornus and Sagittarius, it is easy to locate M75 (NGC 6864), near 20^h06^m $-22°$ and less than $7°$ southwest of β (beta) Cap (drawn above the top border of the chart). M75 is a small rich globular cluster of 8th magnitude, bright enough to be readily visible in good binoculars when it is low in the southern sky on clear August nights. Since it is one of the most compact globular clusters, it can be resolved only in telescopes of fairly large aperture. M75 lies on the far side of our galaxy, about 60,000 light-years away.

The most prominent globular cluster in Capricorn is M30 (NGC 7099), on the top left side of this chart. It is easily found with binoculars or small telescopes, about $4°$ southeast of ζ (zeta) Cap and slightly northwest of the 5th-magnitude star 41 Cap, near 21^h40^m $-23°$. Like M75, though, M30 is difficult to resolve even with larger telescopes because of its extremely compact center. It lies about 26,000 light-years away.

One of the farthest objects known in the universe, the quasar PKS 2000-330, is at the right middle of the chart. Because of the expansion of the universe, it is receding from us so rapidly that its spectral lines are redshifted by 3.7 times (370 percent of) their original wavelengths. That puts this quasar more than 12 billion light-years away, which means that we are seeing it as it was 12 billion years ago.

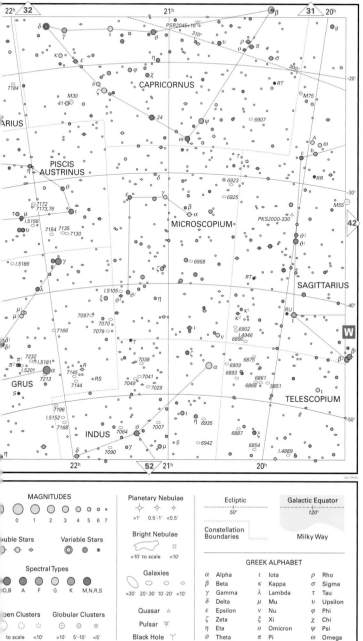

MAGNITUDES								
0	1	2	3	4	5	6	7	

Double Stars	Variable Stars

Spectral Types					
O,B	A	F	G	K	M,N,R,S

Open Clusters		Globular Clusters		
to scale	<10'	>10'	5'-10'	<5'

Planetary Nebulae		
>1'	0.5'-1'	<0.5'

Bright Nebulae
>10' to scale   <10'

Galaxies				
>30'	20'-30'	10'-20'	<10'	

Quasar △
Pulsar ☆
Black Hole

Ecliptic
50°

Constellation
Boundaries

Galactic Equator
120°

Milky Way

	GREEK ALPHABET				
α	Alpha	ι	Iota	ρ	Rho
β	Beta	κ	Kappa	σ	Sigma
γ	Gamma	λ	Lambda	τ	Tau
δ	Delta	μ	Mu	υ	Upsilon
ε	Epsilon	ν	Nu	φ	Phi
ζ	Zeta	ξ	Xi	χ	Chi
η	Eta	ο	Omicron	ψ	Psi
ϑ	Theta	π	Pi	ω	Omega

Fig. 7-44 An amateur photo of the Helix Nebula (NGC 7293), a planetary nebula in Aquarius. See also figures 5-9 and 5-10 for higher-resolution views. (Tim Puckett)

ATLAS CHART 44. FOMALHAUT, HELIX NEBULA This region is uninteresting except for α (alpha) Piscis Austrini, a bright white star whose brilliance is intensified by the comparative darkness of the starless background. α PsA is also called Fomalhaut, a name derived from the Arabic for "mouth of the fish." At magnitude 1.16, Fomalhaut is the eighteenth-brightest star in the sky. To northern-hemisphere observers it is visible in autumn, low above the horizon in an empty region of the southern sky.

Though not readily visible to observers at northern latitudes, an interesting object in this region is the planetary nebula NGC 7293, near $22^h30^m -21°$ in Aquarius; find it by looking 21° south from ζ (zeta) Aqr. NGC 7293 is the Helix Nebula, a huge flattened disk of gas about 3 light-years in diameter surrounding a 13th-magnitude central star (fig. 5-9 and above). A planetary results from the death of a star containing about as much mass as the sun. The Helix is the largest planetary, about half the apparent diameter of the full moon. Although its total brightness is magnitude 6.5, its surface brightness is low because of its large size, making it difficult to detect except with low power on very dark nights. Even on the best nights, the nebula shows only as a faint, circular haze; little structure is visible except in larger telescopes and on long-exposure photographs (fig. 5-10). To observe it, try the technique of averted vision: look not directly at the nebula but toward one side, which brings its image onto a more sensitive area of your retina.

The double star 41 Aqr lies several degrees west and slightly south of the Helix Nebula. Its contrasting components are magnitudes 5.6 and 7.1. Separated by 5.0 arc sec, the components can be resolved in small telescopes.

Southern constellations such as Grus, the Crane, are relatively unfamiliar to most northern-hemisphere observers. θ (theta) Gru, at $23^h07^m -44°$, is another double star that can be separated in a small telescope, with components of magnitudes 4.5 and 7.0.

In the constellation Phoenix, SX Phe, a dwarf Cepheid variable, lies about 7½° west of α (alpha) Phe. SX was originally discovered as a star with large proper motion in 1938; it is about 400 light-years away. Its period is about 79 minutes, ranging in magnitude from about 6.8 to 7.5.

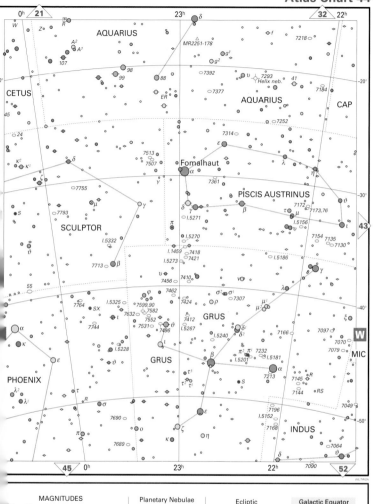

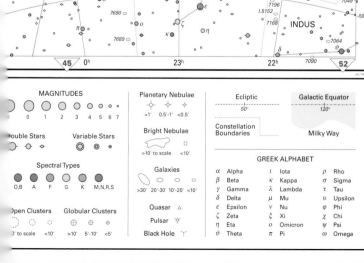

MAGNITUDES								
	0	1	2	3	4	5	6	7

Double Stars   Variable Stars

Spectral Types
O,B   A   F   G   K   M,N,R,S

Open Clusters   Globular Clusters
>10' to scale   <10'   >10'   5'-10'   <5'

Planetary Nebulae
>1'   0.5'-1'   <0.5'

Bright Nebulae
>10' to scale   <10'

Galaxies
>30'   20'-30'   10'-20'   <10'

Quasar △
Pulsar ☿
Black Hole

Ecliptic
50°

Constellation
Boundaries

Galactic Equator
120°

Milky Way

	GREEK ALPHABET				
α	Alpha	ι	Iota	ρ	Rho
β	Beta	κ	Kappa	σ	Sigma
γ	Gamma	λ	Lambda	τ	Tau
δ	Delta	μ	Mu	υ	Upsilon
ε	Epsilon	ν	Nu	φ	Phi
ζ	Zeta	ξ	Xi	χ	Chi
η	Eta	o	Omicron	ψ	Psi
ϑ	Theta	π	Pi	ω	Omega

Fig. 7-45 The Small Magellanic Cloud in Tucana and the globular cluster 47 Tucanae. (Ronald Royer)

ATLAS CHART 45. SMALL MAGELLANIC CLOUD, 47 TUCANAE We begin here a series of eight polar charts showing the region surrounding the south celestial pole. The most interesting objects in this chart are the Small Magellanic Cloud and the bright globular cluster 47 Tucanae (NGC 104). Both can be found slightly below the center of the chart, near 0^h30^m $-72°$. See also fig. 7-53.

The Small Magellanic Cloud and its companion galaxy, the Large Magellanic Cloud (see Atlas Chart 47), are two of the closest galaxies to the Milky Way Galaxy. The SMC and the LMC are separated by about 80,000 light-years.

The Small Magellanic Cloud can be observed with all ranges of telescopic power by observers in the southern hemisphere. To the naked eye it looks like a cloud about $3\frac{1}{2}°$ across, with its hazy outline contrasting with the dark background.

Many open and globular clusters have been found in the Small Magellanic Cloud. Both the Small and the Large Magellanic Clouds also contain many Cepheid variables. In 1917 Harlow Shapley, using the relationship between the periods and luminosities of the Cepheid variables, calculated that the Magellanic Clouds were very distant, with the Small Cloud now known to be about 195,000 light-years away from us. This calculation sparked a tremendous debate about whether the Magellanic Clouds were within our galaxy or were far outside its bounds and therefore were separate galaxies themselves. We now know that they are two of the dozens of members of the "Local Group" of galaxies to which our Galaxy belongs.

Slightly to the west of the Small Magellanic Cloud, though not connected to it, is 47 Tucanae, a large highly concentrated globular cluster that is bright enough (total magnitude 4) to be easily visible to the naked eye. 47 Tuc is one of the nearest globular clusters, only 14,000 light-years away.

In a similar direction is a second globular cluster, NGC 362, about twice the distance of 47 Tuc from us. It consists of stars of magnitudes 13 to 14 and also is highly compact. NGC 362 is barely visible to the naked eye at magnitude 6.6.

The noteworthy double stars in this region include β (beta) Tuc (at 0^h32^m $-63°$), a wide double that can be resolved with low power into a pair of 4.4- and 4.8-magnitude stars. This system is actually a sextuple, though the other components are relatively faint. κ (kappa) Tuc, (near 1^h16^m $-69°$) is a blue-white pair, of magnitudes 5.1 and 7.3, that can be resolved in a small telescope. At least two fainter components are also present.

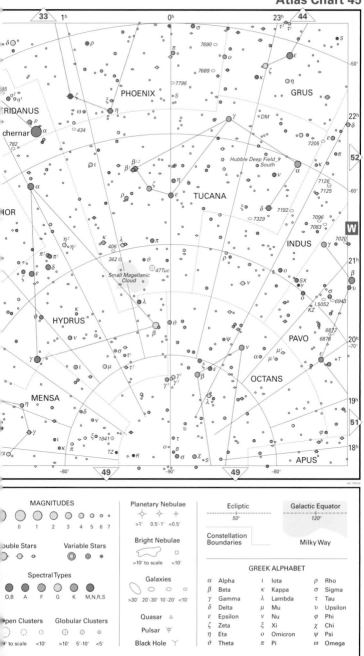

MAGNITUDES							
0	1	2	3	4	5	6	7

Planetary Nebulae
>1' 0.5'-1' <0.5'

Ecliptic
50°

Galactic Equator
120°

Double Stars **Variable Stars**

Bright Nebulae
>10' to scale <10'

Constellation Boundaries

Milky Way

Spectral Types
O,B A F G K M,N,R,S

Galaxies
>30' 20'-30' 10'-20' <10'

GREEK ALPHABET

α	Alpha	ι	Iota	ρ	Rho
β	Beta	κ	Kappa	σ	Sigma
γ	Gamma	λ	Lambda	τ	Tau
δ	Delta	μ	Mu	υ	Upsilon
ε	Epsilon	ν	Nu	φ	Phi
ζ	Zeta	ξ	Xi	χ	Chi
η	Eta	o	Omicron	ψ	Psi
ϑ	Theta	π	Pi	ω	Omega

Open Clusters **Globular Clusters**
' to scale <10' >10' 5'-10' <5'

Quasar △

Pulsar ☼

Black Hole Y

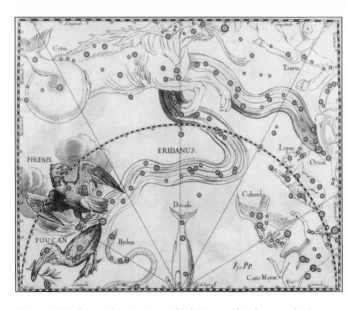

Fig. 7-46 Eridanus, the river into which Cygnus dived in search of a friend. Cygnus was then changed by Apollo into a swan and placed in the sky. The sea-serpent Hydrus is nearby. The drawing of the constellation is backward, as are all drawings from the star atlas of Hevelius. (Jay M. Pasachoff)

ATLAS CHART 46. ACHERNAR This chart shows an area with few bright stars, except for Achernar, α (alpha) Eridani, near 1^{h}38^m −57°. Achernar, whose name means "the star at the end of the river," lies at the southernmost point of Eridanus. At magnitude 0.46, it is the ninth-brightest star in the sky. Though it is never visible from most of the continental U.S., on autumn evenings it can sometimes be seen just above the southern horizon from southern Texas and Florida, and it appears higher in the sky from Hawaii. Achernar is a hot blue giant, about 650 times as bright as the sun and about 120 light-years away. Just north of Achernar is p Eri, a wide orange pair of 6th-magnitude stars.

Below α (alpha) Reticuli (the Net), which is located at 4^{h}14^m −62°, θ (theta) Ret is a double star with components of magnitudes 6.2 and 8.3, separated by 3.9 arc sec. A couple of degrees northeast of θ Ret, R Ret is a long-period variable that ranges in magnitude from 6.5 to 14.0. Its period is 278 days.

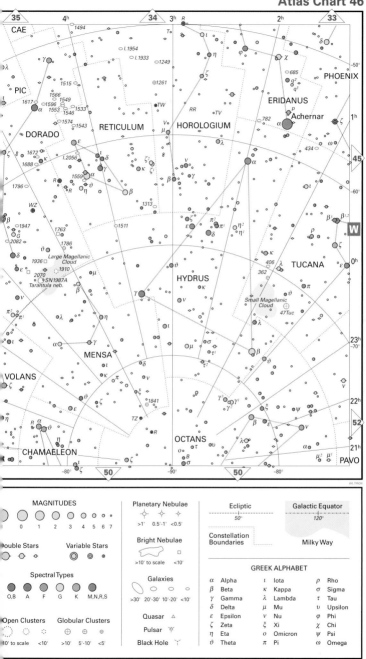

MAGNITUDES								
	●	●	◯	◦	·	·	·	
	0	1	2	3	4	5	6	7

Double Stars ◑ ◔ ◇
Variable Stars ◉ ◎ ●

Spectral Types
● ● ● ◯ ● ●
O,B A F G K M,N,R,S

Open Clusters **Globular Clusters**
⬡ ⬡ ⬡ ⊕ ⊕ ⊜
10' to scale <10'   >10' 5'-10' <5'

Planetary Nebulae ✦ ✧ ✧
>1' 0.5'-1' <0.5'

Bright Nebulae ⬡ ▢
>10' to scale <10'

Galaxies ⬭ ◯ ◯ ◦
>30' 20'-30' 10'-20' <10'

Quasar △
Pulsar ☿
Black Hole ⅄

Ecliptic

50°

Constellation Boundaries ⋯⋯⋯

Galactic Equator
120°

Milky Way

GREEK ALPHABET

α	Alpha	ι	Iota	ρ	Rho
β	Beta	κ	Kappa	σ	Sigma
γ	Gamma	λ	Lambda	τ	Tau
δ	Delta	μ	Mu	υ	Upsilon
ε	Epsilon	ν	Nu	φ	Phi
ζ	Zeta	ξ	Xi	χ	Chi
η	Eta	o	Omicron	ψ	Psi
ϑ	Theta	π	Pi	ω	Omega

Fig. 7-47 The Large Magellanic Cloud in Dorado. (Akira Fujii)

ATLAS CHART 47. LARGE MAGELLANIC CLOUD, TARANTULA NEBULA, CANOPUS In the center of this region is the Large Magellanic Cloud, the huge irregular companion galaxy to the Small Magellanic Cloud (Chart 45). Covering a vast area of the sky, the Large Magellanic Cloud (figs. 5-15, 5-16, and 5-58) can be resolved even with relatively small telescopes. It looks like a hazy cloud to the naked eye and is bright enough to be visible even under a full moon.

The central part of the LMC extends over 20,000 light-years in length, while the whole galaxy extends at least 50,000 light-years. It contains at least 30 billion stars, including some supergiants larger than any in our galaxy. At least 400 planetary nebulae and more than 700 open clusters lie within the Large Magellanic Cloud, along with about 60 globular clusters, most of which are similar to those found in our galaxy.

The most striking areas of the LMC are the huge bright regions of glowing hydrogen where supergiant stars provide the energy for emission nebulae. More than 50 diffuse nebulae are visible with medium-sized telescopes. The most outstanding of these is the Tarantula Nebula (NGC 2070), a looped nebula surrounding the star 30 Doradus (in Dorado, the Swordfish). Visible to the naked eye, the Tarantula Nebula is the largest diffuse nebula known in the universe. Its complex structure shows much detail (fig. 5-16). More than 100 supergiant stars cluster at its center, a region of nebulosity where stars are forming. Nearby, a supernova exploded on February 24, 1987; for a few weeks it was bright enough to be seen with the naked eye. As it fades, astronomers are studying it carefully (chapter 5). It may brighten again any time as the material from the explosion runs into material ejected earlier; it bears monitoring to see if it returns to naked-eye brightness. Studies of the supernova have shown that it is 169,000 light-years away.

The brightest star in the Large Magellanic Cloud is S Doradus in the open cluster NGC 1910. S Dor is an irregular variable star that ranges from magnitude 8.6 to 11.7. Its average luminosity is more than 500,000 times that of the sun.

Near 6^h24^m $-53°$ is α (alpha) Carinae, also known as Canopus. At magnitude −0.72, Canopus is the second-brightest star in the sky, surpassed in brightness only by Sirius. From the southern half of the U.S. it can be seen during the winter months low on the southern horizon, passing south of us about 20 minutes before Sirius does. Because it is so bright and so isolated from other bright stars in the sky, Canopus is often used for navigation by spacecraft going to the outer planets.

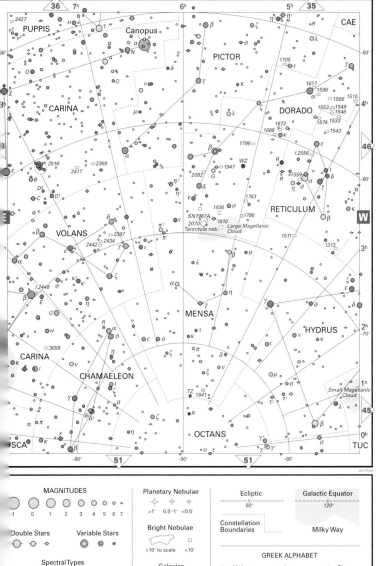

PUPPIS
Canopus
PICTOR
CAE
CARINA
DORADO
RETICULUM
VOLANS
WZ
SN1987A
Tarantula neb.
Large Magellanic Cloud
MENSA
HYDRUS
CHAMAELEON
OCTANS
Small Magellanic Cloud
TUC
MUSCA

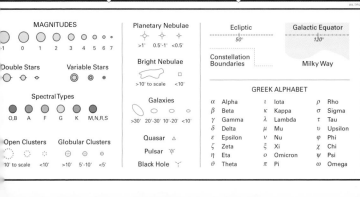

MAGNITUDES								
-1	0	1	2	3	4	5	6	7

Double Stars   Variable Stars

Spectral Types
O,B   A   F   G   K   M,N,R,S

Open Clusters   Globular Clusters
>10' to scale   <10'   >10'   5'-10'   <5'

Planetary Nebulae
>1'   0.5'-1'   <0.5'

Bright Nebulae
>10' to scale   <10'

Galaxies
>30'   20'-30'   10'-20'   <10'

Quasar   △
Pulsar   ☼
Black Hole   ⚉

Ecliptic
50'

Constellation
Boundaries

Galactic Equator
120'

Milky Way

GREEK ALPHABET

α	Alpha	ι	Iota	ρ	Rho
β	Beta	κ	Kappa	σ	Sigma
γ	Gamma	λ	Lambda	τ	Tau
δ	Delta	μ	Mu	υ	Upsilon
ε	Epsilon	ν	Nu	φ	Phi
ζ	Zeta	ξ	Xi	χ	Chi
η	Eta	ο	Omicron	ψ	Psi
ϑ	Theta	π	Pi	ω	Omega

ATLAS CHART 48. SOUTHERN MILKY WAY In Carina, Vela, and Centaurus, the southern Milky Way again comes into view. Here, along the galactic equator, the Milky Way's brilliant star fields are extremely rich, rewarding amateurs for hours of random scanning with binoculars or small telescopes.

The numerous open clusters all merit observation. Among these are NGC 3114 (at 10^h03^m $-60°$), NGC 3293 (at 10^h35^m $-58°$), and NGC 2516 (at 8^h $-61°$). One of the finest open clusters visible to observers in the lower latitudes of the northern hemisphere, NGC 2516 should be viewed with a wide-field telescope because of its great size (about 1° or 15 light-years). It lies about 1,300 light-years away and is older than the double cluster in Perseus (an extremely young cluster, only 10 million years old), and younger than the Pleiades (still fairly young at 100 million years old, less than 1 percent of the age of the universe).

One noteworthy globular cluster in this region is NGC 2808, above the center of the chart, near 9^h47^m $-65°$. It is a large, rich cluster of 13th- to 15th-magnitude stars, covering an area about 5 arc min across. At its center is a brilliant group of loosely packed stars.

Of the double stars, H Vel can be resolved in a small telescope. It is a blue-orange pair with magnitudes 4.9 and 7.7. υ (upsilon) Car (at 9^h47^m $-65°$), at magnitude 3.1, is another double that can be resolved in small telescopes; its 6th-magnitude secondary component is 5 arc sec away from the third-magnitude primary. t^2 Car, near 10^h40^m $-59°$, is a wide orange-green pair.

At 9^h45^m $-63°$, l ("ell") Car is one of the brightest Cepheid variables. It reaches magnitude 3.3 at maximum and is visible to the naked eye throughout its cycle. It is too far south for most U.S. observers, though. This star is one of the largest Cepheids known, with a diameter about 200 times that of the sun. l Car has a period of 35½ days and is more than 3,000 light-years away.

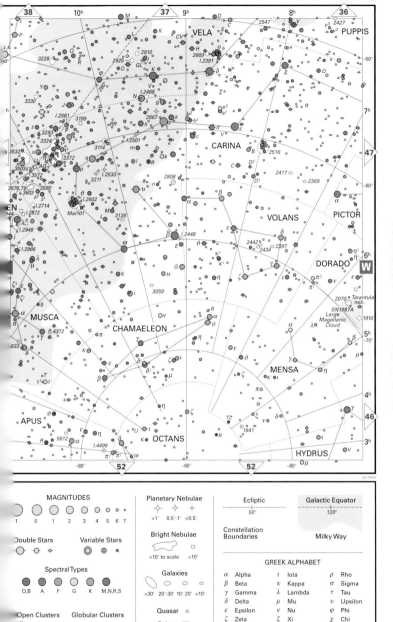

MAGNITUDES

| 1 | 0 | 1 | 2 | 3 | 4 | 5 | 6 | 7 |

Double Stars

Variable Stars

Spectral Types

O,B   A   F   G   K   M,N,R,S

Open Clusters   **Globular Clusters**

10' to scale   <10'   >10'   5'-10'   <5'

Planetary Nebulae

>1'   0.5'-1'   <0.5'

Bright Nebulae

>10' to scale   <10'

Galaxies

>30'   20'-30'   10'-20'   <10'

Quasar   △

Pulsar   ☆

Black Hole

Ecliptic
50°

Galactic Equator
120°

Constellation Boundaries

Milky Way

GREEK ALPHABET

α	Alpha	ι	Iota	ρ	Rho
β	Beta	κ	Kappa	σ	Sigma
γ	Gamma	λ	Lambda	τ	Tau
δ	Delta	μ	Mu	υ	Upsilon
ε	Epsilon	ν	Nu	φ	Phi
ζ	Zeta	ξ	Xi	χ	Chi
η	Eta	ο	Omicron	ψ	Psi
ϑ	Theta	π	Pi	ω	Omega

The Milky Way continues here through Carina, Crux, and Centaurus, offering outstanding sights for observers in southern latitudes. In the upper right-hand corner of the chart, near 10^h45^m $-59°$, the Eta Carinae Nebula (NGC 3372) warrants special attention. Visually the brightest part of the Milky Way, this extended nebula shows a highly complex structure with intermingling dark and light lanes and patches. The dark cloud superimposed on the brightest area at the center of the nebula is the Keyhole Nebula (NGC 3324). The brightest star to the left of the Keyhole is the variable star η (eta) Carinae, lying about 8,000 light-years from us.

η (eta) Carinae itself, 10^h45^m–60, is brightening and could be starting to erupt. η Car is surrounded by a small nebulous shell that is expanding at the rate of about 4 arc sec per century. Although this variable star shows some similarities to most novae, it also shows some peculiar differences: it remained bright for more than a century and was unusually luminous at maximum. First noted by Halley in 1677, it varied in brightness during the nineteenth century, finally reaching its maximum of magnitude –1 in 1843. It faded to 8th magnitude and has since brightened to 5th magnitude, becoming at least four times brighter in the second half of the twentieth century. With a mass approximately 100 times that of the sun, it may explode as a supernova; the presence of some of the heavy elements in unusual amounts, discovered by analysis of its spectrum, shows that this star is in an advanced stage of life.

Slightly north of the Eta Carinae Nebula is the open cluster NGC 3293, surrounded by reflection nebulae. The cluster is only 1,500 light-years away from us and is not physically connected to the Eta Carinae Nebula. About 3° northeast of Eta Carinae is another open cluster, NGC 3532. Since it is very large and elongated, it is best viewed with a wide-field telescope. This superb cluster contains at least 400 members, of which more than 150 stars are 12th magnitude or brighter. About 5° south of the nebula is the scattered open cluster IC 2602. θ (theta) Car is the bright central star of this cluster, which contains 30 stars that are brighter than 9th magnitude and many fainter ones. The cluster is more than 1° across and is best viewed with a rich-field telescope or at low power. With its central star θ Car at a distance of only 750 light-years from earth, IC 2602 may be one of the nearest open clusters to us.

Farther east (left) along the Milky Way is IC 2944 (at 11^h35^m $-63°$), a faint shell around the star λ (lambda) Cen. IC 2944, along with many other regions of nebulosity visible here, lies in a region of ionized hydrogen that marks one of the spiral arms of

Fig. 7-49 *The Southern Milky Way, from the Southern Cross to Carinae. Acrux (alpha Crucis) is at left center, just at the right edge of the dark Coalsack. It is the southernmost point of the Southern Cross, which looks like a kite. Mimosa is the bright star to the upper left of Acrux. The other two stars of the southern cross are fainter and continue around clockwise. The red (hydrogen-alpha) complex of nebulosity around η (eta) Carinae is at right center. This star may go supernova at any time. A meteor also shows. (Daniel Good)*

our galaxy. Slightly east (left) of IC 2944 is the loose open cluster IC 2948, lying behind the densest part of IC 2944. To the north, on the galactic equator (near $11^h35^m -62°$), NGC 3766 is a very rich, concentrated open cluster that contains at least 200 stars from 8th to 15th magnitude; it covers an area about 10 arc min across. Beautiful even through binoculars, it is similar in appearance to M37 in Auriga (see Atlas Chart 11). About 5° to the north is the 11th-magnitude planetary nebula NGC 3918. It appears as a rich blue featureless disk and has been compared to the planet Uranus.

Continuing eastward (to the left) along the Milky Way, we come to Crux, a small constellation nearly surrounded by Centaurus. Crux contains the famous Southern Cross, a diamond-shaped constellation. α (alpha) Cru—Acrux—marks the foot of the cross. β (beta) Cru—Mimosa—to the northeast marks the eastern end of the crosspiece. γ (gamma) Cru—Gacrux—to the northwest (upper right on the chart) marks the head of the cross and is nearly the same brightness as β. δ (delta) Cru, to the southwest of γ, marks the western end of the crosspiece. Three of these stars are among the 30 brightest stars in the sky.

At 1st magnitude, α Cru (Acrux) lies nearly 400 light-years away. It is an easily resolved double, with two blue stars of magnitudes 1.4 and 1.9, separated by 4½ arc sec. Each of these components is also a spectroscopic double.

Fig. 7-50 The Jewel Box (NGC 4755), an open cluster surrounding the red supergiant star κ (kappa) Crucis, the leftmost star of the Southern Cross, above the Coalsack. It was named from the comment of Sir John Herschel, who mapped the southern sky in the early nineteenth century, that in a telescope the stars looked like precious stones. The eight brightest stars range from about 6th to 10th magnitude and are much bluer than κ Crucis; the cluster is about 10 arc min across. (Gabriel Martin, National Optical Astronomy Observatories/CTIO)

β Cru (Mimosa), 20 percent farther away than α Cru, is the nineteenth-brightest star at magnitude 1.25. Nearly as bright is γ Cru, only about 90 light-years away. Like Antares, it is a red giant, though not so large and bright. An optical double, its components show a striking contrast in both color and magnitude.

Just to the south of β Cru is a magnificent open cluster called the Jewel Box (NGC 4755). It is one of the finest objects in the southern Milky Way. This cluster contains more than 100 stars scattered over an area 50 light-years across. About 50 of its brightest stars, most notably the 6th-magnitude star κ (kappa) Cru, are concentrated at its center. Many of the cluster's stars are supergiants, including some of the brightest ones known in our galaxy. Most of its members are bluish white or white, contrasting with a single red supergiant—like diamonds and a ruby. Not more than a few million years old, the Jewel Box is one of the youngest open clusters known. Visible with the naked eye, it should be viewed with low power because of the richness of the surrounding star field.

South of κ Cru and east of α Cru is an almost starless area, caused by a vast dust cloud about 7° by 5° across (60–70 light-years in diameter) that obscures the Milky Way in the background. This is the Coalsack, the most famous of the naked-eye dark nebulae. The few stars visible through telescopes are foreground objects lying between us and the obscuring dust. The Coalsack is one of the nearest dark nebulae, only 500–600 light-years away. It is easily apparent to the naked eye.

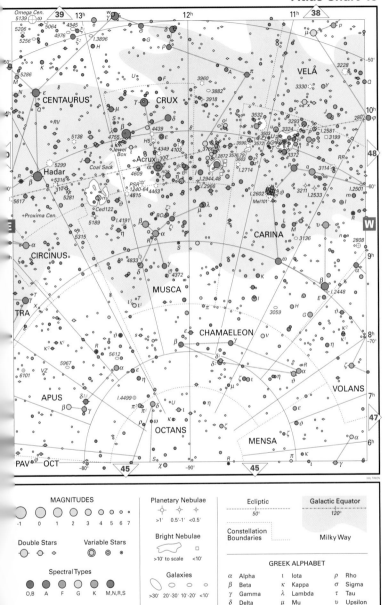

MAGNITUDES								
-1	0	1	2	3	4	5	6	7

Double Stars

Variable Stars

Spectral Types

O,B A F G K M,N,R,S

Open Clusters

>10' to scale   <10'

Globular Clusters

>10'   5'-10'   <5'

Planetary Nebulae

>1'   0.5'-1'   <0.5'

Bright Nebulae

>10' to scale   <10'

Galaxies

>30'   20'-30'   10'-20'   <10'

Quasar △

Pulsar ☆

Black Hole

Ecliptic

50°

Constellation Boundaries

Galactic Equator

120°

Milky Way

GREEK ALPHABET

α	Alpha	ι	Iota	ρ	Rho
β	Beta	κ	Kappa	σ	Sigma
γ	Gamma	λ	Lambda	τ	Tau
δ	Delta	μ	Mu	υ	Upsilon
ε	Epsilon	ν	Nu	φ	Phi
ζ	Zeta	ξ	Xi	χ	Chi
η	Eta	ο	Omicron	ψ	Psi
ϑ	Theta	π	Pi	ω	Omega

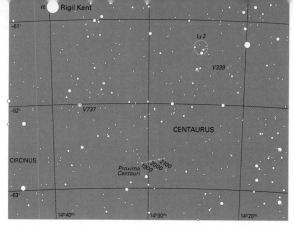

Fig. 7-51 A chart showing the region near Proxima Centauri, the closest star to the sun. Because it is so close, over a period of years we can see it move with respect to more distant stars, as marked. (Wil Tirion)

ATLAS CHART 50. ALPHA CENTAURI Through Centaurus and Norma, the brightness of the Milky Way is divided by clouds of obscuring dust. The most distinctive features in this region are the two bright stars α (alpha) Centauri and β (beta) Centauri (magnitudes 0.0 and 0.6, respectively), located above the center of the chart. The former is known in navigational circles as Rigil Kent (short for Kentaurus), while the latter is sometimes called Hadar or Wazn. A bright pair of stars in the southern sky separated by 4½° (two fingers' width) often turns out to be α and β Centauri (Cen).

The fourth-brightest star in the night sky at magnitude 0.0, α Cen A is in the nearest star system to the earth. Alpha is 4.4 light-years away from us. It is a multiple system; its primary star is almost three times as bright as its secondary. A third, fainter companion lies farther away from the primary and secondary stars, in a direction closer to the sun. This third component is Proxima Centauri (see fig. 7-51), 4.2 light-years from us. Proxima ("the nearest") is a red dwarf with a diameter about ½₀ that of the sun. One of the smallest stars known, it is one-third the size of Jupiter, though its mass is much greater.

β Centauri is the eleventh-brightest star at magnitude 0.61. It also has a companion star only 1.3 arc sec away; the pair is difficult to resolve because of the primary's brightness. β Cen is a blue-giant star, about 10,000 times as bright as the sun.

The Hourglass Nebula (fig. 5-8), also known as MyCn18 (the 18th object in a catalog by Mayall and Cannon), is too faint to show on the Atlas Chart. It is a planetary nebula at 13ʰ40ᵐ −67°23' in Musca.

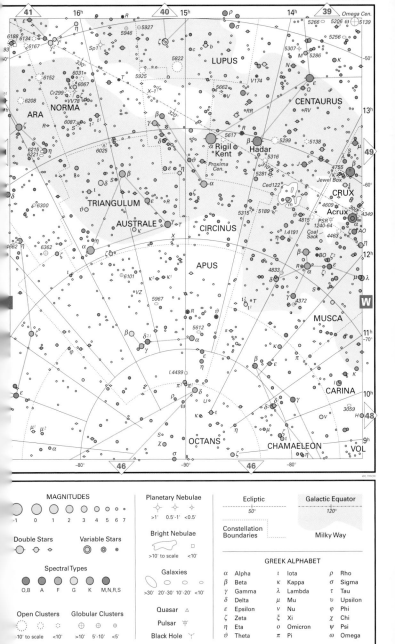

MAGNITUDES

-1 0 1 2 3 4 5 6 7

Double Stars

Variable Stars

Spectral Types

O,B A F G K M,N,R,S

Open Clusters

10' to scale <10'

Globular Clusters

⊕ >10' ⊕ 5'-10' ⊕ <5'

Planetary Nebulae

>1' 0.5'-1' <0.5'

Bright Nebulae

>10' to scale <10'

Galaxies

>30' 20'-30' 10'-20' <10'

Quasar △

Pulsar ☆

Black Hole ˚·˚

Ecliptic

50°

Galactic Equator

120°

Constellation Boundaries

Milky Way

GREEK ALPHABET

α	Alpha	ι	Iota	ρ	Rho
β	Beta	κ	Kappa	σ	Sigma
γ	Gamma	λ	Lambda	τ	Tau
δ	Delta	μ	Mu	υ	Upsilon
ε	Epsilon	ν	Nu	φ	Phi
ζ	Zeta	ξ	Xi	χ	Chi
η	Eta	ο	Omicron	ψ	Psi
ϑ	Theta	π	Pi	ω	Omega

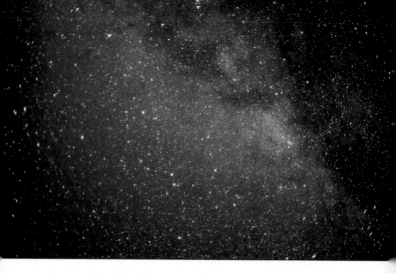

Fig. 7-52 The Milky Way in Ara and Norma (Ronald Royer).

ATLAS CHART 51. OCTANS, PAVO, APUS, ARA As we move southeast of the Milky Way in this chart, the star fields diminish in richness, though globular clusters become more apparent. NGC 6397 (at $17^h40^m -54°$) in Ara, the Altar, is a naked-eye globular cluster of exceptional brilliance. It can be found by looking toward the eastern (left) edge of the Milky Way, about $10\frac{1}{2}°$ south of θ (theta) Sco, though it can't be seen from most of the U.S. It is a loose scattered group with a total magnitude of 5.7, easily resolved in small telescopes. About 7,200 light-years distant, NGC 6397 may be the nearest globular cluster to our solar system.

Another exceptionally bright cluster, NGC 6752 (at $19^h10^m -60°$), has a total magnitude of 5.4, placing it among the half-dozen brightest globular clusters. The third-largest in apparent size, exceeded only by ω (omega) Cen and 47 Tuc, NGC 6752 is also one of the nearest globular clusters, about 12,700 light-years away.

A noteworthy double star in this region is ξ (xi) Pav (at $18^h23^m -62°$), an orange-green pair with a primary component of magnitude 4.4 and a 9th-magnitude secondary.

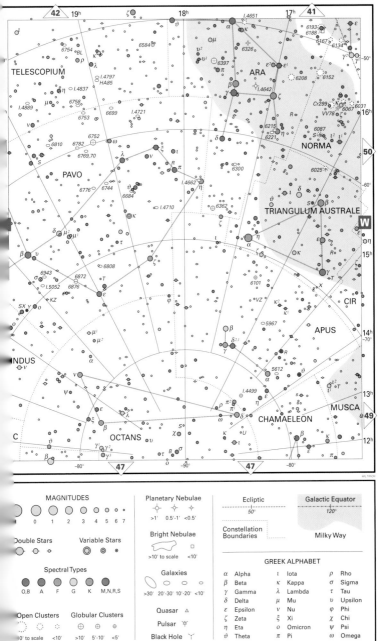

MAGNITUDES

○ ○ ○ ○ ○ ○ ○ ∘ ·
0 1 2 3 4 5 6 7

Double Stars
○⊙ ○⊙ ◇

Variable Stars
◎ ⊙ ⊙

Spectral Types
● ● ● ● ●
O,B A F G K M,N,R,S

Open Clusters
10' to scale <10'

Globular Clusters
⊕ ⊕ ⊙
>10' 5'-10' <5'

Planetary Nebulae
✧ ✧ ✧
>1' 0.5'-1' <0.5'

Bright Nebulae
>10' to scale □ <10'

Galaxies
◯ ◯ ◯ ○
>30' 20'-30' 10'-20' <10'

Quasar △

Pulsar ⚡

Black Hole ˇ

Ecliptic
- - - - - - -
50°

Constellation
Boundaries

Galactic Equator
120°

Milky Way

GREEK ALPHABET

α	Alpha	ι	Iota	ρ	Rho
β	Beta	κ	Kappa	σ	Sigma
γ	Gamma	λ	Lambda	τ	Tau
δ	Delta	μ	Mu	υ	Upsilon
ε	Epsilon	ν	Nu	φ	Phi
ζ	Zeta	ξ	Xi	χ	Chi
η	Eta	ο	Omicron	ψ	Psi
ϑ	Theta	π	Pi	ω	Omega

Fig. 7-53 The Small Magellanic Cloud and the globular cluster 47 Tucanae. (Akira Fujii)

ATLAS CHART 52. HUBBLE DEEP FIELD-SOUTH The constellations in this area contain relatively few stars. Here we find only two bright stars, α (alpha) Pavonis, the Peacock Star, and α (alpha) Gruis, Al Na'ir, in Grus, the Crane. Indus, the Indian, is the other constellation that takes up most of this chart.

The region near the south celestial pole is almost barren. Although the number of background stars here does not differ greatly from the number near the north celestial pole, that pole has enough fairly bright stars to at least make the surrounding constellations look interesting. The 5th-magnitude star σ (sigma) Octantis, lying within 1° of the south celestial pole, is the southern pole star. Though it is very inconspicuous compared with Polaris and lacks the Pointers and Little Dipper that make Polaris easy to find, σ Oct is easily visible to the naked eye in a dark sky.

λ (lambda) Oct (at $21^h 51^m -83°$) is a double star easily resolved in a small telescope. It shows contrasting orange and green components of magnitudes 5.4 and 7.7. The Hubble Deep Field-South is near $22^h -60°$.

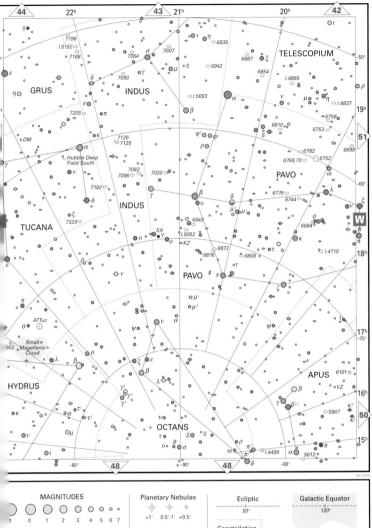

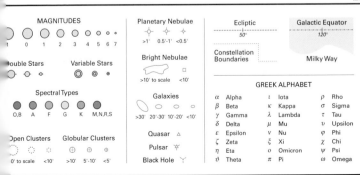

THE MOON

The moon is often the most prominent object in the nighttime sky. The moon is somewhat more than one-quarter the diameter of the earth. This makes it the largest substantial satellite (moon) in the solar system in comparison to its parent planet. (Three moons of Jupiter and one each of Neptune and Saturn are physically larger than our moon; Pluto's small moon Charon is nearly half Pluto's size.)

The moon orbits the earth every $27\frac{1}{3}$ days with respect to the stars. But during that time, the earth and moon have moved as a system about $\frac{1}{12}$ of the way in their yearly orbit around the sun. So if the moon at a certain point in its orbit is directly between the earth and the sun, $27\frac{1}{3}$ days later it has not quite returned to that point directly between the earth and the sun. The moon must orbit the earth a bit farther to get back to the same place with respect to the line between the earth and the sun. The moon reaches this point in a couple of days, making the *synodic period* of the moon equal to $29\frac{1}{2}$ days. (The synodic period is the interval between two successive conjunctions — coming to the same celestial longitude — of two celestial bodies, in this case conjunctions of the moon and sun as observed from the earth.) It is the synodic months that are taken into account in lunar calendars.

THE PHASES OF THE MOON

The phases of the moon repeat with this $29\frac{1}{2}$-day period, since the phases simply depend on the angle between the earth, sun, and moon. One-half of the moon is always illuminated by the sun. To us on earth, the moon appears to go through phases because in the course of the month we see different fractions of its lighted half. When the sun and moon are on opposite sides of the earth, the moon is *full* (fig. 8-1). Everyone on earth who can see the

Fig. 8-1. A full moon. North on the moon is up, as it would be for a northern-hemisphere observer viewing the moon with the naked eye or with binoculars. West is at left. Mare Crisium is the small mare on the northeast limb (just above center at the right). Tycho is the crater at lower center with prominent rays emanating from it. (Daniel Good)

moon above the horizon sees the same phase of the moon at the same time.

When the sun and moon are on the same side of the earth, we are looking past the moon at the sun. At this time, the far side of the moon is illuminated by the sun, but the side that faces us receives no sunlight. We say that the moon is *new*. Actually, we can often dimly see the dark face of the moon by *earthshine* — sunlight that bounces off the earth and back up to the moon.

In between the new moon and the full moon, the moon is a *crescent* (less than half-illuminated, fig. 8-2), a *half-moon* (half the face we see illuminated, fig. 8-3), or is *gibbous* (more than half-illuminated. (Gibbous comes from the Latin word that means "a hump," or "bulging.") The first half-moon after the new moon is called the first quarter, since we are one-quarter of the way through the monthly cycle of phases. The half-moon after the moon is full is called the third quarter. While the phases are changing from new to full, with more of the illuminated side of the moon becoming visible, we say the moon is *waxing*. Between full and new phases the moon is *waning*.

Table 14 lists dates for the phases of the moon. Since the moon's orbit is elliptical, the moon's speed around the earth varies, making the period between its phases vary slightly.

The angle between the earth, the moon, and the sun determines which phase will appear, so the time of night the moon rises depends on its phase. A full moon rises when the sun sets. A new moon rises when the sun rises and sets when the sun sets. The moon rises about 50 minutes later each night. A first-quarter moon thus rises at about noon and sets around midnight, and so is at its highest point each day at sunset. An interesting project is to observe the moon on as many days and nights of the month as possible, and to sketch each phase and its orientation with respect to the sun. You can often see the moon in the daytime, especially if you first figure out approximately where to look for it.

TABLE 14. PHASES OF THE MOON

NEW MOON D H M (U.T.)	FIRST QUARTER D H M (U.T.)	FULL MOON D H M (U.T.)	LAST QUARTER D H M (U.T.)
		2000	
Jan. 6 18 14	Jan. 14 13 35	Jan. 21 04 41	Jan. 28 07 57
Feb. 5 13 04	Feb. 12 23 22	Feb. 19 16 27	Feb. 27 03 54
Mar. 6 05 17	Mar. 13 06 59	Mar. 20 04 45	Mar. 28 00 21
Apr. 4 18 13	Apr. 11 13 31	Apr. 18 17 42	Apr. 26 19 31
May 4 04 13	May 10 21 01	May 18 07 35	May 26 11 56
June 2 12 15	June 9 03 30	June 16 22 28	June 25 01 01
July 1 19 20	July 8 12 53	July 16 13 56	July 24 11 03
July 31 02 26	Aug. 7 01 02	Aug. 15 05 13	Aug. 22 18 51
Aug. 29 10 20	Sept. 5 16 28	Sept.13 19 37	Sept.21 01 29
Sept.27 19 54	Oct. 5 11 00	Oct. 13 08 54	Oct. 20 08 00
Oct. 27 07 59	Nov. 4 07 27	Nov. 11 21 15	Nov. 18 15 25
Nov. 25 23 12	Dec. 4 03 56	Dec. 11 09 03	Dec. 18 00 42
Dec. 25 17 22			
		2001	
	Jan. 2 22 32	Jan. 9 20 25	Jan. 16 12 35
Jan. 24 13 07	Feb. 1 14 03	Feb. 8 07 12	Feb. 15 03 24
Feb. 23 08 22	Mar. 3 02 04	Mar. 9 17 24	Mar. 16 20 46
Mar. 25 01 22	Apr. 1 10 50	Apr. 8 03 22	Apr. 15 15 32
Apr. 23 15 26	Apr. 30 17 08	May 7 13 53	May 15 10 11
May 23 02 47	May 29 22 10	June 6 01 40	June 14 03 29
June 21 11 58	June 28 03 20	July 5 15 04	July 13 18 46
July 20 19 45	July 27 10 09	Aug. 4 05 56	Aug. 12 07 54
Aug. 19 02 56	Aug. 25 19 55	Sept. 2 21 44	Sept.10 19 00
Sept.17 10 28	Sept.24 09 31	Oct. 2 13 49	Oct. 10 04 20
Oct. 16 19 24	Oct. 24 02 59	Nov. 1 05 42	Nov. 8 12 22
Nov. 15 06 41	Nov. 22 23 21	Nov. 30 20 50	Dec. 7 19 52
Dec. 14 20 48	Dec. 22 20 57	Dec. 30 10 41	
		2002	
			Jan. 6 03 55
Jan. 13 13 29	Jan. 21 17 47	Jan. 28 22 51	Feb. 4 13 34
Feb. 12 07 41	Feb. 20 12 02	Feb. 27 09 17	Mar. 6 01 25

NEW MOON D H M (U.T.)	FIRST QUARTER D H M (U.T.)	FULL MOON D H M (U.T.)	LAST QUARTER D H M (U.T.)
Mar. 14 02 03	Mar. 22 02 29	Mar. 28 18 25	Apr. 4 15 30
Apr. 12 19 22	Apr. 20 12 49	Apr. 27 03 00	May 4 07 17
May 12 10 46	May 19 19 43	May 26 11 52	June 3 00 06
June 10 23 47	June 18 00 30	June 24 21 43	July 2 17 20
July 10 23 47	July 17 04 48	July 24 09 08	Aug. 1 10 23
Aug. 8 19 16	Aug. 15 10 13	Aug. 22 22 30	Aug. 31 02 32
Sept. 7 03 11	Sept. 13 18 09	Sept. 21 14 00	Sept. 29 05 28
Oct. 6 11 18	Oct. 13 05 34	Oct. 21 07 21	Oct. 29 05 28
Nov. 4 07 35	Nov. 11 20 53	Nov. 20 01 34	Nov. 27 15 47
Dec. 4 07 35	Dec. 11 15 49	Dec. 19 19 11	Dec. 27 00 32

2003

NEW MOON	FIRST QUARTER	FULL MOON	LAST QUARTER
Jan. 2 20 23	Jan. 10 13 15	Jan. 18 10 48	Jan. 25 08 34
Feb. 1 10 49	Feb. 9 11 12	Feb. 16 23 52	Feb. 23 16 46
Mar. 3 02 36	Mar. 11 07 16	Mar. 18 10 35	Mar. 25 01 52
Apr. 1 19 19	Apr. 9 23 41	Apr. 16 19 36	Apr. 23 12 19
May 1 12 15	May 9 11 54	May 16 03 37	May 23 00 31
May 31 04 20	June 7 20 28	June 14 11 16	June 21 14 46
June 29 18 39	July 7 02 33	July 13 19 22	July 21 14 46
July 29 06 53	Aug. 5 07 28	Aug. 12 04 49	Aug. 20 00 49
Aug. 27 17 27	Sept. 3 12 35	Sept. 10 16 37	Sept. 18 19 03
Sept. 26 03 10	Oct. 2 19 10	Oct. 10 07 28	Oct. 18 12 32
Oct. 25 12 51	Nov. 1 04 25	Nov. 9 01 14	Nov. 17 04 15
Nov. 23 23 00	Nov. 30 17 17	Dec. 8 20 37	Dec. 16 17 43
Dec. 23 09 44	Dec. 30 19 04		

2004

NEW MOON	FIRST QUARTER	FULL MOON	LAST QUARTER
		Jan. 7 15 41	Jan. 15 04 46
Jan. 21 21 05	Jan. 29 06 04	Feb. 6 08 47	Feb. 13 13 40
Feb. 20 09 18	Feb. 28 03 25	Mar. 6 23 15	Mar. 13 21 01
Mar. 20 22 42	Mar. 28 23 48	Apr. 5 11 03	Apr. 12 03 47
Apr. 19 13 22	Apr. 27 17 33	May 4 20 34	May 11 11 05
May 19 04 53	May 27 07 58	June 3 04 20	June 9 20 03
June 17 20 27	June 25 19 09	July 2 11 09	July 9 07 34
July 17 20 27	July 25 03 38	July 31 18 06	Aug. 7 22 02
Aug. 16 01 24	Aug. 23 10 12	Aug. 30 02 23	Sept. 6 15 11
Sept. 14 14 30	Sept. 21 15 54	Sept. 28 13 10	Oct. 6 10 12
Oct. 14 02 49	Oct. 20 21 59	Oct. 28 03 08	Nov. 5 05 54
Nov. 12 14 28	Nov. 19 05 51	Nov. 26 20 08	Dec. 5 00 53
Dec. 12 01 30	Dec. 18 16 40	Dec. 26 15 07	

2005

NEW MOON	FIRST QUARTER	FULL MOON	LAST QUARTER
			Jan. 3 17 46
Jan. 10 12 03	Jan. 17 06 58	Jan. 25 10 33	Feb. 2 07 27
Feb. 8 22 29	Feb. 16 00 17	Feb. 24 04 54	Mar. 3 17 37
Mar. 10 09 11	Mar. 17 19 20	Mar. 25 20 59	Apr. 2 00 51
Apr. 8 20 33	Apr. 16 14 38	Apr. 24 10 07	May 1 06 25
May 8 08 46	May 16 08 57	May 23 20 19	May 30 11 48

NEW MOON D H M (U.T.)	FIRST QUARTER D H M (U.T.)	FULL MOON D H M (U.T.)	LAST QUARTER D H M (U.T.)
June 6 21 56	June 15 01 23	June 22 04 14	June 28 18 24
July 6 12 03	July 14 15 20	July 21 11 01	July 28 03 20
Aug. 5 03 05	Aug. 13 02 39	Aug. 19 17 54	Aug. 26 15 19
Sept. 3 18 46	Sept. 11 11 37	Sept. 18 02 01	Sept. 25 06 41
Oct. 3 10 28	Oct. 10 19 01	Oct. 17 12 14	Oct. 25 01 17
Nov. 2 01 25	Nov. 9 01 58	Nov. 16 00 58	Nov. 23 22 12
Dec. 1 15 01	Dec. 8 09 37	Dec. 15 16 16	Dec. 23 19 37
Dec. 31 03 12			

2006

	Jan. 6 18 57	Jan. 14 09 49	Jan. 22 15 14
Jan. 29 14 15	Feb. 5 06 29	Feb. 13 04 45	Feb. 21 07 17
Feb. 28 00 31	Mar. 6 20 16	Mar. 14 23 36	Mar. 22 19 11
Mar. 29 10 16	Apr. 5 12 01	Apr. 13 16 41	Apr. 21 03 29
Apr. 27 19 44	May 5 05 14	May 13 06 52	May 20 09 21
May 27 05 26	June 3 23 06	June 11 18 04	June 18 14 09
June 25 16 06	July 3 16 37	July 11 03 02	July 17 19 13
July 25 04 31	Aug. 2 08 46	Aug. 9 10 55	Aug. 16 01 51
Aug. 23 19 10	Aug. 31 22 57	Sept. 7 18 43	Sept. 14 11 16
Sept. 22 11 46	Sept. 30 11 04	Oct. 7 03 13	Oct. 14 00 26
Oct. 22 05 15	Oct. 29 21 26	Nov. 5 12 59	Nov. 12 17 46
Nov. 20 22 19	Nov. 28 06 30	Dec. 5 00 25	Dec. 12 14 32
Dec. 20 14 01	Dec. 27 14 48		

2007

		Jan. 3 13 58	Jan. 11 12 45
Jan. 19 4 01	Jan. 25 23 02	Feb. 2 05 46	Feb. 10 09 52
Feb. 17 16 15	Feb. 24 07 56	Mar. 3 23 18	Mar. 12 03 55
Mar. 19 02 43	Mar. 25 18 17	Apr. 2 17 16	Apr. 10 18 05
Apr. 17 11 37	Apr. 24 06 36	May 2 10 10	May 10 04 28
May 16 19 28	May 23 21 03	June 1 01 04	June 8 11 43
June 15 03 14	June 22 13 16	June 30 13 49	July 7 16 54
July 14 12 04	July 22 06 30	July 30 00 48	Aug. 5 21 20
Aug. 12 23 03	Aug. 20 23 55	Aug. 28 10 36	Sept. 4 02 33
Sept. 11 12 45	Sept. 19 16 48	Sept. 26 19 46	Oct. 3 10 06
Oct. 11 05 01	Oct. 19 08 34	Oct. 26 04 52	Nov. 1 21 19
Nov. 9 23 04	Nov. 17 22 33	Nov. 24 14 30	Dec. 1 12 45
Dec. 9 17 41	Dec. 17 10 18	Dec. 24 01 16	Dec. 31 07 51

2008

Jan. 8 11 38	Jan. 15 19 46	Jan. 22 13 35	Jan. 30 05 03
Feb. 7 03 45	Feb. 14 03 34	Feb. 21 03 31	Feb. 29 02 19
Mar. 7 17 15	Mar. 14 10 46	Mar. 21 18 41	Mar. 29 21 48
Apr. 6 03 56	Apr. 12 18 32	Apr. 20 10 26	Apr. 28 14 13
May 5 12 19	May 12 03 48	May 20 02 12	May 28 02 57
June 3 19 23	June 10 15 04	June 18 17 31	June 26 12 10
July 3 02 19	July 10 04 35	July 18 08 00	July 25 18 42
Aug. 1 10 13	Aug. 8 20 21	Aug. 16 21 17	Aug. 23 23 50

NEW MOON D H M (U.T.)	FIRST QUARTER D H M (U.T.)	FULL MOON D H M (U.T.)	LAST QUARTER D H M (U.T.)
Aug. 30 19 59	Sept. 7 14 05	Sept. 15 09 14	Sept. 22 05 05
Sept. 29 08 13	Oct. 7 09 05	Oct. 14 20 03	Oct. 21 11 55
Oct. 28 23 14	Nov. 6 04 04	Nov. 13 06 18	Nov. 19 21 31
Nov. 27 16 55	Dec. 5 21 26	Dec. 12 16 38	Dec. 19 10 30
Dec. 27 12 23			

2009

NEW MOON	FIRST QUARTER	FULL MOON	LAST QUARTER
	Jan. 4 11 57	Jan. 11 03 27	Jan. 18 02 46
Jan. 26 07 56	Feb. 2 23 14	Feb. 9 14 50	Feb. 16 21 38
Feb. 25 01 36	Mar. 4 07 46	Mar. 11 02 38	Mar. 18 17 48
Mar. 26 16 07	Apr. 2 14 34	Apr. 9 14 56	Apr. 17 13 37
Apr. 25 03 23	May 1 20 45	May 9 04 02	May 17 07 27
May 24 12 12	May 31 03 23	June 7 18 12	June 15 22 15
June 22 19 36	June 29 11 29	July 7 09 22	July 15 09 54
July 22 02 35	July 28 22 00	Aug. 6 00 55	Aug. 13 18 56
Aug. 20 10 02	Aug. 27 11 43	Sept. 4 16 03	Sept. 12 02 16
Sept. 18 18 45	Sept. 26 04 50	Oct. 4 06 11	Oct. 11 08 56
Oct. 18 05 34	Oct. 26 00 43	Nov. 2 19 14	Nov. 9 15 56
Nov. 16 19 14	Nov. 24 21 40	Dec. 2 07 31	Dec. 9 00 14
Dec. 16 12 03	Dec. 24 17 37	Dec. 31 19 13	

2010

NEW MOON	FIRST QUARTER	FULL MOON	LAST QUARTER
			Jan. 7 10 40
Jan. 15 07 12	Jan. 23 10 54	Jan. 30 06 18	Feb. 5 23 49
Feb. 14 02 52	Feb. 22 00 43	Feb. 28 16 38	Mar. 7 15 42
Mar. 15 21 02	Mar. 23 11 01	Mar. 30 02 26	Apr. 6 09 37
Apr. 14 12 30	Apr. 21 18 20	Apr. 28 12 19	May 6 04 16
May 14 01 05	May 20 23 43	May 27 23 08	June 4 22 14
June 12 11 15	June 19 04 30	June 26 11 31	July 4 14 36
July 11 19 41	July 18 10 11	July 26 11 31	Aug. 3 04 59
Aug. 10 03 09	Aug. 16 18 15	Aug. 24 17 05	Sept. 1 17 22
Sept. 8 10 30	Sept. 15 05 50	Sept. 23 09 18	Oct. 1 03 53
Oct. 7 18 45	Oct. 14 21 28	Oct. 23 01 37	Oct. 30 12 46
Nov. 6 04 52	Nov. 13 16 39	Nov. 21 17 28	Nov. 28 20 37
Dec. 5 17 36	Dec. 13 13 59	Dec. 21 08 41	Dec. 28 04 19

Notes: d = day, h = hour, m = minute. To convert Universal Time (U.T.) to time in U.S. time zones, subtract:

5 hours to get E.S.T., 4 hours to get E.D.T.
6 hours to get C.S.T., 5 hours to get C.D.T.
7 hours to get M.S.T., 6 hours to get M.D.T
8 hours to get P.S.T., 7 hours to get P.D.T.
9 hours to get Alaska S.T., 8 hours to get Alaska D.T.
10 hours to get Hawaii S.T.

Waxing crescent · Nearly first quarter · Waxing gibbous

Full moon

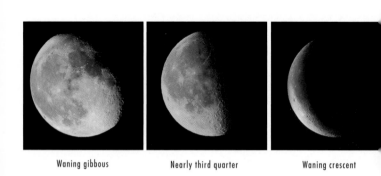

Waning gibbous · Nearly third quarter · Waning crescent

Fig. 8-2 Phases of the moon. (Hopkins Observatory, Williams College)

Fig. 8-3 Maria show prominently on this third-quarter moon, 21 days old (that is, 21 days past new). Note how the lighting angle makes many features look different from their appearance at the full moon. (Daniel Good)

1E MOON'S SURFACE

Even with the naked eye, we can see that the surface of the moon is varied in structure. The fact that the moon has large flat areas called *maria* (pronounced "MAR ee uh"; singular, *mare,* pronounced "MA ray") and craters was discovered by Galileo when he first turned his small telescope to look at the moon in 1609. Maria means "seas," though there is no water in these lunar seas.

The view of the moon through even a small telescope can be breathtaking. You can see that the maria are very flat and that there are other regions, the *highlands,* that are covered with craters. There are relatively few craters on the maria. These "seas" have, instead, been made flat by volcanic material—lava—that flowed from beneath the lunar surface more than 3 billion years ago. This lava covered whatever craters existed at that time, so the craters now visible in the maria were formed by the impact of interplanetary rocks—meteorites—that have hit the moon since then.

When observing the moon with binoculars or with a telescope, you will find it most interesting to observe the *terminator,* the line separating light from dark (day from night, on the moon). At the terminator, we are seeing the region where the sun's rays are hitting the moon at the most oblique angle, and therefore shadows are longest there. This increases the contrast of the surface features. One can even calculate the height of lunar mountains by measuring the lengths of their shadows.

In this chapter, we provide detailed maps of the lunar surface. They are artist's drawings based on National Aeronautics and Space Administration (NASA) photographs, which are able to bring out the structure of features of all parts of the moon. The maps are drawn in a projection that enlarges the areas near the edges (called *limbs*) of the moon, which are otherwise too foreshortened to see clearly. In this projection, a dime held over any portion of the map covers the same area on the moon.

The moon keeps essentially the same face toward the earth as it orbits the earth, presumably because a slight bulge of matter exists in the side of the moon that faces the earth, allowing earth's gravity to capture the moon's rotation. (As a result, the moon turns once on its axis with respect to the stars during each of its orbits around the earth.)

We are not, however, completely limited to seeing only 50 percent of the moon. The moon is sometimes turned slightly one way and sometimes slightly the other. These variations, called *librations,* enable us to see about ⅗ of the surface. Because of this fact, the special projection of the lunar maps in this guide may be more useful to an observer than a photograph taken at a certain instant when the moon may have been turned at a different libration angle. The librations result from several causes, the chief of which is the fact that the moon is turning at a constant rate, while it is moving in its elliptical orbit at slightly different speeds, depending on how far away it is from the earth.

When we observe the moon with the naked eye or through a telescope, we see various maria (fig. 8-3). The largest mare is actually called an ocean, Oceanus Procellarum. Other maria have names such as Palus, signifying a marsh, and Lacus, meaning a lake, though we now know that there is certainly no water on the moon. At the edges of some of the maria we find indentations, each known as a bay—a *sinus.*

Craters of all sizes can be found, ranging up to hundreds of kilometers across. Many of the craters have *central peaks*—raised regions at their centers. Scientists have been able to duplicate these central peaks in laboratories on earth by shooting bullets at sand; the central peaks arise from a rebound of material. The craters, for the most part, have been formed where meteorites

Fig. 8-4 Huge lava flows near the crater Herschel, with the Lunar Module, photographed from Apollo 1 2 in orbit around the moon. (NASA)

have hit the moon. When the meteorites crash, they release their energy explosively, as though an atomic bomb or tons of TNT have gone off. It is this energy that forms each crater, which explains why the craters on the moon are relatively round, even though the meteorites may have hit at an angle.

The moon has *mountain peaks* and *mountain ranges,* along with smaller raised *ridges.* A few *valleys* can be seen on the lunar surface. Some small valleys are known as *rilles;* rilles often seem sinuous in form, winding their ways for hundreds of kilometers across the lunar surface.

The photographs show that at certain places, one can see *rays* that appear to radiate outward from a few craters, most prominently from Tycho. These rays consist of material ejected when the craters were formed. Over millions of years, the material in the rays darkens until it matches the surrounding surface. The fact that the rays extending from Tycho are still clearly visible tells us that this is one of the youngest of the large craters on the moon.

Fig. 8-5 The moon with Earthrise, photographed from Apollo 16. (NASA)

For thousands of years, humans had to guess what the moon was like. But in 1969, with the first Apollo landings of U.S. astronauts, we began to acquire first-hand knowledge of the lunar surface. Even such basic knowledge as the fact that the moon has a hard surface with a thin coating of dust had to await the first landings of unmanned spacecraft on the moon and then observations that the Apollo astronauts didn't sink into the surface. Twelve astronauts on six Apollo missions landed on the moon (table 15) and carried out many types of experiments there. The astronauts measured the flow of heat from beneath the lunar surface, collected particles from the sun that hit the moon, and measured the hardness of rocks on the surface. Of course, the collecting of moon rocks and moon dust for detailed analysis back on earth was the most widely known of the experiments. Three un-crewed Soviet landers also brought back to earth a few samples of moon rock and dust, though in far smaller quantities than the 382 kilograms (843 pounds) of material brought back by the American astronauts. Many of the lunar rocks are basalts, a common type of rock on earth formed by vulcanism.

As a result of the analysis of moon rocks on earth, scientists were able to work out an accurate chronology of the moon's history. The oldest rocks were 4.42 billion years old, which presumably represents the time when the moon's surface began to solidify, shortly after the overall formation of the moon 4.6 billion years ago. Then, from about 4.2 to 3.9 billion years ago, the surface was bombarded heavily by meteorites. A hundred thousand years later, radioactive elements inside the moon had generated so much heat that lava flowed onto the moon's surface, filling the largest basins and making the maria that we see today. This era of lunar vulcanism ended about 3.1 billion years ago, and the surface of the moon has changed little since then.

Though pictures of the moon taken from earth are often spectacular, whether taken with small telescopes or with some of the largest ones, any view of the moon from earth is always somewhat blurred by the effect of looking through the earth's atmosphere. The photographs from the Apollo spacecraft therefore can show much finer detail (fig. 8-5). And from the earth, we see only the moon's near side. The far side was completely unknown to us until spacecraft circled the moon. In this chapter, in addition to providing eight detailed maps of the moon's near side, we present a map of the far side, on a less expanded scale. These maps show some of the names that have been assigned to craters and other features by the International Astronomical Union, which has the responsibility for naming astronomical features in the universe.

TABLE 15. CREWED MISSIONS TO THE MOON

MISSION	YEAR	LANDING SITE — SEE MOON MAP
Apollo 11	1969	6
Apollo 12	1969	3
Apollo 14	1971	3
Apollo 15	1971	6
Apollo 16	1972	2
Apollo 17	1972	6

LUNAR ECLIPSES

The plane of the moon's orbit is tilted 5° with respect to the earth's. Thus as the moon orbits the earth each month, it usually passes above or below the earth's shadow. (The earth's shadow is a cone that—at the moon's distance from the earth—is about twice as broad as the moon's diameter.) But every few months, the moon passes either partially or entirely into the earth's shadow, and we have a partial or total *lunar eclipse* (fig. 8-6).

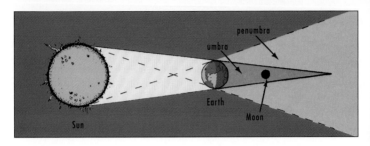

Fig. 8-6 A drawing of a lunar eclipse, showing the umbra of the earth's shadow (the area defined by solid lines) and the penumbra (defined by dotted lines).

The earth's shadow has a darker inner cone, the *umbra*, from which the sun cannot be seen at all, and an outer cone, the *penumbra*, from which part of the sun can be seen. Most of the penumbra is not dark enough to have a noticeable effect on the moon's brightness, so penumbral eclipses are largely ignored.

During a partial lunar eclipse, the umbra of the earth's shadow advances onto part of the moon's surface and then moves off the other side. No other effect is seen.

During a total lunar eclipse (table 16), the umbra hits the moon (first contact) and then gradually covers the moon until the

TABLE 16. LUNAR ECLIPSES

DATE	MOON ENTERS UMBRA	TOTALITY BEGINS	TOTALITY ENDS	MOON LEAVES UMBRA	MAXIMUM UMBRAL MAGNITUDE	TOTAL/ *PARTIAL*
1/21/00	3:01	4:04	5:23	6:26	1.33	T
totality: Americas, western Europe, West Africa						
7/16/00	11:57	13:02	14:49	15:54	1.77	T
totality: Australia, New Zealand, eastern Indonesia, Papua New Guinea						
1/9/01	18:42	19:50	20:52	21:59	1.19	T
totality: Europe, Africa, Asia as far east as mid-Indonesia and omitting Japan; *partial phases visible from eastern U.S. and Canada*						
7/5/01	13:35			16:15	0.50	P
partial phases: Australia, New Zealand, eastern Asia, Japan						
5/16/03	2:03	3:14	4:06	5:18	1.13	T
totality: eastern North America, South America						
11/9/03	(23:32)	1:07	1:30	3:05	1.02	T
totality: eastern North and South America, Europe, Africa						
5/4/04	18:48	19:52	21:08	22:12	1.361	T
totality: eastern Europe, all but West Africa, western Asia						
10/28/04	1:14	2:23	3:45	4:54	1.31	T
totality: Americas excluding west coast, western Europe and Africa						

TABLE 16. LUNAR ECLIPSES

DATE	MOON ENTERS UMBRA	TOTALITY BEGINS	TOTALITY ENDS	MOON LEAVES UMBRA	MAXIMUM UMBRAL MAGNITUDE	TOTAL/ PARTIAL
10/17/05	11:34			12:32	0.07	P

partial phases: Hawaii, Alaska, northwestern U.S. and Canada, Australia, New Zealand

| 9/7/06 | 18:05 | | | 19:37 | 0.19 | P |

partial phases: Eastern Africa, Asia, western Australia; some partials in Europe

| 3/3/07 | 21:30 | 22:44 | 23:58 | (1:12) | 1.24 | T |

totality: Europe, Africa; *partial phases: Americas, Asia, western Australia*

| 8/28/07 | 8:51 | 9:52 | 11:23 | 12:24 | 1.48 | T |

West coast of U.S., Alaska, Hawaii, New Zealand, eastern Australia; *partial phases visible throughout the U.S. and Canada*

| 2/21/08 | 1:43 | 3:01 | 3:51 | 5:09 | 1.11 | T |

totality: Americas except for the west coast of U.S. and Canada, Europe, West Africa; *partial phases visible from the west coast of U.S. and Canada*

| 8/16/08 | 19:36 | | | 22:45 | 0.81 | P |

Europe, Africa

| 12/31/09 | 18:52 | | | 19:53 | 0.08 | P |

Europe, Africa, Asia, Australia; some partial phases visible from New England and eastern Canada

| 6/26/10 | 10:16 | | | 13:00 | 0.54 | P |

Hawaii, Australia, New Zealand, Papua New Guinea, Pacific Ocean

| 12/21/10 | 6:32 | 7:40 | 8:54 | 10:01 | 1.26 | T |

totality: North America, Central America

NOTES: Entries in **boldface** = totality visible in at least some of the U.S. and Canada. Partial eclipses are in italics. Times in parentheses are on the adjacent day.

All times are given in Universal Time (U.T.). To convert Universal time to time in U.S. time zones, subtract:

5 hours to get E.S.T., 4 hours to get E.D.T.

6 hours to get C.S.T., 5 hours to get C.D.T.

7 hours to get M.S.T., 6 hours to get M.D.T.

8 hours to get P.S.T., 7 hours to get P.D.T.

9 hours to get Alaska S.T., 8 hours to get Alaska D.T.

10 hours to get Hawaii S.T.

Total lunar eclipses visible in the U.S. and Canada are in boldface. The maximum umbral magnitude is the fraction of the moon's diameter that is in the umbra at the deepest part of the eclipse. The whole moon is barely in the umbra for maximum magnitude of 1, and is farther in the umbra for maximum magnitude greater than 1. Penumbral eclipses are not listed.

(Based on *Canon of Lunar Eclipses* by Jean Meeus and Hermann Mucke and *Fifty Year Canon of Lunar Eclipses* by Fred Espenak. See also maps in Philip Harrington's *Eclipse!*.)

moon is entirely covered (second contact). Then, for an hour or two, the moon is immersed in the earth's shadow (fig. 8-7). Though no sunlight falls directly on the totally eclipsed moon, some sunlight is bent around the earth's atmosphere. From this refracted sunlight, the blue light is removed when the light is scattered in the earth's atmosphere. This creates blue skies overhead for people on earth; only the red light gets through to the moon. Thus the totally eclipsed moon appears reddish (fig. 8-8). Just how reddish it looks depends on whether volcanic dust is present in the earth's atmosphere; the dust makes the moon look darker and less reddish. Even giant storms or cloudy regions on earth can affect the earth's shadow, perhaps making the darkness of the shadow appear uneven on the moon. The moon will also appear less evenly illuminated if it passes closer to the sides of the umbra instead of through the umbra's center. The shape of the umbra can be photographed (fig. 8-9).

To photograph a lunar eclipse, use a telephoto lens with a focal length of at least 200 mm for close-ups, though wide-angle views can also be interesting. During the partial phases, the bright part of the moon is illuminated by the sun; since it is about the same distance from the sun as the earth is, use the same exposure you would use to photograph people outdoors on a sunny day on earth. (For example, a lens setting of $f/11$ at $\frac{1}{125}$ second for ISO 100 film would be typical.) During the total part of the eclipse, however, the moon is much darker, and time exposures of a minute or more with your lens wide open may be necessary. These will blur because of the moon's motion, unless your camera or telescope is on a tracking mount. For best results, take a wide range of exposures.

Fig. 8-7 (left) *A series on a single piece of film of the stages of a lunar eclipse, over a period of hours. (Akira Fujii)*
Fig. 8-8 (right) *The reddish tinge of a totally eclipsed moon; the reddishness remains visible even when the moon is only partly eclipsed. (Jay M. Pasachoff)*

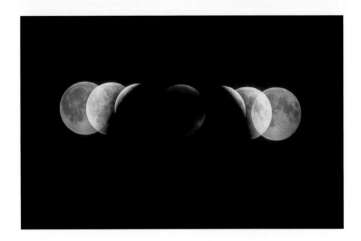

Fig. 8-9 Images of the moon over the interval of a total eclipse, tracking with the earth's shadow, reveals the roundness of the shadow and therefore of the earth itself. (Akira Fujii)

THE MOON MAPS

DESCRIPTIONS ACCOMPANYING MAPS ARE BY EWEN A. WHITAKER

MAP PROJECTION

As seen from earth, the limb (outer) regions of the moon appear very crowded together (foreshortened) because of the viewing angle; formations situated within 10° of the map edges can be observed only with difficulty, and then only when illumination and libration conditions are favorable. The lunar maps used here (which were prepared in a cooperative effort of the National Geographic Society and the U.S. Geological Survey) are drawn in a projection that spreads out the limb regions and makes them easier to see (fig. 8-10). The projection preserves the relative areas that features cover on the lunar surface but slightly alters their true dimensions.

DIRECTIONS ON THE MOON

Since the moon has north and south poles, these directions are unambiguous. Prior to 1961, east and west on the moon were reckoned in the same directions as the sky, i.e., for an observer in the earth's northern hemisphere facing south (toward the moon,

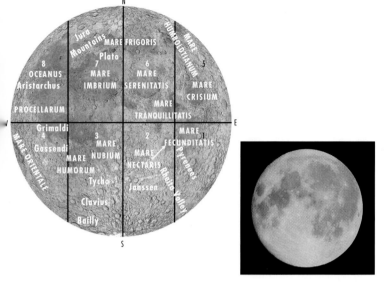

Fig. 8-10 *The moon's near side (North is up).* (left) *This orientation map shows the location of each of the moon maps that follow. (Astrogeology Team, U.S. Geological Survey, Flagstaff, Arizona)* (right) *A photograph of the moon, taken with a telescope that has a diagonal mirror. The addition of such a mirror to the optical path flips the image from side to side, reversing the orientation of what you see. Note, for example, the relation of Mare Fecunditatis and Mare Tranquillitatis. (Daniel Good)*

from U.S. and Canadian latitudes), east was to the left and west to the right. However, these directions are opposite to those used universally in terrestrial mapping, and the advent of the Space Age required that a common convention be adopted. Thus east and west in the accompanying moon charts and descriptions, as with all professional documents published since 1961, conform to these true lunar directions. Readers who may wish to consult some of the more detailed descriptive works and maps of the moon published before 1961 should be aware that "east" and "west" in these publications refer to directions in the sky. All common astronomical telescopes invert the moon's image, so the enlarged maps of the moon's near side in this guide are oriented with south at the top and lunar east to the left, to make the maps easy to use at the telescope. The orientation chart of the near side, however, is not inverted; it directly corresponds with the view seen with the naked eye or through binoculars.

Fig. 8-11 An overexposed crescent moon, with the rest of the moon illuminated by earthshine, sunlight reflected off the Earth. Jupiter and its moons also show. (Michael Myer)

MOON MAP 1. This area consists largely of the lunar highlands, which are the highly cratered and lighter-toned areas of the moon, although most of Mare Fecunditatis lies within the boundaries of this map, as do portions of Mare Smythii and Mare Australe. The major craters are Humboldt (toward the left edge), Furnerius, Petavius, Vendelinus, and Langrenus.

Humboldt, a relatively fresh crater with a diameter of 125 miles (200 km), would be a very impressive feature if it were situated closer to the center of the moon's face (disk). It is best viewed one night after full moon, but favorable librations are essential. Vendelinus and Furnerius are better placed for good viewing but are older, as evidenced by the number of more recent craters that have overlapped them and by their generally smoother appearance. Petavius, 110 miles (177 km) in diameter, is somewhat younger and is interesting because of a major rille that runs from its west wall to the imposing group of central peaks. Langrenus, named for (and by!) Van Langren, who produced the first lunar map, is also fairly fresh and clear-cut in appearance, with a small central peak and finely terraced walls. To the southwest (upper right) of Petavius are situated the sources of two extensive systems of bright surface streaks called rays. These are best seen when the terminator lies some distance away, i.e., during the period toward full moon. The rays are remarkable in that their source craters, Furnerius A and Stevinus A, are so small.

Mare Fecunditatis consists of two contiguous, nearly round areas of dark basaltic lavas; the northern part has about three times the diameter of the southern. The larger part of this "sea" displays a typical system of fairly prominent ridges, the result of compressional forces. The small craters Messier and Messier A, situated near the lunar equator, are interesting both because of their somewhat anomalous shapes and their unusual pair of rays, resembling searchlight beams, that stretch westward from Messier A. These craters were studied extensively in the past because of suspected changes in their appearance, which are now known to be the result of changes in illumination and viewing angles.

S

Jeans

Lyot

Brisbane

Peirescius

Hamilton

Gem

Vega

Oken

MARE AUSTRALE

Marinus

VALLIS RHEITA

Abel

Fraunhofer

Furnerius

Barnard

Adams

Furnerius A

Rheita

Stevinus

Legendre

Hase

Stevinus A

Phillips

Palitzsch

Humboldt

Snellius

Reichenbach

Hecataeus

Petavius

Schorr

Wrottesley

Borda

Gibbs

Biot

Santbech

Balmer

Holden

Monge

Behaim

Cook

Montes Pyrenaeus

Lamé

Vendelinus

Colombo

La Pérouse

McClure

Kapteyn

Crazier

Ansgarius

Lohse

Bellot

Magelhaens

Langrenus

Goclenius

Kästner

FECUNDITATIS

MARE

Gilbert

Lubbock

ss

Von Vleck

Maclaurin

Messier A

Weierstrass

Webb

Messier

TH

E

Luna 16

5

Fig. 8-12 Craters from Hipparchus through Jacobi. (Stephan Martin, Hopkins Observatory, Williams College)

MOON MAP 2. This area also consists largely of highlands, but is interesting because of the presence of Mare Nectaris (lower left) and its associated basin, and of the Rheita Valley (Vallis Rheita, upper left). This lunar valley, which is best viewed when the moon is either about 4–5 days old (during the crescent phase), or 2–3 days after the full phase, is about 300 miles (480 km) long. However, there is some evidence that the valley is not a single structure but consists of two segments of different origin. The 200-mile (320-km) northern segment appears to have been caused by a subsidence of the moon's crust; the southern segment, which lies at an angle to the northern one, resembles many of the valleys that appear to radiate from the Imbrium basin (Map 7), such as those southeast of the crater Hipparchus (lower right of this map; fig. 8-12).

Janssen is one of the largest craters (120 miles—190 km—in diameter) shown on this map (upper left center); most of its floor is very rough. This floor also contains a prominent rille, with many lesser rilles and a clifflike fault branching from it. Maurolycus (70 miles—112 km—in diameter), at center, is one of the deepest (15,000 ft.—4570 m) of the larger craters; it looks very impressive when situated near the terminator. Theophilus (bottom center), with a diameter of 62 miles (100 km), looks somewhat fresher (sharper-edged). It is well situated for observations of part of its *ejecta blanket*—the blanket of material ejected from the crater—that extends in a northerly direction.

A medium-sized telescope will reveal the roughness of this area compared with the fairly smooth surface of Mare Nectaris. Mare Nectaris is a prominent, somewhat circular area of dark, basaltic lavas some 220 miles (354 km) in diameter that displays some concentric ridges on the eastern part. This mare represents a filling by lavas of a pre-existing central depression caused by the impact that formed the Nectaris basin. The outer boundary of this basin is marked, on the southwest, by the fault line named the Altai Scarp (Rupes Altai) and on the east, by the Pyrenees Mountains. These and other segments of scarps and mountain blocks form a circle that is concentric with and has twice the diameter of the mare's surface—a characteristic shared by all the larger impact structures on the moon.

Amundsen→

Demonax→ Scott→ Malapert→

←Wexler

Neumayer→

Helmholtz→ Boguslawsky→

Gill→ Simpelius→

Pantecoulant→ Boussingault→ Curtius

Manzinus→

Mutus→ Pentland→

←Nearch Kinau→ Zach→

Hanno→ Hagecius→

Biela→ Rosenberger→ Tannerus→ ←Jacobi

Asclepi→ Lilius→

Hommel Cuvier→

←Reimarus Pitiscus→ Ideler→ ←Baco Clairaut→ Heraclitus

Watt→ Vlacq Lockyer→ Dove Breislak→ Licetus→

Steinheil→ Spallanzani→ Barocius→

←Mallet Young→ Nicolai→ Faraday→

Janssen Maurolycus→

Fabricius→ Stöfler→ ←Nasireddin

Metius→ Wöhler→ Büsching→ ←Buch ←Pernelius

←Rheita Stiborius→ Gemma Frisius→ Kaiser→

Rabbi Levi→ Celsius→ Nonius→ Walter

Neander→ Lindenau→ Zagut Goodacre→ Aliacensis→

Piccolomini→ Rothmann→ ←Wilkins Poisson→ Werner→

Weinek→ Pons Pohtanus→ Krusenstern→ Blanchinus

Sacrobosco Playfair→

Santbech→ Azophi→ ←Abenezra Delaunay→

Fracastorius→ Geber→ Airy→ ←Faye

Beaumont→ Argelander→ Vogel Parrot→

MARE Almanon→ Abulfeda→ Burnham→ ←Klein

Cyrillus Descartes→ ←Andel Albategnius

NECTARIS Kant→ Halley→

Gaudibert→ Mädler→ Apollo 16 Hipparchus→

←Gutenberg Theophilus→ Zöllner→ Taylor Hind

Capella→ Isidorus→ Alfraganus→ Horrocks→ Pickering→ Gylden→

Torricelli→ Hypatia→ Seeliger→ ←Reaumur

Censorinus→ Delambre→ SINUS

Moltke→ MEDII

←6

E ←1 3→

MOON MAP 3. For an earth-based observer, the area covered by this map is dominated by the dark expanses of maria—Mare Nubium, Mare Cognitum, Mare Humorum, and a portion of Oceanus Procellarum (see also Map 8). The highlands area to the south also displays some interesting formations, including the 55-mile (90-km) diameter crater Tycho (left center). When situated near the terminator, this crater appears both rougher and deeper than its neighbors, with strongly terraced inner walls and a well-formed central peak. At phases near full moon, you can see that Tycho is the center of a very extensive system of bright rays.

There are several large craters to the south of Tycho, but Clavius (140 miles—225 km—in diameter) is undoubtedly the most eye-catching, with an arc of smaller craters spanning its level floor. A smaller crater to its west, Schiller (upper right of center), is unique because it is noticeably elongated, a condition that is further enhanced by the foreshortening that occurs near the moon's limbs. Nearby (farther west) are the large crater Schickard and the unique crater Wargentin. Wargentin's interior has been filled to the rim with material. The elevated floor is smooth and flat, with a few low, curving ridges.

Nearer to the center of the moon's disk, at the lower left of this map, we find the large craters Ptolemaeus, Alphonsus, and Arzachel, which gradually decrease in diameter and increase in depth and freshness of appearance. Southwest of Arzachel, near the eastern edge of Mare Nubium, lies the Straight Wall (Rupes Recta), a 70-mile-long (113-km-long) fault in the mare surface that averages almost 1000 ft. (300 m) in height, with a slope of about 40°. When the moon is 8–9 days old, the Wall is impressive at sunrise, for it casts a wide wedge-shaped shadow.

To the west, Mare Humorum (at right) is a reasonably circular dark area similar in size to Mare Nectaris (Map 2), but the surroundings look more degraded; as a result the edge of the basin is scarcely discernible in a telescopic view. The terrain between the two rings that surround Mare Humorum has been largely flooded with lavas (in Palus Epidemiarum, for example); the area between the rings surrounding Mare Nectaris is largely highlands material. The area lying between Mare Humorum and Palus Epidemiarum includes three parallel, curving rilles that are concentric with the Mare. These rilles are true *graben*—areas of subsidence between parallel faults in the lunar crust. The crater Gassendi, at the northern end of the Mare, is a much-observed and interesting formation. It is about 70 miles (100 km) in diameter, with a cluster of central peaks and a rough floor that displays a complicated network of rilles. Transient Lunar Phenomena (TLPs)—temporary appearances such as apparent obscurations, colorations, glows, etc.—have been reported in this crater.

S

Drygalski

Malapert · Cabeus

Le Genfil

Newton

Hausen

Casatus

Short · Klaproth

Bailly

Maretus

Wilson

Gruemberger

Kircher

Cysatus

Blancanus →

Bettinus

← Zucchius

Pingré →

Rutherfurd →

Scheiner

← Segner

Clavius

Rost

← Weigel

Phocylides →

Porter

Nasmyth

Deluc

Schiller

Wargentin

Bayer →

Maginus

Nöggerath

Longomontanus

Schickard

Proctor →

Tycho

Mee

← Drebbel

Pictet →

Saussure →

Wilhelm

Hainzel

LACUS

Huggins →

Epimenides

EXCELLENTIAE

Surveyor 7

Hemsius

Clausius

Orontius

Sasserides

Elger

Wurzelbauer

Capuanus

Ball →

Lexell →

Walter

Cichus

PALUS

Palmieri →

Heil

Weiss

EPIDEMIARUM

Deslandres

Pitatus →

Promontorium Kelvin →

Hesiodus

Purbach

Kies

Campanus

MARE

König →

MARE

Thebit →

Hippalus →

HUMORUM

chel →

NUBIUM

Bullialdus

Rupes Recta

Ranger 9

Alpetragius

Promontorium Taenarium →

Gassendi →

Lubiniezky

MARE

Davy

OCEANUS

Alphonsus

Ranger 7 ·

COGNITUM

PROCELLARUM

Parry →

erschel →

Ptolemaeus

Euclides

pörer →

Fra Mauro

mmarion →

Apollo 14 · Apollo 12 & Surveyor 3

SINUS

· Luna 5

MEDII

MARE INSULARUM

· Lansberg

W
4

7

MOON MAP 4. Besides Oceanus Procellarum (lower left; see also Map 8), this area contains the eastern half of Mare Orientale, the Orientale basin, and its extensive ejecta blanket (fig. 8-10). However, these latter formations are too close to the limb to permit good observations from earth; lunar features that are located farther from the limb command more interest in the telescope. In the

Fig. 8-13 The dark central area is Mare Orientale (Eastern Sea), on the moon's edge as seen from earth. The basin was caused by the impact of a body that was probably about 30 miles (50 km) across. At right, we see part of the moon's front side. Grimaldi is the small dark region to the right (east) of the Orientale basin, with Oceanus Procellarum above (northeast of) it. The regions on the left are from the moon's far side, which is never visible from earth. North is up in this photo taken in 1990 by NASA's Galileo spacecraft. (NASA)

highlands areas these include the crater Grimaldi (lower right), the Sirsalis-Darwin rille system, and the crater Mersenius and its environs (left center). Grimaldi is really a small basin, since it has the characteristic inner and outer concentric scarps. Its floor has also been flooded with mare lavas except toward the north, where they have not reached the crater wall.

The Sirsalis Rille (Rima Sirsalis) is one of the longest graben on the moon and is contiguous with the Darwin system. Where the rille crosses crater walls, the walls have been cut by the parallel faults, as has the lower terrain.

The crater Mersenius, 51 miles (83 km) in diameter, is unusual in that its floor is noticeably convex—it is more convex than the moon's surface. This bulge is noticeable at sunrise, when the eastern part of the floor is well illuminated while the western part is still largely in shadow.

The western "shores" of Mare Humorum display some interesting features, such as the rough surface northeast of Mersenius, the clifflike faults southeast of that crater, and a small crater (Liebig F) that sits squarely on one of these faults. The flooded crater Doppelmayer is interesting for its concentric structure.

The dark-floored crater Billy and its companion crater Hansteen, which is of similar size but quite different appearance, are situated near the shoreline of Oceanus Procellarum. Between these craters lies the exceptionally bright mountain Mons Hansteen; special photography has shown it to be notably redder than the highlands surface nearby.

Catalan

Inghirami

Vallis Baade

Baade

Graff

Vallis Bouvard

Schickard

Vallis Inghirami

Shaler

Wright

Lehmann

Piazzi

Lagrange

Lacroix

Pettit

Krasnov

Nicholson

Fourier

Vieta

MONTES

MARE ORIENTALE

LACUS VERIS

Palmieri

Byrgius

Doppelmayer

Cavendish

Eichstadt

Lamarck

MONTES

MARE

Liebig F

Liebig

Henry Frères

MONTES CORDILLERA

ROOK

HUMORUM

Henry

Kopff

Mersenius

De Vico

Darwin

Gassendi

Rima Sirsalis

LACUS AUTUMNI

Zupus

Fontana

Crüger

Billy

Rocca

Sirsalis

Hansteen

Mons Hansteen

Hartwig

Grimaldi

Schlüter

OCEANUS

Flamsteed

Damoiseau

Surveyor 1

Orbiter 5

Hermann

Lohrmann

Riccioli

PROCELLARUM

Hevelius

Fig. 8-14 *The earth is rising over the lunar limb as the Apollo 11 Lunar Module rises to rendezvous with the orbiter. In this westward-looking view, the large dark area in the background is Smyth's Sea, which is on the moon's east limb as seen from earth. (NASA)*

MOON MAP 5. The most notable feature on this map is Mare Crisium, an oval dark spot that is readily visible to the naked eye. This lunar sea was once named "The Caspian." It is comparable in size with both Mare Nectaris and Mare Humorum, but unlike those two seas, it is totally isolated from the other major maria. The outer ring of the Crisium basin, twice again the diameter of the mare, is visible mostly on the northern (lower) side when the sun is setting on this region, about two days after the full moon. The area between the rings contains several patches of dark lavas, which have been named Mare Anguis (Serpent Sea), Mare Undarum (Sea of Waves), and more recently, Lacus Bonitatis (Lake of Goodness). Although it is only 17½ miles (28 km) in diameter, the crater Proclus, to the west (right) of Mare Crisium, catches the eye during most phases because of its extreme brilliance. As the moon approaches the full phase, a striking ray system, which extends from all but the southwest sector of the crater, also becomes apparent. Similar asymmetrical ray patterns can be reproduced in the laboratory by using low-angle impacts.

At the north end of Mare Fecunditatis lies the crater Taruntius, which displays a weak ray system but has a flooded floor with concentric rilles and a small central peak.

Of the larger craters on the map, Neper (named for Napier, the inventor of logarithms), with its great terraced walls and prominent central peak, would be very impressive if it were situated nearer the central regions. However, it is on the lunar limb and can only be seen either soon after new moon or full moon, and then only if the moon's libration is favorable. Nearby are the dark areas of Mare Marginis and Mare Smythii (Smyth's Sea, fig. 8-14), equally difficult to observe. The crater Gauss (110 miles — 177 km — in diameter), at lower left, is somewhat easier to observe; its floor exhibits some smaller craters and other detail. Burckhardt, north of Mare Crisium, is interesting in that it has partially obliterated two older, slightly smaller craters, which now protrude like ears on opposite sides of it.

MARE

MARE FECUNDITATIS

MARE SMYTHII

Schubert

Dubyago

MARE SPUMANS

MARE UNDARUM

Secchi

TRANQUILLITATIS

Luna 18 & Luna 20

Tar- untius

Apollonius

Firmicus

Lawrence

MARE

Banachiewicz

Shapley

Da Vinci

Neper

Auzout

Lick Glaisher

Lyell

Jansky

Condorcet

Luna 24 * Luna 23

Picard

Yerkes

PALUS SOMNI

Promontorium Agarum

MARE

Proclus

Hansen

CRISIUM

* Luna 15

Peirce

MARE MARGINIS

Alhazen

Macrobius

Goddard

Tisserand

LACUS BONITATIS

6

Cannon

MARE ANGUIS

Eimmart

Cleomedes

Hubble

Delmotte

Tralles

Plutarch

Debes

Seneca

Burckhardt

Geminus

Hahn

Berosus

Berzelius

Bernouilli

Messala

Rayleigh

Gauss

Hooke

Riemann

Schumacher

Carrington

Zeno

Mercurius

Endymion

Boss

MARE HUMBOLDTIANUM

MOON MAP 6. As seen from earth, this area consists mostly of mare surfaces, including Mare Serenitatis and its adjacent Lacus Mortis and Lacus Somniorum, Mare Tranquillitatis and Mare Vaporum, Sinus Medii, and Mare Frigoris. The only reasonably pristine highlands areas on this map are those lying well north of Mare Frigoris. The area toward the center of the moon's face, which lies in Sinus Medii at upper right, abounds with interesting formations. Triesnecker marks the approximate center of a branching system of rilles.

North of this system lies the more prominent Hyginus Rille, which cuts right through the small crater Hyginus.

East of Hyginus lies the Ariadaeus Rille, a great linear graben that cuts through lowlands and mountain ridges alike. The area immediately north of the Hyginus Rille is one of the darkest spots on the moon; from lunar satellites and data acquired from other ground-based studies, we know that this darkness is largely due to a higher than average content of the mineral ilmenite (ferrous titanate) in the soil; in this area the soil consists mostly of the lunar equivalent of volcanic ash (mainly small glass beads).

The full moon appears almost silvery white in a dark sky, but its true color is mostly brownish gray. You can see color variations, however, especially if you view Mare Serenitatis and Mare Tranquillitatis in a telescope at low power when the terminator is not too close by. The rather warm, brownish tone of Serenitatis contrasts noticeably with the more steely gray tones of Tranquillitatis; the grayish color results from the unusually high ilmenite content of the basaltic lavas of Mare Tranquillitatis. This mare has a very strong system of ridges that traverse its western half.

Mare Serenitatis displays the typical concentric ridges of a flooded basin. The landing site of Apollo 17 is situated just off the easternmost corner of the mare, in a patch of noticeably darker material. Lunokhod 2, carried by the Soviet Luna 21 spacecraft, landed and operated near Le Monnier crater 100 miles (160 km) farther north. Slightly farther north is Posidonius, a prominent crater on the shore of Mare Serenitatis. Both sinuous and graben-type rilles cross the floor of Posidonius. The sinuous rilles follow tortuous paths over the lunar surface and are true lava-flow channels or collapsed lava tubes.

South of Mare Frigoris, craters Aristoteles, Eudoxus, Atlas, and Hercules are the remaining large easily observable craters in this map area. Atlas (55 miles — 88 km — in diameter) is unusual because of the rilles and dark spots on its floor.

The Clementine and Lunar Orbiter spacecraft in the 1990s reported the discovery of ice in the permanently shadowed craters near the lunar poles. The water and the component hydrogen and oxygen could be very useful for lunar exploration.

Apollo 11

SINUS
MEDII

Menzel

Sabine · Ritter

Surveyor 5

Ranger 8

Ariadaeus

Hyginus

Triesnecker

MARE

Lamont

Rima Ariadaeus

Rima

Hyginus

TRANQUILLITATIS

Julius
Caesar

Ranger 6

Ross

MARE

Lyell

Jansen

Plinius

Manilius

VAPORUM

Marco
Polo

Franz

Promontorium Archerusia

MONTES HAEMUS

Yangel

MONTES

APENNINUS

Vitruvius

Carmichael

Hill

MARE

Apollo 17

Franck

Littrow

Mons Hadley

Apollo 15

Römer

Le Monnier

Santos-Dumont

PALUS

MONTES

Luna 21

SERENITATIS

PUTREDINIS

TAURUS

Luna 2

Posidonius

Chacornac

Autolycus

G. Bond

Luther

Aristillus

Hall

Daniell

Theaetetus

LACUS

Calippus

Cassini

Maury

SOMNIORUM

Franklin

Grove

Mason

Alexander

Cepheus

Williams

Plana

Lamech

MONTES ALPES

Shuckburgh

Oersted

Bürg

Eudoxus

Chevallier

LACUS

Egede

Vallis Alpes

Hercules

MORTIS

Mitchell

Atlas

Baily

Aristoteles

MARE

FRIGORIS

Galle

Protagoras

Archytas

Endymion

Gärtner

Sheepshanks

De La Rue

Democritus

Kane

C. Mayer

Timaeus

Thales

W.

Strabo

Schwabe

Maigno

Bond

Arnold

Neison

Barrow

Peters

Meton

Bel'kovich

Hayn

Euctemon

Cusanus

Baillaud

Scoresby

Challis

Petermann

De Sitter

Main

Byrd

Nansen

Peary

MONTES

CAUCASUS

N

This area is dominated by Mare Imbrium, with the mare areas of Mare Insularum, Sinus Aestuum, Mare Frigoris, and Sinus Roris occupying much of the remainder. The Imbrium basin (fig. 8-15) is second only to the Orientale basin in freshness, with an age (it is believed) of about 3.9 billion years. On the moon's earthside face, only the much older Procellarum basin is larger than this huge Imbrium basin. Its formation therefore had a profound effect on the pre-existing lunar surface, and all craters situated south of Mare Frigoris (on this map) necessarily postdate that event. The mare lavas must also be more recent than the basin. There is strong evidence that all lunar vulcanism that produced dark mare lavas occurred no earlier than about 3.7 billion years ago. The surface of Mare Imbrium displays the usual concentric ridge systems, but under good conditions, lava-flow fronts —sharp demarcations—can be detected. Isolated mountain ranges and peaks, such as the Straight Range (Montes Recti), Montes Teneriffe, Mons Pico, Montes Spitzbergen, and Mons La Hire, lie on a circle that is half the diameter of the circle marked by the Carpathians (Montes Carpatus), Apennines (Montes Apenninus), and Caucasus (Montes Caucasus, Map 6). These inner ranges thus mark the ring that is equivalent to the general shoreline for Mare Crisium, Mare Nectaris, and Mare Humorum.

The craters Archimedes (on the east—to the left—of Mare Imbrium), Aristillus, and Autolycus form a very well-known landmark group. The first crater is flooded with mare lavas and thus is older than the lava flows that filled it, but the other two are clearly more recent than the lavas, displaying typical unflooded impact-crater floors and rough, radial ejecta deposits on top of the mare surface.

The Apennine Scarp (or Mountains, Montes Apenninus) is extremely impressive when viewed just after first quarter or near the last quarter phase. The highest peaks rise some 16,000 ft. (5 km) above the mare surface, but the pinnacle-like drawings of them in older books stem from a misinterpretation of the shapes of their shadows; photographs taken by the Apollo 15 astronauts revealed the generally smooth, rolling nature of the mountains that comprise this range. The Hadley Rille, which was visited by those astronauts, can be glimpsed with small telescopes under steady atmospheric conditions when the moon is about nine days old (past new moon).

The north shore of Mare Imbrium is marked by a strip of highlands that contains the lunar Alps (Montes Alpes), the crater Plato, and the spectacular Sinus Iridum (Bay of Rainbows). The location of this strip is quite anomalous with respect to the rest of the Imbrium basin; there is mounting evidence that the strip is

SINUS
MEDII
Surveyor 4
Surveyor 6
Schröter
Reinhold
Kunowsky
Murchison
Pallas
Surveyor 2
Encke
Maestlin
Bode
Copernicus
MARE
INSULARUM
Kepler
Milichius
SINUS
AESTUUM
Stadius
MONTES CARPATUS
OCEANUS
Eratosthenes
Gay-Lussac
Mons Wolff
Draper
T.Mayer
Mons Ampère
Wallace
MONTES
Mons Huygens
Pytheas
Mons Bradley
Apollo 15
MARE
Euler
Aristarchus
APENNINUS
PALUS
Lambert
Montes
Prinz
PUTREDINIS
Timocharis
Diophantus
Harbinger
Archimedes
Mons La Hire
Caventou
Krieger
olycus
Luna 2
IMBRIUM
Delisle
Angström
W
Aristillus
Montes
Heis
Gruithuisen
PROCELLARUM
Spitzbergen
C.Herschel
Mons
Kirch
Luna 17
8
Piton
Piazzi Smyth
LeVerrier
Helicon
Promontorium Heraclides
Mont Blanc
Promontorium
SINUS
Mairan
MONTES
Mons
Laplace
IRIDUM
Pico
Montes
Montes
Louville
Tenerife
Recti
JURA
Sharp
Vallis Alpes
Maupertuis
MONTES
Bianchini
Plato
Bouguer
Foucault
SINUS RORIS
La Condamine
MARE
Harpalus
FRIGORIS
Hornsbow
Robinson
Markov
Timaeus
Fontenelle
South
Birmingham
J.Herschel
Oenopides
Epigenes
Babbage
Anaximander
Pythagoras
Goldschmidt
Philolaus
Cleostratus
Anaxagoras
Carpenter
Mouchez
Anaximenes
Desargues
Boole
Poncelet
Cremona
Gioja
Pascal
Sylvester
Brianchon
Hermite

N

Fig. 8-15 The northeast portion of Mare Imbrium, showing Montes Caucasus, Montes Alpes (the Alps), and Vallis Alpes (the Alpine Valley). (Stephan Martin, Hopkins Observatory, Williams College)

actually a coalesced pair of crustal islands that were torn from their original locations some 100 miles (160 km) to the northeast, during the refilling of the Imbrium crater by magma flowing underneath the moon's crust. The generally parallel-sided shape of Mare Frigoris, the "wrong" placement of the Alps, and other bits of evidence support this conclusion. The Alps, a group of isolated mountains and hills, contain the famous Alpine Valley (Vallis Alpes), a roughly parallel-sided "gash" more than 100 miles (160 km) long and averaging 6 miles (10 km) in width (fig. 8-12). It was once surmised that this valley was formed either by a tangentially traveling meteorite or else by a flying fragment from the Imbrium basin–forming impact, but this idea has long been discounted. It appears to be a grabenlike feature, possibly caused by the crustal movements noted above.

Plato, 63 miles (101 km) in diameter, is prominent because of its smooth, level, dark floor, which contrasts strongly with the surrounding lighter highlands materials. The crater's floor contains craters 1½ miles (2½ km) in diameter and smaller; the largest one is located near the center. A medium-sized telescope is needed to show these objects as recognizable craters. Sinus Iridum looks beautiful when the sun is rising on it and its illuminated western edge protrudes far into the unlit surrounding area. This occurs when the moon is about 10 days old.

The crater Copernicus (57 miles — 92 km — in diameter), situated at the top of the map in Mare Insularum, is at least as famous and frequently observed as Plato. It is an excellent example of a large comparatively fresh impact crater. It has a level but somewhat rough floor, with a group of small central peaks. The strongly terraced walls, a result of post-impact slumping, rise some 15,000 ft. (5 km) above the floor. Copernicus is the center of one of the most extensive ray systems anywhere on the moon. The smaller crater Kepler (20 miles — 32 km — in diameter), at the top right of the map, also has a notable ray system, the rays of which mingle with those of Copernicus.

MOON MAP 8. This chart is taken up almost entirely by the gray expanse of Oceanus Procellarum, whose western shore is never far removed from the lunar limb. In addition to the usual ridge systems and sprinkling of small craters, this area displays some unique features and interesting formations. The showpiece is undoubtedly the crater Aristarchus, plus the whole Aristarchus Plateau region to the northwest of the crater itself.

Aristarchus is one of the brightest craters of its size anywhere on the moon. It is a relatively fresh impact crater, 28 miles (45 km) in diameter, with a rim rising about 9,000 ft. (2,700 m) above the floor. It is the center of a major ray system, as are all relatively fresh impact craters. More Transient Lunar Phenomena (TLPs) have been reported in and near Aristarchus than at any other lunar location. In the past, many of these reports undoubtedly stemmed directly or indirectly from effects of the crater's brightness. Thus when the moon is less than a week old, the crater can be seen glowing in the earthlit part of the disk; observed rather low in the sky, the sunlit crater is bright enough against the dark mare background to produce a spectrum as the earth's atmosphere spreads out the lunar light, causing some of the reported color phenomena. However, dedicated observers are now aware of these "red herrings," and not all reports can be dismissed so easily. The whole area is unusual from several standpoints, and rare phenomena, such as emissions of gas, may occur here. Indeed, the orbiting Apollo spacecraft detected more radon gas here than anywhere else on the moon.

Close to the northwest lies the moon's largest volcano and associated lava-flow channel, visible in the smallest telescopes. Although it was discovered by Huygens, this channel is known as Schröter's Valley (Vallis Schröteri), after the German amateur astronomer who first observed it. The volcano itself is the rather unassuming hill immediately northwest of Aristarchus, and the valley's northern flank starts with a depression nicknamed the "Cobra Head," because of the valley's resemblance to the head and body of a snake. The valley cuts through a rough area that is mostly darker than the surrounding mare, despite the presence of a concentration of rays from Aristarchus. Evidence from other investigations indicates that the whole area is covered with a blanket of lunar-type volcanic ash with a high ilmenite content. Although dark, the area is distinctly browner than the surrounding mare, or even other dark, ash-blanketed areas, which are all grayer than their surroundings. It is interesting that this volcanic soil with high ilmenite content can exist in two forms of identical composition but different color. Thus the black and orange soils collected near Shorty crater (which is not shown because of its small size) by the Apollo 17 astronauts are identical apart from color. The difference in color may be due to differing cooling rates, which determine whether the black ilmenite crystallizes out or stays dispersed as an orange coloration in the glass beads in the soil.

The bright marking known as Reiner Gamma is an example of a rare type of surface marking; other examples are confined almost entirely to the moon's far side. It is not connected with any visible topography, nor does it have any relief of its own. It does, however, exhibit a magnetic field; this field may shield the surface from the darkening effects of the solar wind and thus preserve the area's brightness. To the northeast lies a large field of rather small volcanic structures known familiarly as the Marius Hills, but the details of their topography only begin to become visible in large telescopes.

In the highlands region at upper right, the crater Hevelius—74 miles (119 km) in diameter—commands the most attention. Its floor is crisscrossed by a lattice of linear rilles, two of which cut through the crater wall and continue into the adjacent terrain.

The crater Lichtenberg, situated in the mare much farther to the north, has a ray system that extends only to the north, thus superficially resembling the Proclus system. However, the cause in this case is quite different—the southern portion of the Lichtenberg ray system was covered up by a later lava flow.

Fig. 8-16 Lunar images from the Clementine spacecraft. (top) Tycho. (Naval Research Laboratory) (bottom) Copernicus. (Lawrence Livermore National Laboratory)

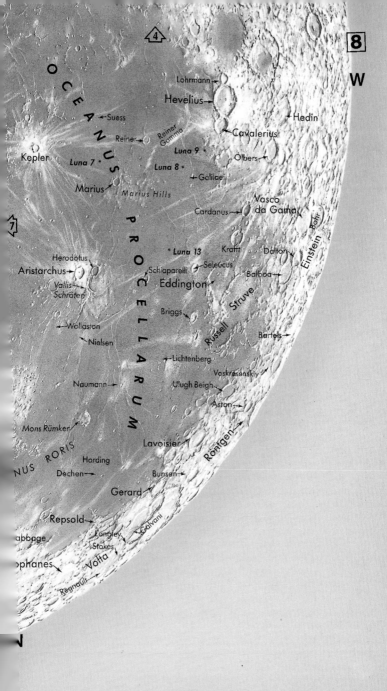

W

O C E A N U S

Lohrmann

Hevelius

Hedin

Suess

Cavalerius

Reiner

Reiner
Gamma

Kepler

Luna 9

Luna 7

Olbers

Luna 8

Galilaei

Marius

Marius Hills

Vasco
da Gama

Cardanus

Bohr

P R O C E L L A R U M

Luna 13

Krafft

Dalton

Herodotus

Aristarchus

Schiaparelli

Seleucus

Einstein

Vallis-
Schröten

Eddington

Balboa

Struve

Russell

Briggs

Wollaston

Bartels

Nielsen

Lichtenberg

Voskresensky

Naumann

Ulugh Beigh

Aston

Mons Rümker

NUS RORIS

Lavoisier

Röntgen

Harding

Dechen

Bunsen

Gerard

Repsold

Galvani

bbage

Longley

Stokes

ophanes

Volta

Regnault

N

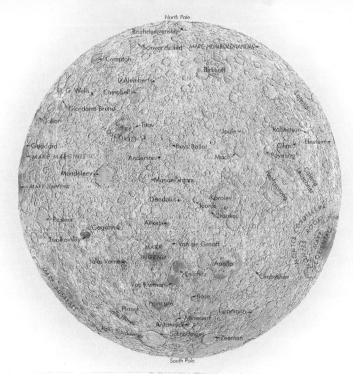

Fig. 8-17 The moon's far side. (North is up.)

THE MOON'S FAR SIDE

The far side of the moon has been seen only by Apollo astronauts, but it has been photographed both by them and by uncrewed spacecraft (fig. 8-13), so its topography is quite well documented. In 1990, the Galileo spacecraft en route to Jupiter also photographed the moon's far side. The Lunar Orbiter that reached the moon in 1997 has no camera, but is compiling a complete map and mapping the relative compositions of minerals.

The most obvious difference from the near side is the scarcity of dark maria. The far side also displays some remnants of larger basins. The cratered areas of the far side resemble the heavily cratered areas toward the south on the moon's near side, with a range of diameters and degrees of apparent freshness.

The paucity of maria (lava-flooded areas) on the far side means an almost complete absence of the volcanic-related features such as rilles, domes, vents, etc., that diversify the near-side topography.

9

2004/9

FINDING THE PLANETS

The brightest planets are quite easy to recognize. They often stand out because they are so bright and shine so steadily instead of twinkling. The reddish color of Mars also makes that planet distinctive. When we are observing the sky, these characteristics can help us recognize that we are observing planets instead of stars. The fact that the planets all lie close to the *ecliptic*—an imaginary line drawn across the sky to mark the yearly path of the sun among the stars—also helps us pick out the planets. The ecliptic is marked with a dotted line on the Monthly Sky Maps in chapter 3 and on the Atlas Charts in chapter 7.

Most of the planets can be seen with the naked eye; Venus, for example, is often conspicuous as a bright object in the sky near sunrise or sunset—the "morning star" or the "evening star," respectively. The planets are most enjoyable to observe, however, when you turn a telescope on them. In the next chapter, I will describe what each planet looks like in a telescope, and what some of the latest observations from space have revealed about the planets. In this chapter, I will show you how to locate the naked-eye planets in the sky, and how to know in advance which planets will be "up"—visible above the horizon—when you plan to observe. Also, if you see a bright object in the sky, you can use the information in this chapter to determine whether it is a planet.

As they orbit the sun, the planets change their positions with respect to the stars. The positions of the planets in the sky from night to night are displayed in a series of Graphic Timetables of the Heavens at the end of this chapter. The Graphic Timetables are a compact way of displaying the rising, setting, and transit times of the planets during the night; they show the times for sunrise, sunset, and twilight as well. (An object *transits* when it crosses your *meridian*, the line drawn from the north celestial pole through your zenith and then through the point due south on your horizon to the south celestial pole; thus a planet "transits" when it is passing due north or south of you, and is then at its highest in the sky.) Once you have become familiar with the Graphic Timetables, which takes a little practice, you will then be able to determine quickly which objects are well suited for observing at any given time of the year and at any time of the night; you will also be able to identify the planets you are seeing in the sky. You can also locate the planets by using the information in appendix 11 to plot their positions along the ecliptic.

I have provided one Graphic Timetable for each year. To use any of the Timetables, scan down the left-hand axis to find the date of the year. Then scan across the top or bottom axis to find the time of night in local time for your time zone. An extra line across the bottom shows the time in daylight saving time.

Look along the horizontal line that corresponds to your date to see which planets can be seen. The time of a planet's transit tells you the time when it will appear due south of you. Before that time, the planet will be to the east along the ecliptic; after that time, it will be to the west along the ecliptic. If a planet sets during the night, it will generally be visible from dusk until it sets. If a planet rises during the night, it will generally remain visible until morning twilight.

The Graphic Timetables in this guide are drawn for observers at latitude 40°N; the correction for observers at different latitudes amounts to only a few minutes, so we do not give it here. The Timetables give rising and setting times for objects assuming a flat horizon; if you have mountains on your horizon, of course, planets will become visible later or will disappear from your view earlier.

Each year's planetary curves are different from those of other years, yet distinct patterns appear. The Timetables show the curves even before sunset and after sunrise, when the planets aren't usually visible (though the brighter ones can sometimes be seen), in order to display the patterns better.

The regular cycling of Mercury's rising and setting times as it rapidly orbits the sun is displayed as distorted sine waves oscillating on either side of the sunrise and sunset lines. Venus displays a similar pattern but over a much longer period. Notice that the rising and setting curves for Venus recur every 19.2 months, the period with which Venus returns to successive *superior conjunctions* —when Venus lines up on the far side of the sun for us on earth. (This interval is Venus's *synodic period,* as opposed to its *sidereal period*—the time it takes to return to the same position in the sky with respect to the stars.) Whenever Venus sets more than 45 minutes after sunset, you should be able to spot it easily with the naked eye. Mercury and Venus have orbits smaller than earth's and are known as *inferior planets.*

The *superior planets* have orbits outside the earth's. They can be seen rising and setting far from the sun in the sky, and the curves showing their positions on the Graphic Timetables are more linear. Mars's pattern repeats every 25–27 months, which means that oppositions of Mars—when Mars is opposite the sun in our sky, transiting in the middle of the night, 12 hours after the sun has transited—occur at intervals of about two years, two months. During an opposition, Mars is on your meridian at midnight, when the sun is as far on the other side of the earth (and thus invisible) as it gets. Jupiter returns to opposition every 13 months and Saturn every 12.4 months; in other words, in the 12 months the earth takes to complete an orbit of the sun, Jupiter and Saturn have moved ahead a little and it takes us a few more weeks to catch up. Oppositions of Mars, Jupiter, and Saturn are listed in the next chapter.

For practice in using the Graphic Timetables, let us pick an example. Look at the night of June 1, 2005. Follow across the Timetable with your finger from left to right. First you find that Mercury sets just as twilight starts, which means that you can't see it. Then, in twilight, Jupiter transits. Thus it is high in the south as it grows dark and will be well placed for viewing in the evening hours. Later in twilight, Venus sets. Venus is so bright that it may be visible low in the sky, though you can see by following Venus's green line to later in the year that it will be up after dark in the fall. The change in shading shows that twilight ends after 10 P.M. at that time of year, using the daylight time axis at the bottom instead of the standard time axes that appear at both top and bottom. At about 11:30, daylight-saving time, Saturn sets, so it has been low in the west before that. Mars rises shortly after 2:00 A.M., but will take an hour or more to become high in the sky for most people to view it. Before 3 A.M., Jupiter sets. At the end of twilight, too late to be visible, Mercury rises. Venus rises in daylight, also too late to be seen. The times of Mars's transit and Saturn's rise are also not sigificant for observing them on this day.

You can use a Graphic Timetable in reverse to identify an object in the sky. You must check both the Graphic Timetables of the planets to see if the object is a planet and the Graphic Timetables of the Brightest Stars (fig. 1-3, p. 31) to see if it is a star. If the object is in the east, check the appropriate Timetable in this chapter for a curve representing a recent rising time of a planet. If it is to the south, then search for a transit curve and compare it with those for the brightest stars, in fig. 1-3. If the object is to the west, look to see what bright planets are about to set at the time when you are observing.

The positions of Mars, Jupiter, and Saturn in the sky are shown in a series of charts.

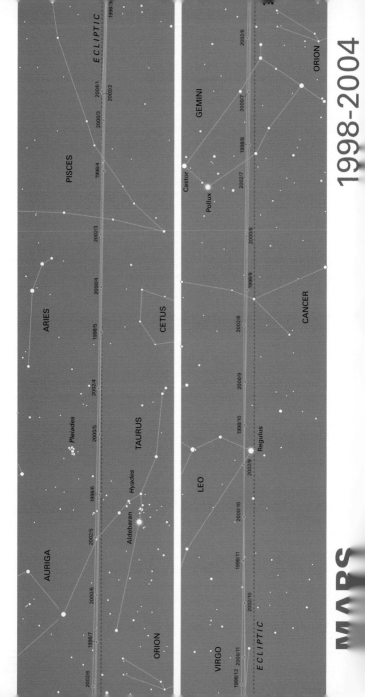

1998-2004

MARS

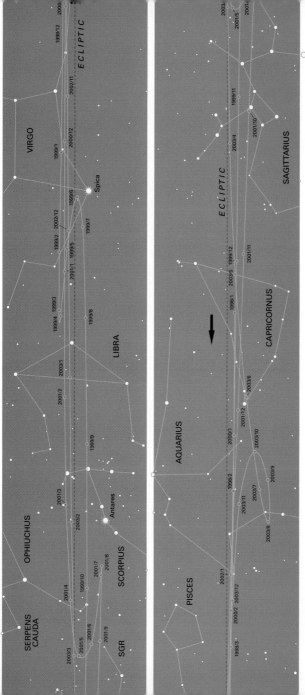

MARS

1998-2004

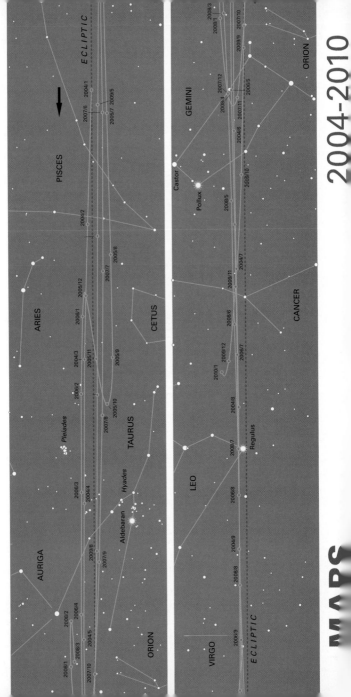

2004-2010

MAPS

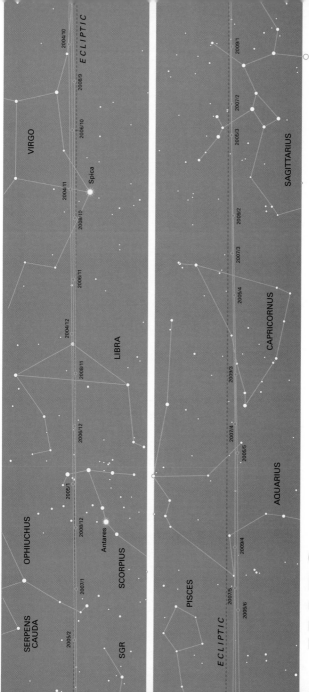

MARS

2004-2010

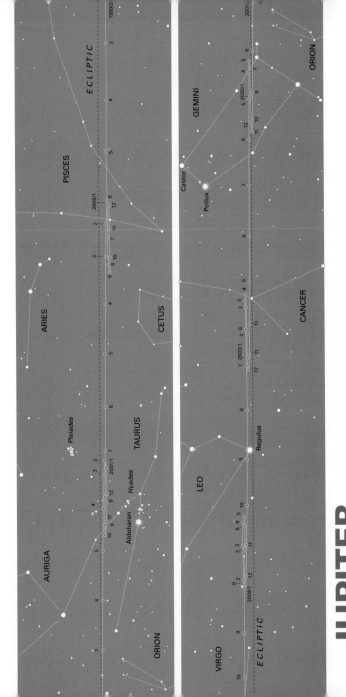

JUPITER

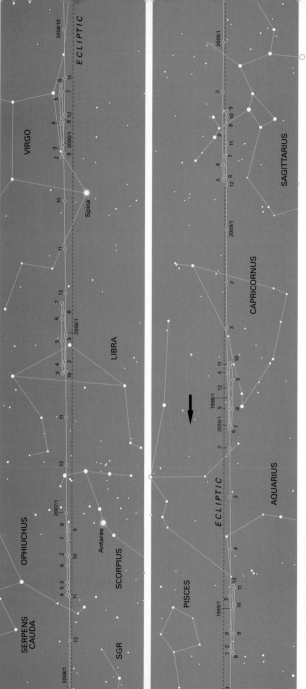

JUPITER

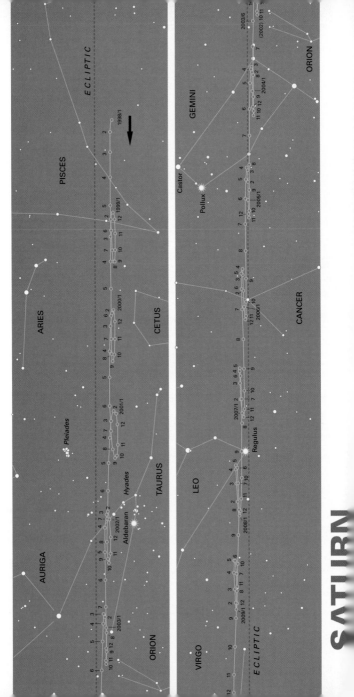
SATURN

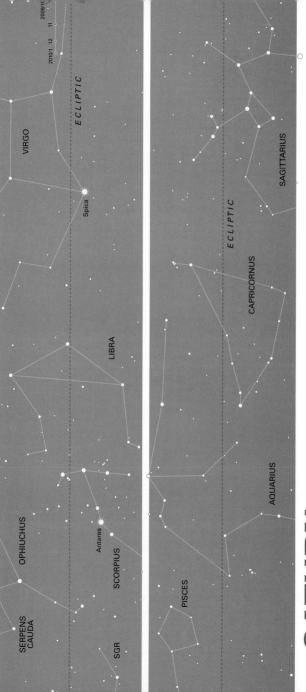

SATURN

TOTAL LUNAR ECLIPSE on January 20/21 (visible in North America and western Europe, among other places) and a total lunar eclipse on July 16 visible in Australia and vicinity.

SUN has no total eclipses. Partial solar eclipses on February 5 (Antarctica only); July 1 (southern tips of Chile and Argentina); July 30 in Pacific time (nw U.S. and Canada, northern Russia); and December 25 (all of continental U.S. and most of Canada).

MERCURY has its best evening viewing in mid-February and early June, when it does not set until the end of twilight. It might also be seen as an evening star low in the west in September and early October. Mercury is most visible in the morning sky in mid-November and will also be visible in twilight from mid-March through mid-April and in late July and early August. Look for the Mercury-Mars conjunction in morning twilight on August 10 and their closeness on surrounding days.

VENUS begins the year as the morning star, gleaming brightly in the east before morning twilight begins through mid-February. Look at sunrise for the Venus-Jupiter conjunction of May 17, when these two planets are only 42 arc sec apart, and for their closeness in the sky for days before and after. From July on, it is the evening star, shining brightly through the end of the year.

MARS sets two hours after evening twilight ends as the year begins and sets earlier through mid-April, when it sets at sunset near Jupiter and Saturn. Mars rises at the beginning of morning twilight in late August and is increasingly early until it rises two hours before morning twilight begins at the end of the year. Look for the Mercury-Mars conjunction in morning twilight on August 10 and their closeness on surrounding days.

JUPITER is near its highest due south as the year begins, setting increasingly early through mid-April. Look at sunrise for the Venus-Jupiter conjunction of May 17, when they are only 42 arc sec apart, and for their closeness in the sky for days before and after. By July 1, Jupiter rises before morning twilight begins. In mid-July, look for Jupiter and Saturn near Aldebaran in the morning sky. Jupiter then rises increasingly early until it rises at sunset in late October. It passes the Hyades in August and September and is at opposition on November 28. It is well placed for viewing all night as the year ends.

SATURN is near Jupiter in the sky all year. Both are well placed for evening viewing from January through March, when Saturn trails Jupiter by an hour, and from mid-October through the end of the year, when Saturn leads Jupiter by a half hour. Find Saturn near the moon on October 16, November 12, and December 9. Saturn is at opposition on November 19.

GRAPHIC TIMETABLE OF THE HEAVENS
2000

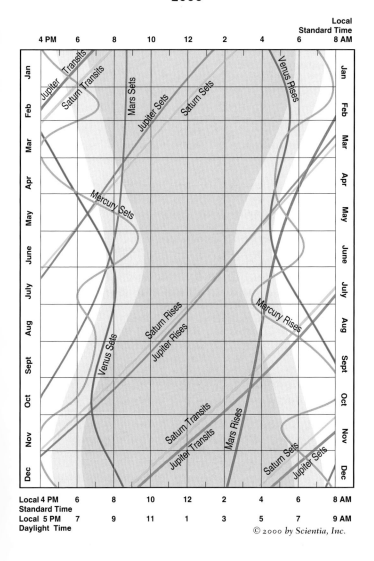

Local Standard Time

| 4 PM | 6 | 8 | 10 | 12 | 2 | 4 | 6 | 8 AM |

Jupiter Transits
Saturn Transits
Mars Sets
Jupiter Sets
Saturn Sets
Venus Rises
Mercury Sets
Saturn Rises
Jupiter Rises
Venus Sets
Mercury Rises
Saturn Transits
Jupiter Transits
Mars Rises
Saturn Sets
Jupiter Sets

Jan, Feb, Mar, Apr, May, June, July, Aug, Sept, Oct, Nov, Dec

Local 4 PM	6	8	10	12	2	4	6	8 AM
Standard Time								
Local 5 PM	7	9	11	1	3	5	7	9 AM
Daylight Time								

© 2000 by Scientia, Inc.

TOTAL LUNAR ECLIPSE on January 9 (visible from Europe), with partial phases visible from the eastern U.S.; and a partial eclipse visible from Australia on July 5.

TOTAL SOLAR ECLIPSE on June 21 in Africa; annular solar eclipse on December 14 in Central America with partial phases over most of the U.S. (excluding the East Coast).

MERCURY sets at the end of evening twilight in mid-May and earlier than that the rest of the year. It rises at the beginning of morning twilight in late October and closer to sunrise than that for the rest of the year. Look at the beginning of morning twilight for the close passages of Venus and Mercury, only half a degree apart, on October 29 and on November 4, and their closeness on surrounding days.

VENUS sets two hours after the end of evening twilight as the year begins, diminishing its distance from the sun until it sets at the end of evening twilight in mid-March. It is a morning star from April through the end of the year. On July 15, Venus passes 0.7 degrees south of Saturn. In mid-July, Venus is near the Hyades, and in early September it is south of the open cluster Praesepe. Look at the beginning of morning twilight for the close passages of Venus and Mercury, only half a degree apart, on October 29 and on November 4, and their closeness on surrounding days. These planets will be within a degree for 11 days.

MARS is well placed in the evening sky from June on, transiting at sunset in mid-July. It is in opposition on June 13. It sets at midnight in mid-September and sets about 10:00 P.M. at the end of the year.

JUPITER is well placed in the evening sky as the year begins, setting after 4:00 A.M. It sets earlier each night until it sets at the end of evening twilight in mid-May. In mid-July look for Jupiter and Venus near Aldebaran in Taurus and for Jupiter and Venus to be especially close together in the days around August 6. Jupiter rises at midnight as October begins, and rises earlier each night through the end of the year, when it is well placed in the evening sky.

SATURN precedes Jupiter by about a half-hour in the sky through April, as both are well placed for evening viewing. Saturn rises at midnight in late August and rises earlier each night through the end of the year, when it is about two hours ahead of Jupiter in the east in the evening sky. Find Saturn near the moon on January 5, February 2, and March 1. Look for a grouping of the crescent moon, Venus, and Saturn on July 17. Saturn is at opposition on December 2.

GRAPHIC TIMETABLE OF THE HEAVENS
2001

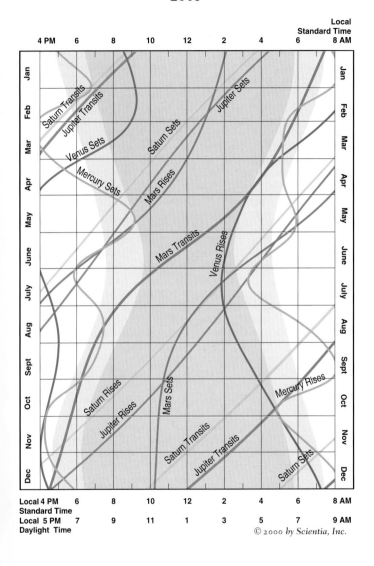

Local Standard Time

| 4 PM | 6 | 8 | 10 | 12 | 2 | 4 | 6 | 8 AM |

Jan
Feb
Mar
Apr
May
June
July
Aug
Sept
Oct
Nov
Dec

Saturn Transits
Jupiter Transits
Venus Sets
Mercury Sets
Mars Rises
Saturn Sets
Jupiter Sets
Mars Transits
Venus Rises
Saturn Rises
Jupiter Rises
Mars Sets
Mercury Rises
Saturn Transits
Jupiter Transits
Saturn Sets

| Local 4 PM Standard Time | 6 | 8 | 10 | 12 | 2 | 4 | 6 | 8 AM |
| Local 5 PM Daylight Time | 7 | 9 | 11 | 1 | 3 | 5 | 7 | 9 AM |

© 2000 *by Scientia, Inc.*

SOLAR ECLIPSES include an annular eclipse on June 10 almost reaching Baja California, with partial phases over the western U.S. and Canada; and a total solar eclipse in Africa on December 4.

PLANETARY CONJUNCTIONS will put all the planets that can be seen with the naked eye in the western sky at dusk during the last week in April and the first two weeks of May.

MERCURY sets as late as the end of evening twilight only at mid-January and at the end of April and the beginning of May. It rises as early as the beginning of morning twilight only in mid-February and mid-October. Look for Mercury and Mars only 3° apart at the beginning of morning twilight on October 10 and close together on surrounding days.

VENUS sets after the end of evening twilight from April through August, so it is the evening star during those times. It is highest in the western sky after sunset in May and June. See Venus near Jupiter in early June and nearest Spica on September 1. It rises as early as the beginning of morning twilight in mid-November and rises earlier through the end of the year, when it rises nearly four hours before sunrise. Look for Venus and Mars only 1½° apart on December 6, rising together in the morning sky a couple of hours before morning twilight.

MARS is in the evening sky as the year begins, setting as evening twilight ends in mid-May. It is then not visible through early October, when it rises as morning twilight begins. It rises as early as sunrise only in mid-October. Look for Mercury and Mars only 3° apart at the beginning of morning twilight on October 10 and close together on surrounding days. Later, look for Venus and Mars only 1½° apart on December 6, rising together in the morning sky a couple of hours before morning twilight. By the end of the year, Mars rises nearly four hours ahead of the sun.

JUPITER is well placed in the evening sky as the year begins, transiting at midnight. It sets earlier each night, until it sets at the end of evening twilight, near Venus, in early June. It becomes visible again in the morning sky in August, rising at 11:00 P.M. at the beginning of November and rising at 7:30 P.M. at the end of the year.

SATURN is well placed in the evening sky as the year begins, transiting before 10:00 P.M., about two hours ahead of Jupiter. See Saturn near the Hyades. It is visible in the evening sky through April. It moves into the morning sky in August, rising at midnight in early September and rising at the end of evening twilight in late November. It ends the year well placed for viewing in the evening sky, reaching opposition on December 16.

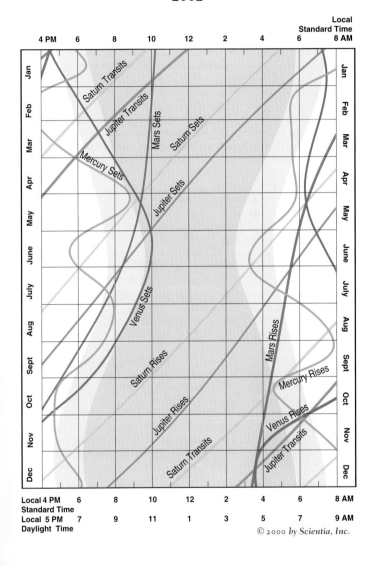

Local Standard Time

4 PM   6   8   10   12   2   4   6   8 AM

Jan · Feb · Mar · Apr · May · June · July · Aug · Sept · Oct · Nov · Dec

Saturn Transits
Jupiter Transits
Mars Sets
Saturn Sets
Mercury Sets
Jupiter Sets
Venus Sets
Saturn Rises
Mars Rises
Mercury Rises
Jupiter Rises
Venus Rises
Saturn Transits
Jupiter Transits

Local	4 PM	6	8	10	12	2	4	6	8 AM
Standard Time									
Local	5 PM	7	9	11	1	3	5	7	9 AM
Daylight Time									

© 2000 by Scientia, Inc.

2003

TOTAL LUNAR ECLIPSE on May 15/16, visible from eastern North America; and a total lunar eclipse on November 8/9 visible from eastern North America and Europe.

SOLAR ECLIPSES include an annular eclipse on May 31 near Greenland with partial phases visible in most of Europe and Asia; total solar eclipse on November 23 visible only from Antarctica (cruises are being arranged) with partial phases reaching Australia.

MERCURY is best visible in the evening sky in April and in evening twilight in mid-July through mid-August. It is best visible in the morning sky in late January and late September and early October and may also be seen in morning twilight in late May and June. On May 7, Mercury goes in transit across the sun, an event visible from most of the world except for the Americas. See the discussion in the next chapter.

VENUS begins the year as a bright morning star, and remains so, rising closer and closer to morning twilight, through March. It sets after sunset from September on, and after the end of evening twilight in December.

MARS rises at 3:00 A.M. at the beginning of the year, rising earlier and earlier until it is up at sunset in early September. It sets during morning twilight in late August and early September and sets at midnight at the end of the year. It is in opposition on August 28, as close to earth and as bright as it gets.

JUPITER is up all night from the beginning of the year through mid-March and sets earlier through the beginning of July, when it sets at the end of evening twilight. It rises at sunrise in late August, rising earlier and earlier, until it rises at 10:30 P.M. at the end of the year.

SATURN is up all night at the beginning of the year, transiting at 11:00 P.M. It sets earlier and earlier until it sets at the end of evening twilight in late May. It then rises at dawn in late July, rising at midnight daylight-saving time in early October and rising at sunset, the beginning of evening twilight, at the end of the year.

URANUS can be found close to Venus on March 28 and surrounding days.

Graphic Timetable of the Heavens
2003

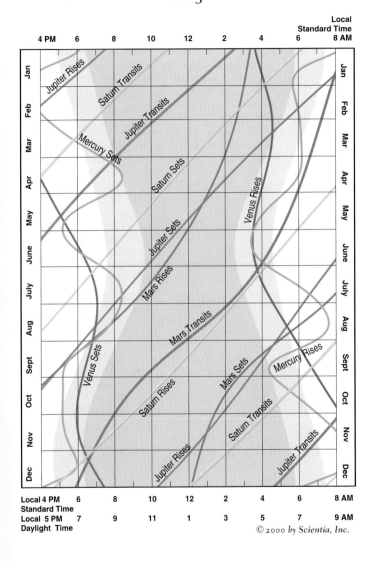

Local Standard Time

4 PM 6 8 10 12 2 4 6 8 AM

Jan · Feb · Mar · Apr · May · June · July · Aug · Sept · Oct · Nov · Dec

Jupiter Rises
Saturn Transits
Jupiter Transits
Mercury Sets
Saturn Sets
Venus Rises
Jupiter Sets
Mars Rises
Mars Transits
Venus Sets
Mercury Rises
Saturn Rises
Mars Sets
Jupiter Rises
Saturn Transits
Jupiter Transits

| Local Standard Time | 4 PM | 6 | 8 | 10 | 12 | 2 | 4 | 6 | 8 AM |
| Local Daylight Time | 5 PM | 7 | 9 | 11 | 1 | 3 | 5 | 7 | 9 AM |

© 2000 by Scientia, Inc.

TRANSIT OF VENUS, June 8, the first since 1882. Entirely visible from Europe to India with the end barely visible in the easternmost U.S. Worth a trip. See the discussion in the following chapter.

TOTAL LUNAR ECLIPSE on May 4 visible from eastern Europe and western Asia; and a total lunar eclipse on October 27/28 visible from the Americas (except the west coast) and from western Europe.

SOLAR ECLIPSES include a partial eclipse on April 19 visible in southern Africa; and a partial eclipse on October 14 visible in Japan and Alaska, and barely reaching western Hawaii.

MERCURY is best visible in the evening during twilight in late March–early April, July, and November. In July through mid-August, look for Mercury and Mars close together, closest on July 10 and August 16. Mercury rises at the beginning of morning twilight in mid-January and early September. It rises close to Jupiter and Mars on and around September 28. Mercury and Venus rise together as morning twilight begins in the last week of December.

VENUS is visible after sunset in the evening sky through May and in the morning sky before sunrise from July through the end of the year. Venus and Jupiter are close together in the morning sky on and around November 4. See above for the transit of Venus.

MARS is an evening object through the winter and spring, setting at midnight as the year begins and as evening twilight ends in June. It becomes visible in the morning sky in twilight in October and before morning twilight in mid-November through the end of the year. It is closest to Mercury on July 10 and August 16.

JUPITER rises at 10:00 P.M. at the beginning of the year and at sunset at the end of February. It is then up all night through mid-April, setting earlier and earlier until it sets at the end of evening twilight at the end of July. It rises before the beginning of morning twilight between mid-October and the end of the year. Venus and Jupiter are close together in the morning sky on and around November 4. An occultation of Jupiter by the moon will be visible from the eastern U.S. on December 7.

SATURN transits at midnight as the year begins, transiting at sunset and setting at 2:00 A.M. in late March. It sets earlier and earlier until it sets at the end of evening twilight in late May. It rises before beginning of morning twilight from August on, rising at the end of evening twilight and remaining visible through the night at the end of the year.

Graphic Timetable of the Heavens
2004

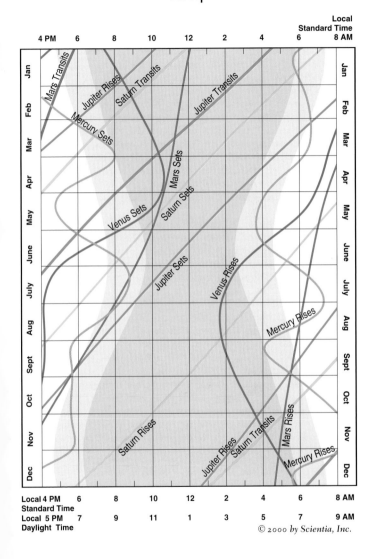

Local Standard Time

4 PM — 6 — 8 — 10 — 12 — 2 — 4 — 6 — 8 AM

Mars Transits
Jupiter Rises
Saturn Transits
Jupiter Transits
Mercury Sets
Mars Sets
Venus Sets
Saturn Sets
Jupiter Sets
Venus Rises
Mercury Rises
Saturn Rises
Jupiter Rises
Saturn Transits
Mars Rises
Mercury Rises

Jan, Feb, Mar, Apr, May, June, July, Aug, Sept, Oct, Nov, Dec

| Local Standard Time | 4 PM | 6 | 8 | 10 | 12 | 2 | 4 | 6 | 8 AM |
| Local Daylight Time | 5 PM | 7 | 9 | 11 | 1 | 3 | 5 | 7 | 9 AM |

© 2000 by Scientia, Inc.

PARTIAL LUNAR ECLIPSE on October 17 visible from Hawaii, Alaska, and the northwestern U.S. and Canada.

ANNULAR/TOTAL SOLAR ECLIPSE on April 8 over the Pacific, with annularity reaching Panama and partial phases over the southern U.S.; annular solar eclipse on October 3 crossing Portugal, Spain, and eastern Africa.

MERCURY is best visible in the evening sky in mid-March and in the morning sky in late August and early December. Look for its conjunction with Venus on June 27, when the two planets are 4 arc min (less than 0.1°) apart in evening twilight. Saturn is close by.

VENUS sets in twilight from April through August and after sunset through the end of the year. Look for its conjunction with Mercury on June 27, when the two planets are 4 arc min (less than 0.1°) apart in evening twilight. Saturn is close by. Venus and Jupiter are close together, setting as evening twilight ends, on September 2.

MARS rises an hour or two before sunrise through mid-May, and earlier and earlier until it rises at midnight daylight-saving time in late July and at sunset in October. Mars is then visible all or most of the night through the end of the year. It is in opposition on November 7, gleaming at magnitude -2.3.

JUPITER rises at about midnight as the year begins, and at the end of evening twilight in mid-March, passing opposition on April 2. It is then well placed for evening viewing. It sets at midnight daylight-saving time in mid-July and at sunset by September. It is visible in the eastern sky before sunrise in December.

SATURN transits at about the same time as Jupiter rises through March. It is well placed for evening viewing through the spring. Saturn sets at midnight daylight-saving time in mid-May and at the end of evening twilight in mid-June. See Saturn, Venus, and Mercury set together in evening twilight in late June. Saturn enters the morning sky in mid-August, rising earlier and earlier until it rises an hour after the end of evening twilight at the end of the year.

GRAPHIC TIMETABLE OF THE HEAVENS
2005

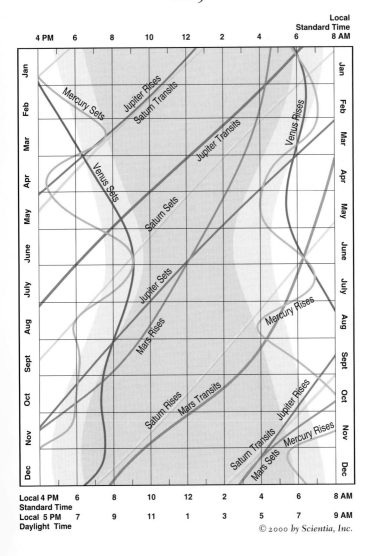

Local Standard Time

| 4 PM | 6 | 8 | 10 | 12 | 2 | 4 | 6 | 8 AM |

Jan
Feb
Mar
Apr
May
June
July
Aug
Sept
Oct
Nov
Dec

Mercury Sets
Jupiter Rises
Saturn Transits
Jupiter Transits
Venus Rises
Venus Sets
Saturn Sets
Jupiter Sets
Mars Rises
Mercury Rises
Saturn Rises
Mars Transits
Saturn Transits
Jupiter Rises
Mars Sets
Mercury Rises

| Local Standard Time | 4 PM | 6 | 8 | 10 | 12 | 2 | 4 | 6 | 8 AM |
| Local Daylight Time | 5 PM | 7 | 9 | 11 | 1 | 3 | 5 | 7 | 9 AM |

© 2000 by Scientia, Inc.

2006

PARTIAL LUNAR ECLIPSE on September 7 partly visible from western Europe and visible in Asia and western Australia.

TOTAL SOLAR ECLIPSE on March 29 visible from Africa, Turkey, and Russia, with partial phases visible throughout Europe; annular eclipse on September 22 from northern South America and the Atlantic Ocean, with partial phases visible in eastern South America and western Africa.

MERCURY On November 8, Mercury goes in transit across the sun, with the beginning visible from North and South America, the peak viewing from Hawaii, and the end visible from Australia as well as barely from Japan and the west coast of Asia. See the discussion in the next chapter.

Mercury is never in the evening sky after the end of evening twilight this year, with its setting closest to that time in late February. Mercury and Venus are only 2° apart in morning twilight on August 10. The next chance of seeing Mercury in the morning sky is in late November.

VENUS is in the morning sky almost all year, moving to the evening sky only for November and especially December. Venus rises near Mercury in early August, passing only 2° from it on August 10, and rises near Saturn, passing only about 4 arc min (less than 0.1°) from it on August 26.

MARS is in the sky at sunset as the year begins, setting at midnight daylight-saving time in mid-May and at the end of evening twilight as July begins. Mars and Saturn are close to each other as they set on June 17 and surrounding days. Mercury and Mars are close together as they rise in morning twilight on December 9 and surrounding days.

JUPITER doesn't rise until 2:30 A.M. as the year begins, but rises at the end of evening twilight by the end of April. Then it is well placed for observing the whole night through mid-June, setting at midnight daylight-saving time about August 1 and at the end of evening twilight in early October.

SATURN is in the sky at sunset as the year begins, setting at midnight daylight-saving time in early June and at sunset in late June. It moves into the morning sky in August, rising with Venus and only about 4 arc min (less than 0.1°) from it on August 26. Saturn rises earlier each night, rising at midnight by November and by 8:00 P.M. at the end of December.

URANUS is less than 1¼ arc min from Mercury in evening twilight on February 14.

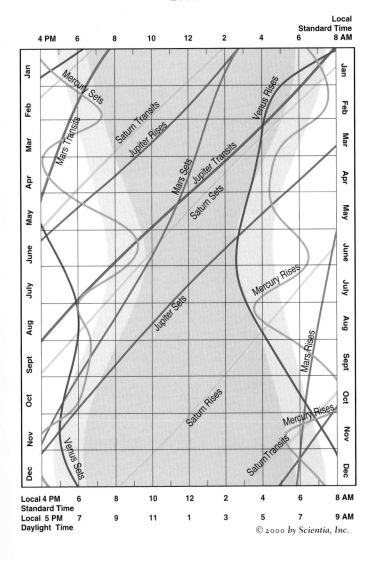

Local Standard Time

4 PM 6 8 10 12 2 4 6 8 AM

Mercury Sets
Mars Transits
Saturn Transits
Jupiter Rises
Venus Rises
Mars Sets
Jupiter Transits
Saturn Sets
Mercury Rises
Jupiter Sets
Mars Rises
Saturn Rises
Mercury Rises
Venus Sets
Saturn Transits

Local Standard Time 4 PM 6 8 10 12 2 4 6 8 AM
Local Daylight Time 5 PM 7 9 11 1 3 5 7 9 AM

© 2000 *by Scientia, Inc.*

2007

TOTAL LUNAR ECLIPSE on March 3 visible in Europe, Africa, and the eastern U.S. and Canada with partial phases visible throughout the U.S. and Canada; total lunar eclipse on August 28 visible from the U.S. West Coast, Hawaii, Alaska, New Zealand, and eastern Australia, with partial phases visible throughout the U.S. and Canada.

PARTIAL SOLAR ECLIPSE on March 19 visible in Asia and, barely, Alaska; partial solar eclipse on September 11 visible in South America.

MERCURY sets before the end of evening twilight all year, with the best chance of seeing it in the evening sky in late May. The best chance of seeing it in the morning sky is in early November, when it rises just before the beginning of morning twilight.

VENUS sets after sunset through June, and will be very visible in the evening sky from March through May. Look for Venus and the crescent moon close together on May 19. Venus will be in the morning sky before the beginning of morning twilight from September through the end of the year. Venus and Saturn are within 3° of each other on October 15 in the pre-dawn sky.

MARS doesn't rise appreciably before the beginning of morning twilight until June. It rises at midnight daylight saving time in late August, and then earlier and earlier until it rises at the end of evening twilight at the beginning of December. November and December, then, are the only opportunity to see it in the evening this year. Mars is in opposition on December 24.

JUPITER rises an hour before the beginning of morning twilight on January 1, rising earlier and earlier until it rises at the end of evening twilight in mid-May. It is nicely placed for observation in the evening all summer and into the fall, setting at midnight daylight saving time in late August and at the end of evening twilight in late November.

SATURN rises soon after sunset in late January and earlier thereafter, so it is well placed for viewing in the evening and nighttime hours through the spring. It sets at midnight daylight saving time in mid-June. It is in the morning sky as of mid-September. Venus and Saturn are within 3° of each other on October 15 in the pre-dawn sky. Saturn rises at about 9:00 P.M. as the year ends.

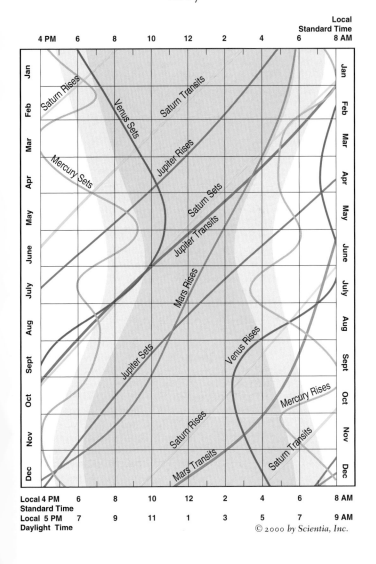

Local
Standard Time

| 4 PM | 6 | 8 | 10 | 12 | 2 | 4 | 6 | 8 AM |

Jan
Feb
Mar
Apr
May
June
July
Aug
Sept
Oct
Nov
Dec

Saturn Rises
Venus Sets
Saturn Transits
Mercury Sets
Jupiter Rises
Saturn Sets
Jupiter Transits
Mars Rises
Jupiter Sets
Venus Rises
Mercury Rises
Saturn Rises
Saturn Transits
Mars Transits

Local 4 PM	6	8	10	12	2	4	6	8 AM
Standard Time								
Local 5 PM	7	9	11	1	3	5	7	9 AM
Daylight Time								

© 2000 by Scientia, Inc.

2008

TOTAL LUNAR ECLIPSE on February 20/21 visible in the Americas (except for the west coast of the U.S. and Canada, where partial phases are visible) and in Europe; partial lunar eclipse on August 16 visible in Europe and Africa.

ANNULAR SOLAR ECLIPSE on February 7 visible only in Antarctica, with partial phases best in New Zealand and reaching eastern Australia; total solar eclipse on August 1 across northern Canadian islands, northern Greenland, Siberia, Mongolia, and China, with partial phases over Europe.

MERCURY is best visible in the evening in early May and in the morning in late October.

VENUS is visible before dawn in the eastern sky in January, and then in evening twilight during the summer. Venus and Saturn set together in evening twilight on August 13. Venus and Mars are within ½° on September 11, with Mercury close by. Venus sets later and later from mid-October on, setting after sunset from mid-October through the end of the year.

MARS is well placed for viewing for the first half of the year. It is up in the sky at sunset, setting at the beginning of morning twilight in mid-January, at midnight daylight-saving time in mid-June, and at the end of evening twilight in late July. Mars is in Praesepe, the Beehive Cluster, on and around May 23, and is only 4° from Saturn on July 11. Mars and Venus are within ½° of each other as they set on September 11, with Mercury close by.

JUPITER is in the morning sky from late January onward, rising as early as midnight daylight-saving time in mid-May and at sunset in early July. It sets at midnight daylight-saving time in early October and at the end of evening twilight at the end of the year. On November 30, Jupiter is only 2° from Venus in the evening western sky. Look for Venus, Jupiter, and the crescent moon close together before dawn on February 4 and at sunset on December 1.

SATURN rises in the early evening in January and is then well placed for viewing all night through early April, setting earlier and earlier until it sets at midnight daylight-saving time in mid-June and at the end of evening twilight in mid-July. Saturn is only 4° from Mars on July 11. It sets together with Venus in evening twilight on August 13. Saturn is in the morning sky from October through the end of the year, rising as early as 10:30 P.M. by December 31.

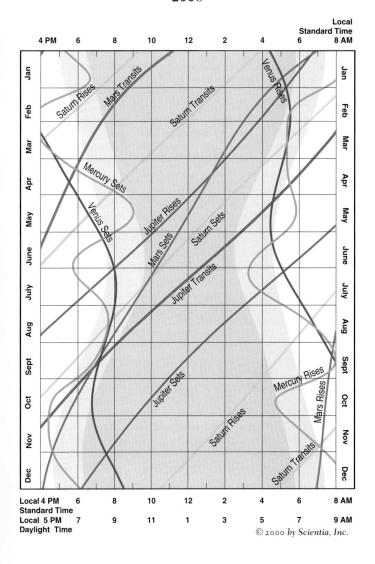

Local Standard Time

4 PM | 6 | 8 | 10 | 12 | 2 | 4 | 6 | 8 AM

Jan
Feb
Mar
Apr
May
June
July
Aug
Sept
Oct
Nov
Dec

Saturn Rises
Mars Transits
Saturn Transits
Venus Rises
Mercury Sets
Venus Sets
Jupiter Rises
Mars Sets
Saturn Sets
Jupiter Transits
Jupiter Sets
Mercury Rises
Mars Rises
Saturn Rises
Saturn Transits

Local Standard Time: 4 PM | 6 | 8 | 10 | 12 | 2 | 4 | 6 | 8 AM
Local Daylight Time: 5 PM | 7 | 9 | 11 | 1 | 3 | 5 | 7 | 9 AM

© 2000 by Scientia, Inc.

2009

PARTIAL LUNAR ECLIPSE on December 31 visible in Europe and Asia; some partial phases visible from New England.

ANNULAR SOLAR ECLIPSE on January 26 in Indonesia; total solar eclipse on July 22 across China.

MERCURY sets as late as the end of evening twilight only at the end of April and is up in evening twilight in April to mid-May, from mid-July through mid-September, and from mid-November to the end of the year. The best chance of seeing it in the morning sky is in early October, when it rises barely before morning twilight begins. Mercury rises close to Saturn on October 8.

VENUS is up in the dark sky before morning twilight from the end of May through the end of October. Venus rises close to Saturn on October 13. Venus is visible in the evening sky after twilight ends from January through mid-March.

MARS is not visible at night in the first part of the year but rises as morning twilight begins at the beginning of June. By mid-October it is rising at midnight daylight-saving time (11:00 P.M. standard time). Mars is in Praesepe, the Beehive Cluster, on and around November 1. At the end of the year, Mars rises just over an hour after evening twilight ends, the only chance this year of seeing it in the evening sky.

JUPITER rises at the beginning of morning twilight in early March, rising at the end of evening twilight in early July. Then it is well placed for viewing all night through the summer. It remains visible for at least the first half of the night for the rest of the year. It sets at the beginning of morning twilight at the beginning of September and sets at 8:30 P.M. at the end of the year.

SATURN rises in the evening at the beginning of the year, transiting at midnight in mid-March. It is well placed for viewing all night and for at least the first half of the night through mid-June and then for another month in the evening. It sets at the end of evening twilight at the end of July. Saturn is very close to Mercury on October 8 and to Venus on October 13, as they rise in the predawn sky. At the end of the year, Saturn is in the morning sky.

GRAPHIC TIMETABLE OF THE HEAVENS
2009

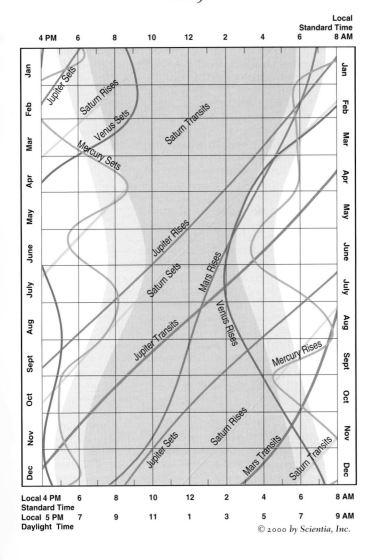

Local Standard Time

4 PM   6   8   10   12   2   4   6   8 AM

Jan
Feb
Mar
Apr
May
June
July
Aug
Sept
Oct
Nov
Dec

Jupiter Sets
Saturn Rises
Venus Sets
Saturn Transits
Mercury Sets
Jupiter Rises
Saturn Sets
Mars Rises
Jupiter Transits
Venus Rises
Mercury Rises
Jupiter Sets
Saturn Rises
Mars Transits
Saturn Transits

Local	4 PM	6	8	10	12	2	4	6	8 AM
Standard Time									
Local	5 PM	7	9	11	1	3	5	7	9 AM
Daylight Time									

© 2000 by Scientia, Inc.

PARTIAL LUNAR ECLIPSE on June 26 visible from Hawaii, Australia, and New Zealand; total lunar eclipse in the early morning of December 21 visible throughout North America.

ANNULAR SOLAR ECLIPSE across Africa, touching southern India and Sri Lanka, and reaching Thailand and China; total solar eclipse on July 11 across the southern Pacific, barely reaching southern Chile.

MERCURY sets barely after evening twilight ends in early April. Mercury and Venus set together, only 3° apart, on April 4. It sets in evening twilight also in July and August and in November through mid-December. The best chance of seeing it in the morning sky is in mid-September.

VENUS sets in or after evening twilight from the beginning of the year through mid-October, setting after evening twilight ends from the beginning of April through the end of July and with the end of twilight through August. Mercury and Venus set together, only 3° apart, on April 4. Look for Regulus near Venus on July 9. Venus is in the morning sky in November and December.

MARS is well placed for viewing in the first half of the year, transiting at 3:00 A.M. as the year begins. Mars is in opposition on January 29, but not an especially close one. It sets at the beginning of morning twilight as March begins, at midnight daylight-saving time in mid-June, and at the end of evening twilight in early September. Look near Mars for Regulus, less than 1° from it, on June 6, and for Spica, less than 2° from it, on September 4.

JUPITER sets shortly after evening twilight ends through January. It moves into the morning sky as of May, rising at midnight daylight-saving time in mid-July and at the end of evening twilight in early August. Then it is well placed for viewing all night through mid-October, when it sets at the beginning of morning twilight. It remains visible in the evening sky, setting at 11:00 P.M. by the end of the year.

SATURN rises before midnight as the year begins and rises as evening twilight ends by the beginning of March. It is well placed for viewing all night, until it sets at the beginning of morning twilight in early May. It remains a good evening object through July, when it sets as evening twilight ends. It is visible in the east before morning twilight begins starting in mid-October, rising shortly after midnight as the year ends.

GRAPHIC TIMETABLE OF THE HEAVENS
2010

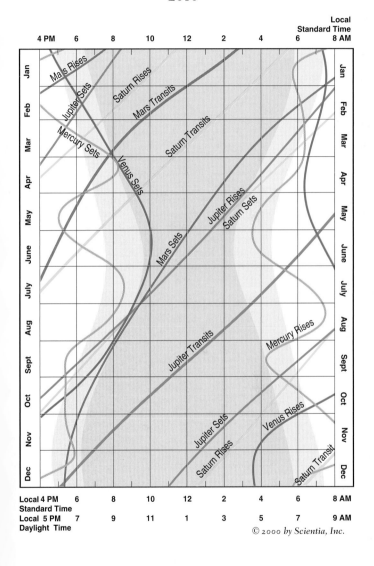

Local Standard Time

4 PM 6 8 10 12 2 4 6 8 AM

Jan · Feb · Mar · Apr · May · June · July · Aug · Sept · Oct · Nov · Dec

Mars Rises
Jupiter Sets
Saturn Rises
Mars Transits
Mercury Sets
Saturn Transits
Venus Sets
Jupiter Rises
Saturn Sets
Mars Sets
Jupiter Transits
Mercury Rises
Jupiter Sets
Saturn Rises
Venus Rises
Saturn Transit

Local Standard Time: 4 PM 6 8 10 12 2 4 6 8 AM
Local Daylight Time: 5 PM 7 9 11 1 3 5 7 9 AM

© 2000 by Scientia, Inc.

OBSERVING THE PLANETS

In the previous chapter, we explained how to find the planets in the sky. This chapter deals with observing the planets themselves. With even a small telescope, you can see that the planets have shapes and that the apparent shapes of some of the planets change over time. Since the easiest and most impressive planets to observe (in order) are Jupiter, Saturn, Venus, and Mars, we describe the planets in that order here, then discuss the planets that are harder to see: Mercury, Uranus, Neptune, and Pluto.

JUPITER

Jupiter, the largest planet in the solar system, is 11 times the diameter and over a thousand times the volume of the earth (fig. 10-1). It is about 5.2 times as far away from the sun as the earth is; since the average distance from the earth to the sun is called 1 Astronomical Unit (1 A.U.), Jupiter averages 5.2 A.U. from the sun. When Jupiter is on the same side of the sun as the earth, it is high in the nighttime sky and is only about 4 A.U. from us (5.2 A.U. minus the earth's orbital radius of 1 A.U. equals 4.2 A.U.). At this distance, the planet covers almost 50 seconds of arc in the sky—$1/40$ the diameter of the full moon—and can be about magnitude −2.5, three times brighter than the brightest star, Sirius.

Jupiter's four largest moons were discovered by Galileo in 1609, when he was the first to observe the sky with a telescope. They are thus known as the Galilean satellites (fig. 10-2). They are easily visible with even the smallest telescope; after all, our small modern telescopes surpass the quality of Galileo's telescopes.

When a planet is on a line with the earth and sun, the planet and sun appear close together in the sky and we say that the planet is in *conjunction*. The actual time of conjunction is when

Fig. 10-1 *Jupiter, photographed from the earth, showing its bright hori-zontal zones and its dark horizontal belts. North is at top. (CCD image by Donald Parker)*

the planet and the sun appear at the same longitude along the ecliptic (see appendix 11). Planets can also be in conjunction with each other or with the moon.

A planet is at *opposition* when it is on the extension of the line of sight from the sun through the earth. Jupiter and the other outer planets are best observed at opposition, when they are above the horizon all night. They then rise at sunset, are bright-

Fig. 10-2 *Jupiter with its four Galilean satellites: Io, Europa, Ganymede, and Callisto. Io is fainter than the others and is barely visible in this im-age, which was exposed to show the features on Jupiter's disk. (Akira Fujii)*

est, and show the largest disks. Jupiter is in conjunction with the sun only when it is on the far side of the sun and consequently appears smaller than it does when it is at opposition.

JUPITER'S DISK

With a small telescope, you can see light and dark bands across Jupiter's disk. The number of bands you see depends not only on how close Jupiter is to earth at the time but also on how steady the atmosphere is for your observing location. The bands represent clouds drawn out into long streaks as Jupiter rotates; the planet, which is made entirely of gas, rotates at different speeds at different latitudes. (Solid bodies like the earth don't do this.) Gas near the pole rotates about five minutes faster than gas near the equator does during each 10-hour rotation period. (Planets rotate —spin—on their axes but revolve—orbit—around the sun.)

A pair of NASA spacecraft called Voyager visited Jupiter in 1979 (fig. 10-3). The Galileo spacecraft reached Jupiter in 1995, dropping a probe into Jupiter's atmosphere and going into orbit to get close-ups of the moons. Even a large telescope on earth does not show much more detail than in figure 10-1, because of the obscuring effects of the earth's atmosphere. The form of the bands changes slowly over long periods of time. The bright bands are called *zones* and the dark bands are called *belts*. The belts and zones are polar, temperate, tropical, and equatorial, respectively,

Fig. 10-3 (left) Jupiter, imaged by the Voyager 1 spacecraft. Io appears as a reddish-brown ball suspended in front of the planet. (Photo NASA/JPL/Caltech) (right) A ground-based CCD image. The Great Red Spot shows, as do Europa and its shadow. (Donald Parker)

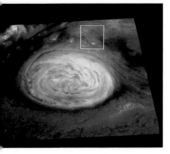

Fig. 10-4 Jupiter's Great Red Spot, imaged with the Galileo spacecraft. The colors in this reproduction are exaggerated. Tiny features in the white box show changes over minutes. We now know of similar storms on Saturn and Neptune. (NASA/JPL/Caltech)

in each hemisphere. The belts and zones show subtle colors.

Jupiter's rotation causes it to bulge at the equator—to be *oblate*—by 7 percent. It is obvious in even a small telescope that Jupiter is out-of-round.

A large reddish region known as the Great Red Spot has been observed on Jupiter for hundreds of years. The Great Red Spot is about 8,400 by 18,000 miles (14,000 by 30,000 km), much larger than the earth. The Great Red Spot is a giant storm in Jupiter's

Fig. 10-5 Jupiter's ring (left), impossible to see from Earth but imaged from the Galileo spacecraft. (NASA/JPL/Caltech)

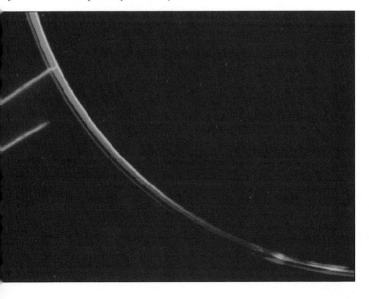

clouds, which is more visible at some times than others. Some-times its shape changes. Close-up views taken from the Voyager spacecraft revealed the rotation of the Great Red Spot (fig. 10-4).

Though the Voyagers discovered a thin ring around Jupiter, later found from Galileo to be made of dust given off by some of Jupiter's moons (fig. 10-5), it is too faint to be detected by ordinary means from earth.

JUPITER'S MOONS

The four brightest moons (satellites) of Jupiter can be seen with even a small telescope and their positions followed from night to night (fig. 10-6). Named Io, Europa, Ganymede, and Callisto, after mythological lovers of Jupiter (as are all of Jupiter's more than 16 moons), they range from 1,860 to 3,180 miles (3,100 to 5,300 km) in diameter. All but one of the four brightest moons are larger than our own moon. The largest of Jupiter's moons, Ganymede, is even larger than the planet Mercury.

Nonetheless, these four "Galilean satellites" (called so since they were discovered by Galileo) were merely points of light in the sky until NASA's Voyager spacecraft radioed back close-up pictures of the moons' surfaces, which have since been seen in even more detail by the Galileo spacecraft (fig. 10-7). Io is covered with volcanoes, of which about 10 are now erupting; it is the most active body in the solar system (fig. 10-8). Europa is covered with ice, which probably covers an ocean (fig. 10-9). Ganymede—which has the largest surface diameter of any moon in the solar system—may contain water and ice surrounding a core of rock; its surface shows craters and strange grooves. Callisto is covered with craters, including a huge bull's-eye called Valhalla.

These four bright moons orbit in the plane of Jupiter's equator, so they always seem stretched out close to a line drawn through Jupiter. The period of Io's orbit is 1 day, 18 hours; Europa's period is 3 days, 13 hours; Ganymede's is 7 days, 3 hours; and Callisto's is 16 days, 16 hours. If you observe these moons in a telescope over a night or from night to night, you can see their relative positions change. We sometimes see the moons move in front of or behind Jupiter; when the moons move in front of Jupiter, their shadows fall on the planet. When the moons are at their brightest, Ganymede can reach magnitude 4.4, and the others are fainter by no more than about a magnitude.

Fig. 10-6 (right) Jupiter's Galilean satellites revolving around Jupiter over a period of a few hours. (Akira Fujii)

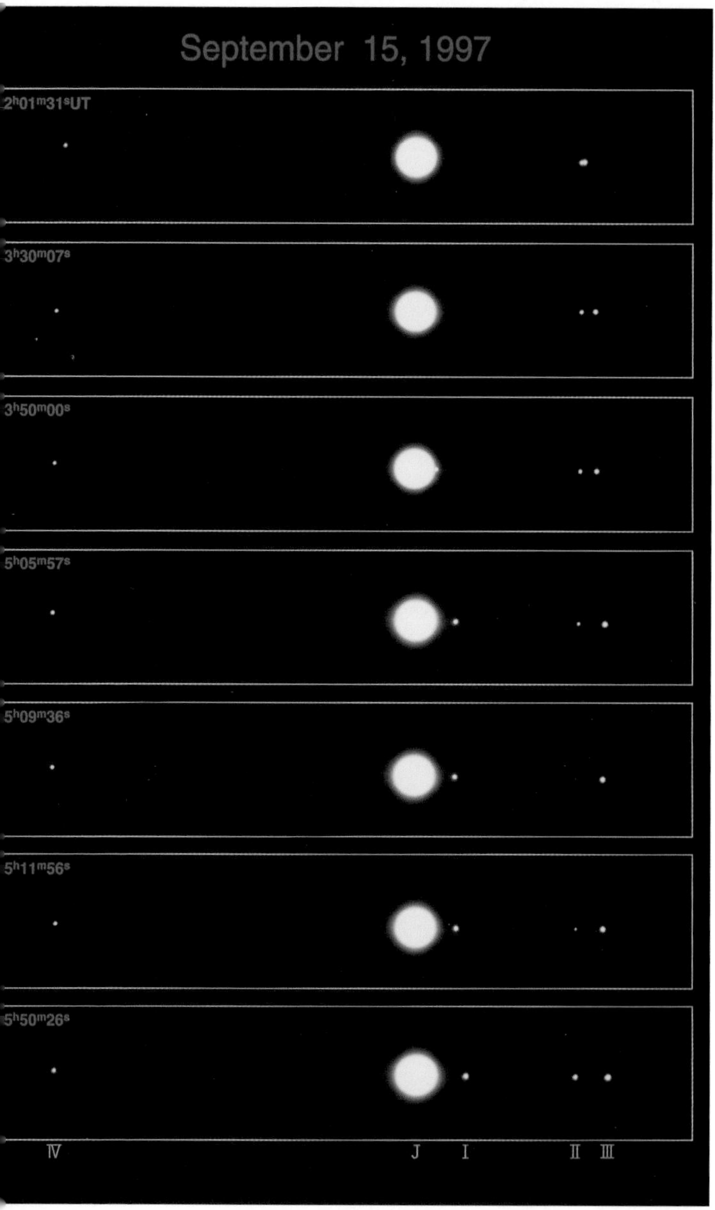

September 15, 1997

2h01m31sUT

3h30m07s

3h50m00s

5h05m57s

5h09m36s

5h11m56s

5h50m26s

IV   J   I   II III

Fig. 10-7 Galileo spacecraft images of Jupiter's Great Red Spot and of its Galilean satellites, to scale. From left to right: Io, made reddish by sulfur from its volcanoes; smooth, icy Europa; Ganymede, the largest solar-system moon; and Callisto, with bright craters. (Photo NASA/JPL/Caltech)

Fig. 10-8 A Galileo close-up of a volcano on Io. Volcanoes erupt to heights of hundreds of kilometers. (NASA/JPL/Caltech)

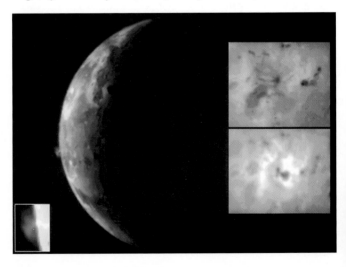

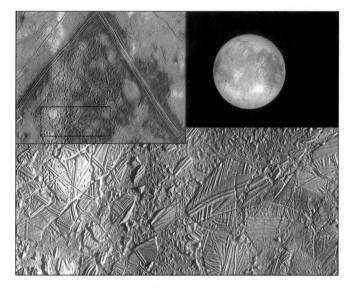

Fig. 10-9 A Galileo close-up of Europa's surface appears to show pack ice such as that on top of an ocean. (NASA/JPL/Caltech)

Each year, the positions of the Galilean moons in relation to the sides of Jupiter are graphed in *The Astronomical Almanac*. They also appear in monthly astronomy magazines. Computer planetarium programs can show their configurations. You can figure out which is which just by watching for even a few hours and remembering their order outward from Jupiter—Io, Europa, Ganymede, and Callisto—and their orbital periods.

OBSERVING JUPITER

Count, sketch, and note the colors of belts and zones on the disk. Observe and sketch the Great Red Spot and perhaps other disturbances in Jupiter's atmosphere; notice the oblateness of Jupiter's disk. Observe and identify the Galilean satellites and how they move; observe their *occultations* by Jupiter (when they pass behind Jupiter's disk), their eclipses, and the transits of the moons and their shadows across Jupiter's disk. Sometimes Jupiter is occulted (hidden) by the moon (fig. 10-11).

Table 17 shows oppositions of Jupiter, when Jupiter is opposite the sun and so is highest in the sky at midnight.

Figure 10-10 Jupiter, imaged from the Hubble Space Telescope in orbit around the earth. Images of this quality can now be obtained from the Hubble Space Telescope whenever the priority for observing Jupiter is sufficiently high. The cloud pattern is entirely different from what it was when the Voyagers passed by Jupiter two decades earlier. Images like this one are being taken on a regular basis to follow how Jupiter's weather evolves. The computer programs used to predict weather on earth will be tested with these changing weather patterns on Jupiter, and we hope that forecasts of weather on earth will improve as a result. At extreme right, we see the Great Red Spot below center and Jupiter's moon Europa disappearing behind the planet above center. (NASA/JPL/Caltech)

Fig. 10-11 A rare double occultation of Jupiter and Venus by the moon had just taken place when this image was taken. Ganymede is visible between Jupiter and the moon. (Olivier Staiger)

TABLE 17. OPPOSITIONS OF JUPITER

| DATE OF OPPOSITION | DIAMETER OF DISK ARC SEC | DISTANCE TO EARTH | | MAG. |
		A.U.	KM (MILLIONS)	
November 28, 2000	49	4.04	606	−2.9
January 1, 2002	47	4.18	626	−2.7
February 2, 2003	46	4.32	647	−2.6
March 4, 2004	45	4.42	662	−2.5
April 3, 2005	44	4.45	667	−2.5
May 4, 2006	45	4.41	660	−2.5
June 6, 2007	46	4.30	644	−2.6
July 9, 2008	47	4.16	622	−2.7
August 14, 2009	49	4.03	602	−2.9
September 21, 2010	50	3.95	592	−2.9

ATURN

Galileo, with his tiny telescope in the early seventeenth century, could see that Saturn wasn't quite round. Some decades later, astronomers realized that Saturn was surrounded by a ring (fig. 10-12). As the decades and centuries passed, we realized that Saturn had more and more rings. In the 1970s, we knew of half a dozen rings; now, with the results from the Voyager flybys, we know of hundreds of thousands of rings.

Saturn is 9.4 times the diameter of earth, almost the size of Jupiter. Its ring extends out much farther, up to 81,000 miles (135,000 km) from Saturn's center.

Saturn is 9.5 A.U. from the sun, which means that at best it is twice as far from earth as Jupiter is. Its maximum size is thus about 20 seconds of arc, and it never goes brighter than about zero magnitude.

When the Voyager spacecraft visited Saturn in 1980 and 1981, they not only observed Saturn's rings and disk in detail but also got close-up views of many of Saturn's moons (fig. 10-13). They even discovered several new moons; a total of 18 moons are now known. NASA's Cassini spacecraft is due at Saturn in 2004 to begin orbiting. It carries the Huygens probe to penetrate the atmosphere of and to land on Saturn's moon Titan.

ATURN'S RINGS

Even small telescopes reveal Saturn's rings. They are often the first objects an amateur looks at, and they are certainly among the first things shown to novices. The rings are so glorious that pro-

Fig. 10-12 Saturn, imaged with a CCD from the ground with a 24-inch (60-cm) telescope. (Donald Parker)

fessionals like to look at them too. When you are observing Saturn (or any of the other planets), you will find that waiting for steady air (good "seeing") will give you better images.

The main division between Saturn's bright middle ring (the *B-ring*) and fainter outer ring (the *A-ring*) is called *Cassini's division*, after its seventeenth-century discoverer. A faint inner *crepe ring* (the *C-ring*) is hard to see. Though Cassini's division looks dark when seen from earth, and therefore resembles a gap in the rings, the Voyagers revealed that it really contains bits of opaque material. A gap called *Encke's division* runs through the A-ring, though it is extremely hard to detect even with large telescopes. The general structure of the rings is visible in images from the Hubble Space Telescope (fig. 10-14).

On close examination from Voyager, Saturn's rings break up into thousands of tiny "ringlets." The rings and ringlets consist of chunks of rock, ranging in size from pebbles to huge boulders, each independently orbiting Saturn.

Saturn's rings and equator are inclined by 27° to Saturn's orbit, so we see them from different aspects at different times. The angle at which we see the rings varies over a 30-year period. The earth last passed through the plane of the rings, when the rings

Fig. 10-13 *A montage of Saturn and some of its moons, photographed from the Voyager 1 and 2 spacecraft. Clockwise from upper left, the moons are Titan (reddish), Iapetus (with one dark side), Tethys, Mimas (a small moon with a giant crater), Enceladus (lower center), Dione, and Rhea. (NASA/JPL/Caltech)*

become invisible, in 1995–96 (fig. 10-15). They are faintest when edge-on.

SATURN'S DISK

Saturn's disk, out-of-round by 10 percent, shows less contrast in its bands than Jupiter's disk does. The effect may be because Saturn is farther from the sun and thus colder; the cooler temperatures may slow down the chemical reactions that create the colors in the clouds.

SATURN'S MOONS

Saturn boasts of the second-largest moon in the solar system, Titan (fig. 10-16). With its clouds included, Titan is 3,625 miles

Fig. 10-14 *Saturn, viewed from the Hubble Space Telescope. (Hubble Heritage Team)*

Fig. 10-15 *Saturn, viewed in the plane of its rings, from the Hubble Space Telescope. (E. Karkoschka, Lunar and Planetary Laboratory; and NASA)*

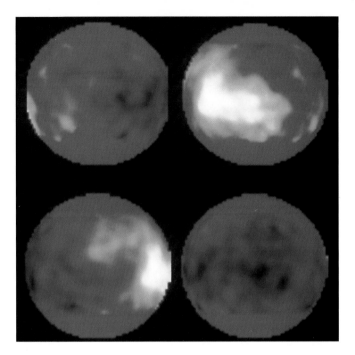

Fig. 10-16 *A glimpse of Titan's surface, obtained by the Hubble Space Telescope's observing at a wavelength in the infrared that penetrates Titan's atmopsheric haze. The large bright region is about the size of the U.S., though we do not know for certain that it is a continent on Titan. It might be a huge cloud or a meteor impact, for example. (Peter H. Smith and Mark T. Lemmon, University of Arizona; and NASA)*

(5,150 km) in diameter, nearly half the size of earth, and is surrounded by an atmosphere that is even thicker than earth's. Titan never becomes brighter than about 8th magnitude; it can, however, be seen as a small starlike point of light with even a small telescope. The Huygens Probe aboard the Cassini spacecraft will drop into its atmosphere in 2004.

Saturn has another dozen moons more than 60 miles (100 km) across, and many smaller ones (appendix 10). A mnemonic for the major moons and their order of distance from Saturn—Mimas, Enceladus, Tethys, Dione, Rhea, Titan, Hyperion, Iapetus, and Phoebe—is "Met Dr. Thip," remembering that Titan is the second T.

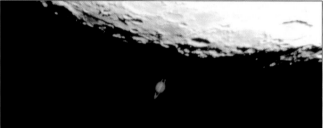

Fig. 10-17 *Saturn, on September 18, 1997, a rare occasion when it was occulted by the moon. The series of photos covers a time span of five minutes. The image was enlarged at the telescope by eyepiece projection.* (© Chris Cook)

OBSERVING SATURN

Observe the rings and their orientation; observe and sketch Cassini's division; try to observe the bands on Saturn's disk. Saturn's brightness varies between magnitudes of about −0.2 and +0.4, so it isn't as bright as the brightest stars or planets.

Observe Titan, Saturn's largest moon, and plot its changes in position during its 16-day orbital period; see how many other satellites you can observe (such as Rhea, Tethys, and Dione) with, say, a 6-inch telescope.

Oppositions of Saturn, when Saturn is opposite the sun and so is highest in the sky at midnight, occur on November 19, 2000 (magnitude −0.4); December 3, 2001 (magnitude −0.4); December 17, 2002 (magnitude −0.4); December 31, 2003 (magnitude −0.4); January 13, 2005 (magnitude −0.4); January 27, 2006 (magnitude −0.4); February 20, 2007 (magnitude −0.3); February 24, 2008 (magnitude −0.3); March 9, 2009 (magnitude −0.3); and March 22, 2010 (magnitude −0.3).

Occasionally, Saturn is occulted by the moon (fig. 10-17).

VENUS

Venus can be the brightest object in the sky besides the sun and moon. Since its orbit is inside that of the earth, we see Venus only when we are looking in the general direction of the sun. This bright planet is visible only during the first few hours after sunset, when it is known as "the evening star," or before sunrise, when it is known as "the morning star." Venus can be brighter than magnitude −4 and can even cast noticeable shadows.

Venus is covered with thick layers of clouds through which we cannot see. From earth, we see no structure, though we do see Venus go through phases as it orbits the sun (fig. 10-18). (Only planets that have orbits smaller than earth's—Venus and Mercury—can go through a crescent phase. The fact that Venus goes through a complete cycle of phases, including the crescent phase, was discovered by Galileo and was major proof of the validity of Copernicus's idea that the sun rather than the earth is at the center of the solar system.)

When Venus is just about to pass between the sun and us, it appears as a crescent and is at its largest. We may even see sunlight bent around toward us through its thick atmosphere. When Venus appears at the same longitude along the ecliptic as the sun, it is at *conjunction*. When Venus lies nearly on a line of sight between us and the sun, it is at *inferior conjunction*. When we can see all of Venus's lighted side (its "full" phase), the planet is on the far side of the sun, at its farthest point from us and therefore at its smallest. It is then at *superior conjunction*.

From spacecraft looking through ultraviolet filters from the vicinity of Venus, we have been able to study the circulation of the planet's clouds (fig. 10-19). The Hubble Space Telescope looks at Venus very rarely, since it is so close to the sun in angle.

Fig. 10-18 The phases of Venus. Note the different size of Venus at its various phases. Whenever we see a crescent, Venus must be on the near side of the sun; as a result, it appears relatively large. (Akira Fujii)

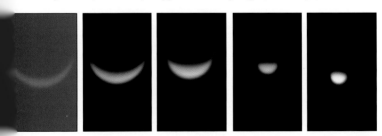

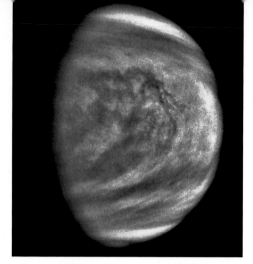

Fig. 1 0-1 9 *The clouds of Venus show when observed from a spacecraft with ultraviolet light of wavelengths slightly too short to come through the earth's atmosphere. This image was taken from the Galileo spacecraft, which got a gravity assist from Venus en route to Jupiter. (NASA/JPL/ Caltech)*

Fig. 1 0-2 0 *The overall radar map of Venus compiled by the Pioneer Venus spacecraft; the radar could penetrate the clouds that enshroud Venus. (NASA/Ames, U.S. Geological Survey, and M.I.T.)*

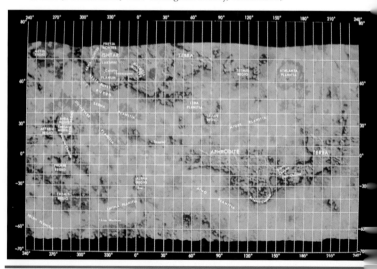

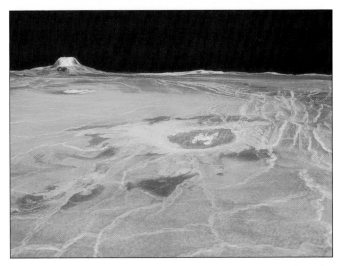

Fig. 10-21 *A computer-generated 3-D perspective view of Venus from the radar on the Magellan spacecraft, with the elevations enhanced by a factor of 10 to make them stand out. The resolution is as small as 100 meters on Venus's surface. The 3-km-high volcano Gula Mons appears on the horizon. (NASA/JPL/Caltech)*

Though the clouds block our view from the earth, the Pioneer Venus Orbiter in 1979 and then the Magellan spacecraft from 1990 to 1995 compiled overall maps of Venus's surface, using radar (fig. 10-20).

NASA's Magellan spacecraft went into orbit around Venus in 1990. Its radar sent back images of Venus's surface much more detailed than those made by earlier spacecraft (fig. 10-21). The Magellan observations have shown details of the geology of Venus's crust and have changed our understanding of the processes that shaped Venus's surface.

A series of Soviet spacecraft have landed on Venus and sent back photographs of its surface (fig. 10-22). From studies of the composition of the surface, we know that the same kind of geological processes that formed the earth's surface also were at work on Venus. The earth, though, has several continents and is mostly covered with deep ocean basins. Venus, on the other hand, has only a few small continents and few deep basins; it is mostly covered with a broad rolling plain.

Fig. 10-22 *The surface of Venus, photographed in 1982 by a Soviet lander, showed flat rocks and soil. The swath photographed is a curved region with horizons appearing diagonally just beyond the upper left and upper right corners. In the center, we look down at the base of the spacecraft. (Image courtesy of Valeriy Barsukov and Arnold Selivanov and the USSR Academy of Sciences)*

Venus's clouds consist primarily of sulfuric acid droplets. Its atmosphere is mostly carbon dioxide, and the surface pressure is 90 times that on earth. The earth contains about the same amount of carbon dioxide, though on earth it is locked up in carbonate rocks formed under the ocean. So the presence of water on earth apparently saved us from undergoing Venus's fate.

The carbon dioxide in Venus's atmosphere traps sunlight, which enters mostly as visible light but is changed to infrared radiation when it heats Venus's surface. The infrared can't escape, mostly because of the carbon dioxide but also somewhat because of other gases and particles, so the atmosphere heats up to a temperature of 900°F (500°C) on Venus's surface. This effect is known as the "greenhouse effect." If we unbalance our earth's atmosphere in some way, perhaps by burning too much fossil fuel and thus putting too much carbon dioxide into the atmosphere, our atmosphere could become as unlivable as Venus's.

Venus's atmosphere also teaches another lesson about air pollution. If we introduce too many fluorocarbons into the earth's atmosphere by using aerosol cans that contain them or by leakage of air-conditioner or refrigerator coolant, we could destroy a lot of the earth's ozone layer, which protects us from the sun's ultraviolet light. The effects of ozone and other gases are now better understood through comparative studies of the atmospheres of the earth, Venus, and other planets.

Only rarely does Venus *transit*—go directly in front of—the sun. It is then visible as a black dot projected on the solar disk; light bent forward to us by Venus's atmosphere makes a bright ring around the planet. Historically, transits of Venus have been important for setting the distance scale in the solar system; we can now find the distance scale more accurately with radar and by tracking spacecraft, so transits of Venus are now merely an observational curiosity.

Transits of Venus come in pairs separated by eight years; the interval between successive pairs is more than 100 years. The last transits were in 1874 and 1882 (fig. 10-23). The next will be on June 8, 2004. In the U.S., only eastern locations will see any of the 2004 transit, with the transit ending just after sunrise on a line from Chicago to Miami and the transit lasting about an hour after sunrise in New York City. The whole 6 hours and 13 minutes of the transit, from 5:13 U.T. (Universal Time) to 11:26 U.T., will be visible from Europe through India. Of course, a solar filter will be necessary to observe the event, which will be barely visible to the unaided eye but which will be clearer in telescopes. The following transit of Venus will be on June 5–6, 2012, when all of the U.S. and Canada will see the beginning but only Asia and the Pacific will see the end. After that, you'd have to wait until December 11, 2117, and December 8, 2125, for the following pair of transits.

OBSERVING VENUS

Venus is the brightest planet in the sky, ranging between magnitudes about −3.9 and −4.5. Observe the phases, especially during the 10 weeks before and after inferior conjunction, when Venus switches from an "evening star" to a "morning star." An ultraviolet filter sometimes adds contrast to photographs of the disk. Fig. 10-11 shows an occultation of Venus by the moon.

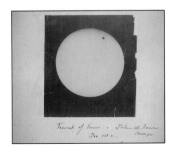

Fig. 10-23 The transit of Venus of 1882. The image was taken by Maria Mitchell and her students. Professor Mitchell became famous in the mid-nineteenth century when she won the King of Denmark's gold medal for discovering a comet; she later was a professor at Vassar College. (Vassar College)

Mars, "the red planet," has long been an object of interest to humanity because of its relatively rapid motion among the stars and because of its reddish color. Over a period of months, Mars's path among the stars apparently reverses itself in a giant loop, as you will see in the chart of Mars's position that precedes the Graphic Timetables in chapter 9. In 1543, Copernicus explained these loops in Mars's orbit, called *retrograde* (backward) *motion*, by showing how it is an effect of perspective. It occurs when the earth passes Mars, as both planets orbit the sun. The orbits of other planets have similar retrograde loops.

Mars's surface, when viewed from earth, shows semipermanent features, which can be followed over the 25-hour period of rotation of Mars, a Martian "day." These features change with the seasons of the year on Mars; the cycle of Martian seasons takes 23 earth months, the period with which Mars orbits the sun. Mars's reddish tinge is obvious even to the naked eye, but a telescope is necessary to see detail on Mars's surface from earth. The seasons change visibly from earth because dust is blown off or settles on Mars's rocky surface features. The features of Mars have long been named (fig. 10-24). CCD images often show more detail than film (fig. 10-25).

Because of Mars's elliptical orbit, at different oppositions Mars is at different distances from us. The angular diameter of its disk

Fig. 10-24 Polar caps and other features on Mars that can be observed from earth. Sinus Meridiani is the horizontal feature on the left photo. Syrtis Major is the triangular feature on the right photo, with the Hellas plain below it. North is at top. These images were made on film. (Donald Parker)

 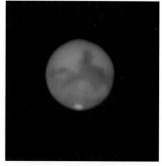

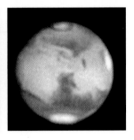

Fig. 10-25 CCD images of Mars, taken on earth with a 24-inch tele-scope. The images show different parts of Mars, as it rotates. (Donald Parker)

at opposition ranges from about 14 arc sec to 25 arc sec, which is ¹/₁₃₀ to ¹/₇₅ the diameter of the full moon and about the same size as Saturn's disk. The closer the planet is to us, the larger it appears. Mars appears still smaller when not at opposition, sometimes as small as 4 arc sec across.

Because of the seasonal changes on Mars, and because some observers thought they saw thin lines crossing Mars's surface, Percival Lowell and others at the turn of the last century suggested that there was life on Mars. The seasonal changes, they thought, were caused by vegetation, and the lines were "canals" dug by Martians to carry water.

But observations from the ground and later from spacecraft have shown that the seasonal changes are the result of blowing dust during huge seasonal storms generated by the effect of solar heating. When dust is blown off certain surfaces, the dark underlying material is revealed. Close-ups have proved that the "canals" don't really exist; they were presumably the effect of the eye and brain, which tend to imagine connections even when they don't exist. In fact, the features that show up in telescopes as bright and dark to earthbound observers don't necessarily correspond to any physical features on Mars.

The polar caps on Mars, which wax and wane with the seasons of each Martian year, turn out to be largely made of frozen carbon dioxide that condenses over a core of frozen water. The existence of large quantities of water is thought to be essential for life as we know it, so the discovery of signs of water on Mars is encouraging for those who hope to find that life has started there independently of life on earth.

The American Viking missions in 1976 carried out close reconnaissance of Mars. Each of the two Viking spacecraft had an or-

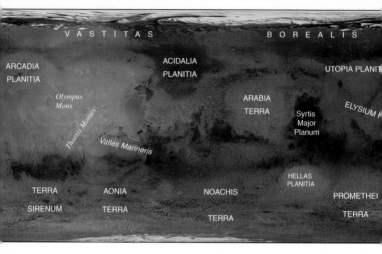

Fig. 10-26 *The surface of Mars from the Viking orbiters. Valles Marineris, the width of the U.S., appears at left center. (NASA/JPL/ USGS processing; labels courtesy of Jennifer Blue and Adrienne Wasserman)*

biter that took photographs of Mars's surface over a period of years. The Viking orbiters' photographs revealed giant volcanoes, a huge canyon larger in diameter than the continental U.S., many craters in certain regions, and many other geological features (fig. 10-26). Branching channels appear like streambeds on earth, indicating that water may have flowed on Mars in the past.

In 1997, the Mars Pathfinder landed on Mars. It carried a small robotic rover, named Sojourner, that wandered around the martian surface. It investigated the rock and soil detail (fig. 10-

Fig. 10-27 *Rocks on Mars, photographed by the Mars Pathfinder lander, the Carl Sagan Memorial Station. (NASA/JPL/Caltech)*

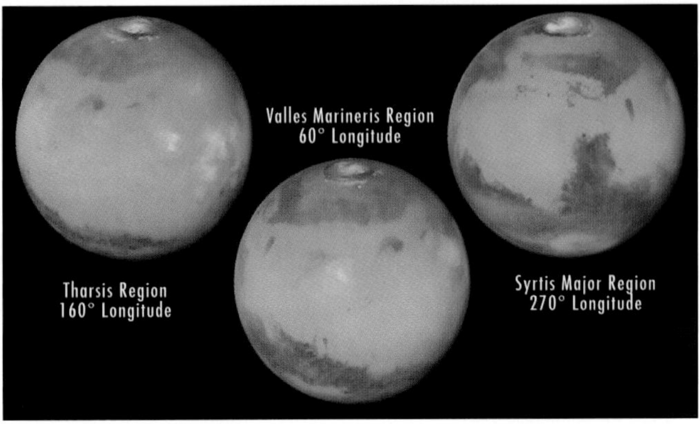
Valles Marineris Region
60° Longitude
Tharsis Region
160° Longitude
Syrtis Major Region
270° Longitude

Fig. 10-28 Views of Mars from the Hubble Space Telescope. The north polar cap shows clearly. At left, a crescent-shaped cloud identifies the volcano Olympus Mons at right center. Farther east (right), a north-south line of clouds marks the volcanoes Ascraeus Mons, Pavonis Mons, and Arsia Mons. At center, the Mariner Valleys (Valles Marineris) span the width of the U.S. at lower left; on the left (west) limb, Ascraeus Mons peers through the clouds. The Chryse basin is at center. At right, the dark Syrtis Major is the giant upward-facing region. The impact basin Hellas is below it. (Philip James, University of Toledo; Steven Lee, University of Colorado; and NASA)

27). The Hubble Space Telescope regularly takes excellent images of Mars (fig. 10-28), showing the changes of season. Starting in 1997, Mars Global Surveyor orbited Mars, sending back detailed images. Launches toward Mars are planned at regular intervals of about 26 months, the time it takes between conjunctions of earth and Mars, corresponding to the shortest travel times.

Mars has two moons, Phobos ("Fear," in Greek) and Deimos ("Terror," in Greek), named after the mythological companions of the Roman war god, Mars. Photographs from the Vikings and from the earlier Mariner 9 spacecraft revealed that both moons are elliptical, with Phobos having a longest diameter of only 16 miles (27 km) and Deimos a longest diameter of only 9 miles (15 km). Phobos was observed most recently in detail by the Soviet Phobos mission in 1989. These moons are not large and are not made round by gravity like our moon, which is really a small planet. The moons of Mars are really only small orbiting chunks of rock, comparable to some of the smaller asteroids. They do not become brighter than 11th and 12th magnitudes and so are not easily observed from earth.

Observe the reddish color and features on the disk (such as Syrtis Major, Hellas, and the polar caps) as the planet rotates; notice how the features change with Mars's seasons. Observe seasonal dust storms. An orange filter is useful to accentuate contrast on Mars's surface; a blue filter may reveal the status of dust storms. Morning clouds show best with a green or yellow filter.

Because of Mars's elliptical orbit around the sun, some oppositions of Mars are more favorable than others for observers on Earth, as shown by Mars being closer and thus larger in angle, also making it brighter (table 18).

TABLE 18. OPPOSITIONS OF MARS

DATE OF OPPOSITION	NEAREST TO EARTH	DIAMETER OF DISK ARC SEC	DISTANCE TO EARTH A.U.	KM (MILLIONS)	MAG.
June 13, 2001	June 21	21	0.45	67	−2.3
August 28, 2003	August 27	25	0.37	56	−2.9
November 7, 2005	October 30	20	0.46	69	−2.3
December 24, 2007	December 18	16	0.59	88	−1.6
January 29, 2010	January 27	14	0.66	99	−1.3

MERCURY

Mercury, the closest planet to the sun, is less than half the size of the earth. It has no moons, no atmosphere, and is an exceedingly inhospitable place. The temperature at midday is over 750°F (400°C).

Mercury never appears too far from the sun in the sky, so it is visible only briefly after sunset or before sunrise, as shown in the Graphic Timetables in chapter 9.

From earth, we can see Mercury's phases, but even the largest telescopes do not reveal any detail on its surface. The surface features are known only from NASA's Mariner 10 spacecraft, which flew close to Mercury three times in 1974 and 1975. The pictures (fig. 10-29) revealed that Mercury has many craters and resembles our moon. The craters are flatter and have thinner rims on Mercury, because of the higher gravity, but those are subtle effects. We deduce that Mercury's interior was once molten from the lava that once flooded the surface. Also, Mariner 10 found a weak magnetic field. We think magnetic fields can remain only if part of the interior is still molten.

Fig. 10-29 A mosaic of Mariner 10 images of Mercury, with an artist's conception of Mercury's interior added. Mercury's core is shown in yellow and may be partly molten. (Courtesy of Robert G. Strom from his book Mercury; *artwork by Karen Denomy)*

TRANSITS OF MERCURY

Transits of Mercury are not as rare as those of Venus; there are about a dozen per century (fig. 10-30). The transit of May 7, 2003, is not visible from the U.S. and Canada, except for a glimpse of the end from the Canadian Maritime provinces. The beginning at 5:14 U.T. is visible from Australia and all of Asia, the middle occurs at 7:53 U.T., and the end at 10:32 U.T. is visible from all of Europe and Africa, with visibility just reaching the Canadian Maritime Provinces and Greenland as well as eastern Brazil. On November 8, 2006, the beginning of the transit at 19:13 U.T. is visible from all of North and South America; midtransit at 21:42 U.T. is visible only from the Pacific Ocean, including Hawaii; and the end of the transit at 00:11 UT the next day is visible from all of Australia and barely reaches the Asian east coast.

Mercury takes about five hours to cross the sun's disk. Observers should note the exact times when Mercury first touches the sun, when it moves entirely within the solar disk, when it first touches the other side of the sun, and when it leaves the solar disk entirely.

Fig. 10-30 *A transit of Mercury across the sun. (Fred Espenak)*

OBSERVING MERCURY

Observe the phases. For best viewing, locate Mercury on a favorable morning when it rises well ahead of the sun, and keep your telescope on it until after sunrise. Mercury can also be seen during transits and during solar eclipses.

URANUS

Uranus reaches magnitude +6 at its brightest, the limit of naked-eye visibility. It is a giant planet, about four times the earth's diameter, and about 15 times its mass. It appears as a tiny greenish or bluish disk in earth-based telescopes because methane in its atmosphere absorbs the other colors, chiefly red. Uranus shows

Fig. 10-31 *Uranus and its five largest moons through a large telescope, in one of the best photographs ever taken from earth. Still, the moons appear only as points of light. (William Liller)*

Fig. 10-32 *Uranus and its five major moons, a montage pasted together from photographs taken with the spacecraft Voyager 2. Uranus is the bluish object at center. Ariel is at lower right foreground, Umbriel is reproduced small to its left, Oberon appears small above it and to the left of Uranus, Titania appears at top left, and Miranda overlaps the top of Uranus. (NASA/JPL/Caltech)*

Fig. 10-33 *(left) The rings of Uranus seen frontlighted, as we see them from earth. (right) The rings of Uranus seen backlighted, after Voyager 2 passed Uranus. Fine dust in the rings shows up especially well. (NASA/JPL/Caltech)*

Fig. 10-34 *Voyager 2's view of a canyon on the edge of Miranda, seen before its photograph of Uranus and with an artist's conception of a ring. (NASA/JPL/Caltech)*

in binoculars or small telescopes as only a dot. Astronomers learned most of what we know about Uranus when the Voyager 2 spacecraft flew by it in 1986.

Uranus's orbit is 19 A.U. in radius, almost twice that of Saturn. Uranus is so distant that it never gets bigger than 4 seconds of arc. Since the best resolution available from earth's surface is about ½ second of arc, even the largest telescopes don't show much detail on its surface.

A mnemonic for the names of Uranus's moons, in order of their distance from the planet—Miranda, Ariel, Umbriel, Titania, and Oberon—is "M-Auto." Uranus's moons are shown in figure 10-31 from earth and in figure 10-32 from space.

When Uranus occulted (that is, passed in front of and therefore hid) a star in 1977, the light from the star winked off and on a few times before and after the star was occulted by the planet itself. This marked the discovery of a set of nine thin rings around Uranus, which have been studied further at subsequent occultations.

Voyager 2 photographed the rings as it passed by. As it approached, it had the same view we have from earth, with the sun over its shoulder. After it passed, it could look through the rings, which stood out when backlighted (fig. 10-33).

Each of Uranus's moons turned out to have a different personality. They are all made out of frozen mixtures of rock and ice, and all are fairly dark, reflecting only a small percentage of the sunlight that hits them. The most interesting is Miranda (fig. 10-34), whose surface is so varied though it is only 300 miles (500 km)

Fig. 10-35 *(top left) Ariel, (top right) Umbriel, (bottom left) Titania, (bottom right) Oberon, (middle) Miranda. All were photographed by Voyager 2 and are shown in their relative sizes and darknesses (a property known as albedo). (NASA and USGS-Flagstaff, courtesy of S. K. Croft and L. Soderblom)*

across; it is like the old joke about the animal that was put together by a committee. Apparently, Miranda was completely broken up in the past in a collision and came together with its parts jumbled.

The surface of Ariel, 720 miles (1,200 km) across, shows few craters or other old features, and so must have been reshaped by geological activity. Umbriel, also 720 miles (1,200 km) across though darker, is covered with craters, and its surface may otherwise be unchanged since it was formed 4.5 billion years ago. Titania and Oberon are each somewhat bigger, 960 miles (1,600 km) across. Titania shows no large craters, which must have been covered over at some time, perhaps by material that flowed out from the interior. Oberon has many large craters. Voyager 2 also discovered 10 additional moons.

OBSERVING URANUS

Uranus varies between magnitudes +5.7 and +5.9, so it is slightly too faint to see with the unaided eye. Use a telescope to observe its greenish disk. Uranus started the millennium in Capricorn as it moved along the ecliptic.

NEPTUNE

Neptune, another of the giant planets, is about the same size as Uranus. It is 30 A.U. from the sun and gets no larger in the sky than 2.5 arc sec. Thus from earth we have been able to see very little detail on its surface.

Until 1999, Neptune was actually the farthest of the nine planets from the sun for a couple of decades, since Pluto's elliptical orbit brings it closer to the sun than Neptune. However, we rank

Fig. 10-36 Neptune, as seen from the Voyager 2 spacecraft. The atmospheric storm known as the Great Dark Spot is clearly visible, as is a bright methane ice cloud. (NASA/JPL/Caltech)

planets by the overall sizes of their orbits, so it was still legitimate to say that Neptune is the eighth planet.

Most of what we have learned about Neptune came from the 1989 flyby by the Voyager 2 spacecraft (fig. 10-36).

Neptune shows a lot of activity on its surface, unlike bland Uranus. Neptune's surface activity results from an internal source of energy. We do not know why Jupiter, Saturn, and Neptune have such additional energy while Uranus does not.

Neptune, like Uranus, gets its color from the traces of methane in its atmosphere, which is mainly made of hydrogen and helium. The methane absorbs the red part of the spectrum, leaving the blue for us to see. The most obvious feature on Neptune's surface was the Great Dark Spot. The Great Dark Spot was a circulating storm, similar in many ways to the Great Red Spot on Jupiter. The Great Dark Spot was about the size of the earth. It was a high-pressure region and rotates every 17 days. It disappeared before the current series of Hubble views.

Once Uranus's rings were discovered from earth when they occulted stars, astronomers started searching for rings around Neptune. Over several years, astronomers sometimes detected drops in the brightness of stars when they passed near Neptune; sometimes they did not detect such drops. They deduced that Neptune might have partial rings, perhaps ring arcs.

Fig. 10-37 The rings of Neptune, photographed by Voyager 2. The left half and the right half were photographed separately. The two major rings plus additional material forming broad rings are visible in these backlighted images. The dark bar in the center provides the appropriate space for the part missing from the photographs. The clumps were on the side of Neptune for each photograph that did not show. (NASA/JPL/Caltech)

Voyager 2 found, however, that Neptune does have complete rings around it (fig. 10-37). The material is not evenly distributed in the rings, though, and in some places the ring material is so thin that it does not noticeably hide a star. Voyager found two clumpy rings at 31,800 and 37,800 miles (53,000 and 63,000 km) from Neptune's center, far above its equator, which is 14,860 miles (24,760 km) from the center. These rings are named Leverrier and Adams, after the two mathematicians who predicted where Neptune would be found. The clumps of the Adams ring are named Liberté, Egalité, and Fraternité, slogans from the French Revolution that were allowed as celestial names because they are universal aspirations. Voyager also photographed a faint, inner ring 25,200 miles (42,000 km) from the center and a broad ring extending from the inner clumpy ring halfway to the outer one. The inner ring is named Galle, after the astronomer who actually saw Neptune first when using the predictions.

From earth, we had known that Neptune had two moons. The largest, Triton, is one of the largest moons of the solar system, somewhat larger than our own moon though not as large as the Galilean satellites of Jupiter. It was named after a sea god, a son of Poseidon. Triton orbits Neptune in the opposite direction from all the other major moons in the solar system around their plan-

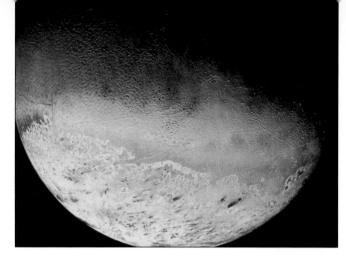

Fig. 10-38 Triton, Neptune's largest moon. We see its southern region. The lowest part of the image is its polar ice cap, probably of nitrogen ice. Ultraviolet sunlight and radiation from Neptune's magnetosphere may have interacted with the ice to turn it pinkish. The image also shows many dark, elongated streaks that are the results of eruptions from under Triton's surface. (NASA/JPL/Caltech)

ets, which indicates that it was formed elsewhere and then was captured by Neptune's gravity.

Voyager 2 was able to pass extremely close to Triton, and the views it sent back were astounding (Figs. 10-38 and 10-39). The broad southern polar cap is pinkish, and plumes of dust were seen erupting. The result of these plumes is the black streaks that are visible on the surface, presumably in the downwind direction. The rest of the surface of Triton that was seen is known as the "cantaloupe" terrain from its puckered appearance. I participated in two expeditions to observe Triton as it occulted a star; from how the star's light dimmed, we analyzed Triton's atmosphere. We found that some global warming is going on there, since the temperature was slightly higher after an interval of several years between our observations and that of Voyager 2.

Nereid, the second moon of Neptune that was known from earth, is much smaller. Its name means "sea nymph" in Greek. Voyager 2 discovered six additional moons of Neptune. One of them is even larger than Nereid's 360-mile (600-km) diameter.

The magnetic fields of both Neptune and Uranus were big surprises. They are both very tilted from the planets' axes of rotation and are offset from the planets' centers. The earth's magnetic

Fig. 10-39. *Neptune and Triton, seen as crescents as Voyager 2 receded into more distant space. (NASA/JPL/Caltech)*

field, on the other hand, is tilted only slightly. We have a lot to learn about how magnetic fields are generated. Probably, those of Uranus and Neptune are formed in shells far from center.

BSERVING NEPTUNE

Neptune varies between magnitudes +7.9 and +8.0. With a telescope of at least moderate size, notice Neptune's blue-greenish color and the fact that it has a slight disk. Neptune, like Uranus, started the millennium in Capricornus as it moved along the ecliptic.

LUTO

Tiny, distant Pluto is so far away and so small that we know little about it. Its orbit is very elliptical and has an average radius of 39 A.U. Pluto was on the part of its orbit that is within the orbit of Neptune for two decades until the year 1999.

Pluto—like Neptune's moon, Triton—has methane ice or snow on its surface and an atmosphere consisting of methane gas plus traces of other substances. Pluto is about magnitude 13.7 and was only slightly brighter in 1989, when it reached *perihelion,* the closest point ("peri-") of its orbit to the sun ("helios"). Though Pluto is as bright as it has been for centuries, it is still

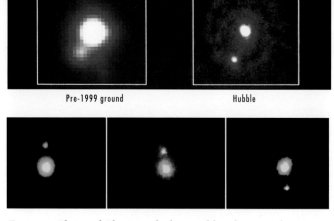

Pre-1999 ground · Hubble

Fig. 10-40 Pluto and Charon, in both ground-based views and an image taken with the Hubble Space Telescope. The Space Telescope was joined in 1999 by the Gemini North telescope in being able to show them as completely distinct objects. (top: NASA and Canada-France-Hawaii Telescope Corp.; bottom: Gemini Observatory, NSF and U. Hawaii)

much too faint to be seen in small telescopes. For keen observers, though, it may be seen as a point of light in 8-inch (20-cm) telescopes. No detail at all is detectable on its surface from earth, even in the largest telescopes.

Photographing Pluto on different nights, though, reveals its slow motion among the stars. Pluto, the only planet to be discovered in modern times, was found from such motion in 1930 by Clyde Tombaugh. Tombaugh studied hundreds of thousands of pairs of star images, comparing one night's observations with another's, before he noticed Pluto. Tombaugh studied the part of the sky where he found Pluto because of predictions that gravity from a planet there could explain why Uranus did not follow exactly in the orbit it otherwise should. (Neptune had not been studied for long enough to use for such a study.) If the discovery of Pluto had really resulted from the prediction, then Pluto would have actually been reponsible for the variations in Uranus's orbit and Pluto would have almost as much mass as the earth.

A ground-based photograph taken in 1978 revealed that the image of Pluto was not quite round. A bulge at the edge of Pluto's image turned out to be a previously unsuspected moon, which was named Charon (fig. 10-40). The third law of Johannes Kepler as emended by Isaac Newton, both in the seventeenth century,

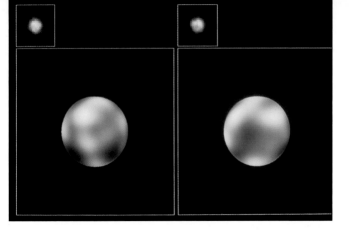

Fig. 10-41 Pluto, imaged with the Hubble Space Telescope, reveals a dozen regions of light and dark on its surface. (Alan Stern, Southwest Research Institute; Marc Buie, Lowell Observatory; and NASA/ESA)

links the period of revolution of a moon with the mass of its central object. Studies of the orbit of Pluto's moon have shown us that Pluto is only ¹⁄₅₀₀ the mass of earth. Thus Pluto could not have been the object that caused the distortions measured in Uranus's orbit. Whether there is still another planet out there, or whether our measurements are too inaccurate to tell, is not known.

It remains quite difficult to observe Charon, and no ground-based image has shown it as an object completely separate from Pluto. Finally, in 1990 the Hubble Space Telescope provided an image that showed Pluto and Charon as separate objects. This and later images are enabling astronomers to measure the size of Charon's orbit and Pluto's mass more accurately.

Other observations indicate that Pluto is less than one-fifth the size of earth, even smaller than our moon. The most interesting of these observations are from mutual occultations of Pluto and Charon, observed as Pluto and Charon passed in front of each other twice every six-hour orbital period from 1985 to 1990. Study of the way in which the total brightness from the pair changes enables not only the sizes of each to be accurately determined but also gives an idea of the darkness of different regions on their surfaces. The Hubble Space Telescope has revealed a dozen different regions of light and dark on Pluto (fig. 10-41). Pluto is only 2,300 km in diameter and Charon is only 1,186 km across. Thus Pluto and Charon are closer in size to each other than any other planet and its moon in the solar system.

Pluto/Charon is therefore often thought of as a double planet.

The best knowledge we have of Pluto's atmosphere is from its occasional occultations of stars. The atmosphere has proved to be more substantial than expected.

Is Pluto a planet? There are those who want to demote it. Proposals have been made to give it a number in the asteroid scheme (a significant number: 10,000) or as one of the trans-Neptunian objects (TNO-1 or even TNO-0). There has not been any official change in Pluto's status by the International Astronomical Union. However, increasing numbers of other trans-Neptunian objects, also known as Kuiper belt objects, are being discovered, and Pluto seems to be merely the largest of such a set of bodies. If some other bodies are discovered out there that are larger than Pluto, Pluto's planetary status will be even more endangered.

OBSERVING PLUTO

Pluto varies between magnitudes of about +13.7 and +13.9. With a sufficiently large telescope, observe Pluto as a dot in the sky, changing position among the stars from night to night.

OBSERVING THE PLANETS

By looking for a long time through a telescope and waiting for instants when the "seeing" is especially good (that is, when the image is constant because the air is especially steady), the human eye can often see more detail than can be recorded with a camera. Thus excellent sketches can be made. You can follow not only the planets' rotation but also seasonal and weather changes on them.

The Association of Lunar and Planetary Observers, c/o Harry D. Jamieson, P.O. Box 171302, Memphis, TN 38187-1302, hjamieson@bellsouth.net, www.lpl.arizona.edu/alpo/ is an amateur society. The British Astronomical Association is at Burlington House, Piccadilly, London W1V 9AG, (011 44 for dialing from the U.S.) 171 734-4145; www.ast.cam.ac.uk/~baa. Regional societies, such as the Amateur Astronomers Association, 1010 Park Avenue, New York, NY 10028, also may have special groups of members who are especially interested in planetary observing. The Astronomical League, www.astroleague.org, maintains a list of amateur astronomy associations across the United States. The International Occultation Timing Association's Web site is www.occultations.org.

COMETS

A bright comet looks spectacular in the sky. Comets are bodies that orbit in our solar system, passing between the planets. Also in the spaces between the nine major planets in our solar system are minor planets known as asteroids. In this chapter we discuss comets and how to observe them.

Fig. 11-1 Comet Hale-Bopp, 1997. Its technical name is C/1995 O1 (Hale-Bopp). Note its whitish dust tail and its bluish gas tail. (Jay M. Pasachoff)

Sometimes a fuzzy object becomes visible in the heavens. It may look like a smudge on the sky. If we are lucky, it will grow brighter over a period of weeks or months and will form a tail. This tail may become so long that it extends across much of the sky (fig. 11-1).

Such bodies are comets, large icy snowballs in the sky. Each comet we see began as one of the hundreds of billions of small bodies in a tremendous cloud surrounding the sun, far beyond the outermost planets (the Oort Cloud) or in a disk of small bodies beyond Neptune (the Kuiper belt). The long-period comets, those with periods of more than 200 years, come from the Oort Cloud; the short-period comets, those with periods of less than 200 years, come from the Kuiper belt. Sometimes a gravitational nudge from a passing star or from the center of our galaxy causes one of these bodies to move closer to the sun. Solar energy heats it, and the body gives off gas and dust that form a tail. The gas is pushed away from the comet by gas flowing outward from the sun —the *solar wind*. The dust is left behind in the comet's orbit by the pressure of the sun's radiation. Thus some comets have two visible tails—a dust tail trailing gracefully behind the comet and a gas tail whose irregularities show the different puffs of the solar wind. The dust tail is yellower, since it is sunlight reflected off dust, and the gas tail is bluer, since we are seeing actual blue emission from gas.

A dozen or more comets can be detected in the sky at a single time, but most are so faint that they can be seen only with large telescopes. Every few years, a comet can be seen with the naked eye, though most appear faint. Bright comets (whose brightness totals 1st or 2nd magnitude) might appear every decade or so. The appearance of a bright comet is usually unpredictable. Of the predictable comets, only the one known as Halley's Comet is spectacular (fig. 11-2). It returns every 76 years or so. Unfortunately, its orbit kept it too far from the earth during its 1986 return for it to appear very bright (fig. 11-3). It won't be back until 2061.

We had enough warning for Halley's Comet that many nations prepared spacecraft to study it up close. The European Space Agency's Giotto spacecraft flew into the comet's head. The head is composed of a nucleus surrounded by a diffuse coma. Giotto showed that the nucleus is a potato-shaped object about 10 miles (16 km) long. The nucleus is an icy snowball covered with a dark crust. The crust is so dark that it reflects only about 2 percent of the light that hits it. Dust spews out from cracks in crust on the

Fig. 11-2. Halley's Comet in 1986, showing its bluish gas tail and its yellowish dust tail. (William Liller)

side that is hit by sunlight. The resulting dust jets reflect the sunlight brightly (fig. 11-4).

A comet bright enough to be seen with the naked eye or with binoculars can be a very beautiful sight (fig. 11-5). Comet Hyakutake in 1996 and Comet Hale-Bopp in 1997 were bright comets that were very widely seen. Such a comet will be one with an orbital period so long that it will arrive unexpectedly. None of the periodic comets expected in the next century will be very bright.

If you plot a comet's position carefully against the background of stars, you will see that it changes by about 1° per day. (Comet Hyakutake exceeded that rate.) A comet's motion, however, is not

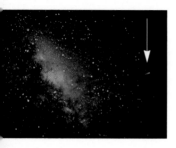

Fig. 11-3. Halley's Comet (extreme right) near the Milky Way. They were photographed together in 1986 after Halley's perihelion. (Akira Fujii)

Figure 11-4 *The nucleus of Halley's Comet and its dust jets. The nucleus, 10 miles (16 km) long by 5 miles (8 km) wide, is seen in silhouette. The dust jets come from the sunlit side of the nucleus.* (© 1986 *Max-Planck-Institut für Aeronomie, Lindau/Hartz, FRG; photographed with the Halley Multicolour Camera aboard the European Space Agency's Giotto spacecraft, courtesy of H. U. Keller*)

apparent to the eye. Though a comet's tail results in part from its motion through the solar system, a comet does not speed across our sky; its tail is not a result of any such apparent motion across the sky.

The time a comet takes to complete its orbit depends on the size of its orbit. Encke's Comet returns every 3.3 years, but is faint. Halley's Comet returns about every 74–79 years, and is always bright enough to be seen with the naked eye. It has been seen on at least 27 occasions since 87 B.C.E., though it wasn't until early in the eighteenth century that Edmond Halley realized that the bright comets reported by previous observers were reappearances of the same comet. However, its 1986 passage near earth was a faint one, and so will be the one in 2061. A comet loses less than 1 percent of its material at each passage near the sun, so it can reappear many times.

The more sensitive your observing equipment, the more comets you can find. More than a dozen comets can actually be studied in the sky at most times, if you have a sensitive electronic imager.

Fig. 11-5 An image of 1997's Comet Hale-Bopp with a wide-field telescope. (Dennis di Cicco)

NAMING COMETS

The method of naming comets changed in the 1990s. Short-period comets like Halley's Comet, whose periods are no more than 30 years, are sequentially numbered. Halley's Comet is the first, 1P, and you can also write 1P/Halley or 1P (Halley). Well over 100 periodic comets have been numbered. Other comets are described by their year of discovery and by the half-month of discovery, following a C/. (The letter I is omitted because it could be confused with J, following an old astronomical tradition also in use for variable stars.) For example, C/1995 O1 was discovered in the 14th half-month of 1995 (O is the 15th letter, the 14th when I is omitted), the second half of July 1995, with the final 1 meaning that it was the first comet discovered in the half-month. It was found independently by Alan Hale and Thomas Bopp, so it is also called Hale-Bopp, and can be written C/1995 O1 (Hale-Bopp). The prefixes D/ and X/ stand for defunct and questionable comets, respectively.

Detailed information on the naming scheme and lists of comets are available at cfa-www.harvard.edu/iau/TheIndex.html.

Fig. 11-6 Comet Hale-Bopp was so bright that it could be seen even from the middle of New York City. Here is a view from Fifth Avenue, looking west over Central Park. To take such images of a bright comet, use a tripod and take a series of images ranging from a few seconds up to 1 or 2 minutes. (Jay M. Pasachoff)

PHOTOGRAPHING A COMET

Occasionally, a comet may be bright enough to be photographed with an ordinary camera (fig. 11-6). No special equipment is required; a sturdy tripod and a lens that can be opened to a wide aperture (preferably about $f/1.4$, and at least as fast as $f/2$) are all that you will need. Use fast film, and be careful not to shake the camera. Use a cable release (or release the shutter as a self-portrait, to eliminate shake from your hand), and, if possible, raise the mirror of a reflex camera before snapping the shutter. Fast color print films, with ISO 800 or greater, give nice images. As always, there is a tradeoff of film speed with film grain; an exposure longer than a minute or so will also show trailing of the stars or comet head.

If you use a normal or wide-angle lens, you may be able to capture the comet at the top of the frame and the horizon or a tree on the bottom. Such a photograph gives a sense of scale and may look more interesting than one in which the image of the comet is centered.

Take a wide range of exposures: try 1, 2, 4, 8, 15, 30, 60, and 120 seconds. Film is cheap compared with the loss of a rare opportunity, so don't hesitate to take more pictures than you usually take at one time.

If you can place your camera piggyback on a telescope that tracks the stars, you will be able to take longer exposures. Again, take exposures over a wide range of times, with your lens wide open or almost wide open.

It can also be very interesting to draw the comet and its tail, instead of photographing it. The eye can detect very interesting detail.

You Discover a Comet

A handful of new comets have been discovered each year by amateur observers, usually with large binoculars (those with huge front lenses) or medium-sized telescopes. The main background you must have to discover a comet is experience in observing what is in the sky, which brings knowledge of what existing nebulae look like, so that you can tell whether an object you see is a new comet or a well-known nebula. The Edgar Wilson award, established in 1998, is divided each year among amateurs who discover comets; with competition from automated searches increasing, it may be that only a couple of comets will be discovered by amateurs each year.

A standard method of searching for comets is to sweep your field of view back and forth at a rate of no more than 1° of sky per second. Start at the horizon after sunset, and sweep back and forth horizontally 45° or more on either side of the sunset point, gradually moving your binoculars or telescope upward so as to slightly overlap your preceding area of observation. Continue until you are about 90° above the sun (technically, an elongation of 90°). If you are searching before sunrise, begin 90° above the sun's position and sweep back and forth while moving your field of view downward.

If you think that you have found a comet, note its position in the sky (right ascension and declination) by finding its position in relation to the stars. The Atlas Charts in chapter 7 are a place to start, though they go only as faint as 7th magnitude, and most comets discovered are a couple of magnitudes fainter than that. For devoted work, therefore, you can go to sources of atlas charts that show fainter objects, both stars and deep-sky objects; some books of atlas charts are listed at the back of this book. If you find a candidate comet, note its brightness (its approximate magnitude) by comparing it with that of plotted stars. Also record the time. Increasingly, such observations are made with photography or with CCD electronic detectors instead of visually. The next step is to watch for a while or to reimage later on in order to note the direction in which the comet is moving and the rate of its motion. It is helpful to plot the position of the comet in the star atlas at different times, so that its motion can be accurately charted; the presence of motion is an important sign that you have in fact found a comet. At a minimum, you should check that the object is present over a period of time, and preferably that an image of that area at another time does not show it. Your checks on the reality of the comet should be made yourself or by consulting an individual you contact directly rather than by a general message sent over the Internet.

If the object passes all the tests, then send that information to the Central Bureau for Astronomical Telegrams of the International Astronomical Union, which is located at the Smithsonian Astrophysical Observatory, 60 Garden Street, Cambridge, MA 02138. The e-mail address is cbat@cfa.harvard.edu. A form for submitting discoveries, along with a readable set of past IAU Circulars with past comet and other announcements, can be found on the Web at cfa-www.harvard.edu/iau/TheIndex.html. As a last resort, you can send a telegram (Western Union is 800-325-6000). Don't forget to include your name, address, and telephone number in the message.

The IAU Central Bureau names each comet after the first people to find it—up to three observers, if there are near-simultaneous independent discoveries.

Fig. 11-7 Comet C/1995 O1 (Hale-Bopp) in 1997 in Vermont seen through an aurora. (Jason Lorentz)

ASTEROIDS

In between the major planets are thousands of minor planets ranging from one kilometer to hundreds of kilometers across; they are called *asteroids* or *minor planets*. The first asteroid, discovered on January 1, 1801, was at first thought to be a new planet; it was named Ceres. The numbers assigned to asteroids (in order of discovery) are now given along with the names, so we call this asteroid 1 Ceres. Within a few years 2 Pallas, 3 Juno, and 4 Vesta were also discovered, and scientists realized that they were dealing with several little planets instead of a few major ones.

Asteroids show up as streaks on astronomical photographs made with telescopes that track the stars (fig. 12-1). On photographs made with telescopes that track the asteroid, the stars are streaks, while the asteroid is a dot.

Many asteroids only 0.6 miles (1 km) in diameter are known. More than 200 of them with diameters greater than 60 miles (100 km) have been discovered, and half a dozen are known to exceed 180 miles (300 km) in diameter. Though most asteroids are in the asteroid belt located between the orbits of Mars and Jupiter, many other asteroids come very close to the earth. The asteroids in one group, the Apollo asteroids (named after the first of the group to be discovered, 1862 Apollo), have orbits that cross earth's orbit. We know of about three dozen asteroids in this group. The orbit of an Aten asteroid (2062 Aten and those with similar orbits) not only crosses the earth's orbit but is also even smaller than the earth's orbit. The orbit of 1566 Icarus, another unusual asteroid, takes it closer to the sun than Mercury. Icarus came within 3.6 million miles (6 million km) of the earth in 1968, a near miss on an astronomical scale.

Lists of asteroids of all types are available on the World Wide Web at cfa-www.harvard.edu/iau/TheIndex.html.

Newly discovered asteroids have even come closer to the earth than the moon, less than 240,000 miles (400,000 km) away. Such an asteroid hitting the earth some 65 million years ago may well have led to so much dust being ejected into the atmosphere that dinosaurs and many other types of animals were pushed into extinction. The study of Near-Earth Objects (NEOs) is attracting widespread interest, including major motion pictures. Astronomers are mounting projects, as funding permits, to allow them to discover all the asteroids that could potentially hit the earth and to chart their orbits. There may be 2,000 such objects larger than half a mile across.

TABLE 19. THE BRIGHTEST ASTEROIDS

ASTEROID		BRIGHTEST VISUAL MAGNITUDE	DIAMETER (KM)	ASTEROID		BRIGHTEST VISUAL MAGNITUDE	DIAMETER (KM)
4	Vesta	5.1	555	15	Eunomia	7.9	261
2	Pallas	6.4	583	8	Flora	7.9	160
1	Ceres	6.7	1025	324	Bamberga	8.0	256
7	Iris	6.7	222	1036	Ganymed	8.1	40
433	Eros	6.8	20	9	Metis	8.1	168
6	Hebe	7.5	206	192	Nausikaa	8.2	99
3	Juno	7.5	249	20	Massalia	8.3	140
18	Melpomene	7.5	164				

NOTE: Brightness data from R. Shuort via J. U. Gunter; diameters based on Tucson Revised Index of Asteroid Data (TRIAD).

Vesta can be brighter than 6th magnitude, the limit of visibility to the naked eye. With a small telescope, we can observe a number of asteroids (table 19). Some five dozen may become brighter than 10th magnitude at some time in their orbits. The motion of these asteroids with respect to the stars can be followed from night to night.

Several spacecraft have traveled close up to asteroids, sending back detailed photographs (fig. 12-2). The Hubble Space Telescope has been used to map the surface of the brightest asteroids (fig. 12-3).

Sometimes an asteroid goes in front of—*occults*—a star, blocking its light from reaching us. Then, by measuring the length of time that the star is hidden, we can calculate the size of the asteroid. The stars are so far away that light rays from them are nearly parallel, so the shadow an asteroid casts on the earth is almost exactly the size of the asteroid itself. Such occultations, for example, have shown us that the size of 2 Pallas is 335 × 316 ×

Fig. 12-1 *An asteroid shows up as a trail in a photograph taken through a telescope that is tracking the stars. The color image of the Helix Nebula (fig. 5-9) was made from three separate exposures each taken through a different colored filter. We see here how the asteroid moved during and between the exposures. (Anglo-Australian Observatory)*

319 miles (558 × 526 × 532 km). To observe these occultations people had to be stationed, on the basis of last-minute observations, in the locations where the asteroid's shadow would most likely pass.

The International Occultation Timing Association coordinates observations of occultations of stars by asteroids. They can be observed, for the brightest stars occulted, with ordinary camcorders. The results allow a profile to be made up that shows the shape and size of the asteroids that are going in front of the stars (fig. 12-4). Their Web site is www.occultations.org.

Fig. 12-2 *Asteroids 253 Mathilde imaged from the Near Earth Asteroid Rendezvous (NEAR) spacecraft in 297; 951 Gaspra, 9.6 × 7.2 miles (16 × 12 km), imaged from the Galileo spacecraft; and 243 Ida, 34 miles (56 km) long, imaged by the Galileo spacecraft. Ida has a satellite Dactyl, 0.9 miles (1.5 km) in diameter and 60 miles (100 km) away. (The Johns Hopkins University Applied Physics Laboratory)*

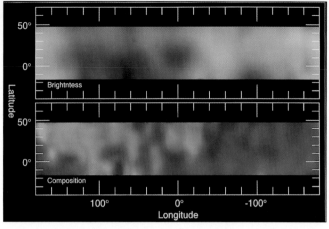

Fig. 12-3 *The asteroid 4 Vesta, 305 miles (510 km) across, imaged with the Hubble Space Telescope. The map of the composition uses false colors to show different kinds of rocks on the surface. The hemisphere shown on the left shows lava that cooled on the surface while the other hemisphere shows lava that cooled below the surface. (Ben Zellner, Georgia Southern University/NASA; STScI)*

Fig. 12-4 *Star trails with the trail of the star 14 Piscium broken for 6 seconds when it was occulted by the asteroid 51 Nemausa. (R. Scott Ireland, Michael Mooney, and Robert Riefer)*

METEORS AND
METEOR SHOWERS

If you are outdoors and looking upward in the late evening around August 12 any year, you will probably see a "shooting star," a *meteor,* travel across the sky in about a second (fig. 13-1). You are actually seeing solid particles from space—larger than atoms but much smaller than asteroids (minor planets)—burn up in the earth's atmosphere. Each August 12, the earth's orbit intersects the center of a stream of particles that makes up the Perseid meteor shower, which in some years consists of meteors that are visible at rates of up to one per minute.

When in space, the bodies are called *meteoroids.* Though the meteoroids from a meteor shower are tiny and always burn up before they reach the ground, some larger meteoroids can survive their fiery passage through the earth's atmosphere. Any part of a meteoroid that reaches the earth (or, indeed, any place where we can examine it) is called a *meteorite.*

On any night with perfect observing conditions, you are likely to see a random meteor in the sky every 10 minutes or so. These are *sporadic meteors.* The brightest meteors are called *fireballs,* and they can take half a minute to cross the sky. They may leave trails *(trains)* that last for minutes, and when noises accompany them they are called *bolides.* Fireballs can be magnitude −5, brighter than Venus.

Many times a year, the earth's orbit crosses a stream of particles, believed to be (in most cases) from a comet. These particles make up meteor showers (table 20). Some showers have more meteors per hour than others. Some, such as the Perseids around August 12, the Geminids near December 14, and the Quadrantids near January 3, are of about the same strength each year, and we can expect to see approximately the same number of meteors at maximum. Others, such as the Leonids around November 17, differ in strength from year to year. Still others, such as the delta

Fig. 13-1 *A Perseid meteor, viewed as always on or about August 12. The meteor, magnitude −4, is entering Taurus. Auriga is at left above the tree. (© James N. Smith)*

TABLE 20. MAJOR METEOR SHOWERS

SHOWER	DATE OF MAXIMUM 2000, 2004, 2008	TIME (E.S.T.)	RADIANT (AT MAXIMUM) R.A. (2000) H M	DEC. °
Quadrantids	Jan. 3	9 P.M.	15 28	+50
Lyrids	Apr. 21	11 P.M.	18 16	+34
Eta Aquarids	May 4	2 A.M.	22 24	0
Delta Aquarids	July 28	5 A.M.	22 36	−17
Perseids	Aug. 11–12	—	03 04	+58
Orionids	Oct. 21	midnight	06 20	+15
Taurids	Nov. 1–7	—	03 32	+14
Leonids	Nov. 17	7 A.M.	10 08	+22
Geminids	Dec. 13	7 P.M.	07 32	+32
Ursids	Dec. 22	1 A.M.	14 28	+76

Notes: Since the year is actually 365¼ days long, we have listed the date and time starting in March for leap years (2000, 2004, 2008, etc.); add 6 hours for each year following in the four-year sequence. Most showers are named after the constellation in which their radiant is located, or after the bright star their radiant is near. The Quadrantids were named after Quadrans Muralis (in the northern part of Boötes), a constellation suggested by J. E. Bode in 1801 that is no longer accepted as a constellation.

Fig. 13-2 A fireball in the sky from the 1998 Leonid meteor shower.
(Pekka Parviainen)

TABLE 20 (CONT.). MAJOR METEOR SHOWERS

ASSOCIATED COMET OR ASTEROID	AVG. HOURLY RATE (FOR A SINGLE OBSERVER)*	AVG. DURATION OF MAXIMUM (BETWEEN DAYS OF 1/4 MAXIMUM)**
	40	1
Comet C/1861 G1 (Thatcher)	10	2
Comet 1P/Halley	20	3
	20	7
Comet 109P/Swift-Tuttle	50	5
Comet 1P/Halley	25	2
Comet 2P/Encke	10	†
Comet 55P/Tempel-Tuttle	15	—
Asteroid 3200 Phaethon	70	3
Comet 8P/Tuttle	15	2

* = Number of meteors (both shower and sporadic meteors) you can expect to see under ideal observing conditions—i.e., no clouds, and a clear, dark sky in which 6th-magnitude stars are visible. Shower strength fluctuates from night to night and from year to year.

** = Background rate of sporadic meteors subtracted.

— = Broad maximum; duration of shower uncertain.

† = Meteors from a large number of small radiants near the constellation Taurus are visible for several weeks in November.

Aquarids from mid-July to late August or the Taurids in the fall, are spread out over many weeks.

The Leonid meteor shower on November 16–17 is particularly spectacular about every 33 years, when it is notable for many fireballs. Now we will have to wait until about 2031 or 2032 for the next such occasion. During a one-hour period of the 1998 Leonids, more than 100 meteors were observed per minute (fig. 13-2). Some of the brightest trains lasted minutes.

The best way to observe a meteor shower is to lie back on a lawn chair or blanket on the grass and enjoy the night. A meteor shower may appear in virtually any part of the sky, so to use a telescope or even binoculars would simply limit your field of view. Instead, just look up, moving your eyes slowly around the sky. You will probably see a meteor out of the corner of your eye. Of course, the farther you are from streetlights, the better. If the moon is up, and particularly if it is more than half full, the sky will be too bright to see the meteor shower at its best. Sometimes you can wait until the moon sets, or observe before it rises. Patience is a watchword—give yourself at least a half hour of watching at a time. Don't just pop outside for five minutes, or you may not see anything—your eyes wouldn't even be dark-adapted.

Try to trace the paths of meteors in a certain shower back across the sky; all the paths will seem to converge in the same part of the sky, called the radiant. This convergence is an effect of perspective, since the meteors are hitting the earth's atmosphere in parallel lines. Just as parallel railroad tracks seem to converge at a distant point on the horizon, the parallel meteor paths seem to come from a convergent point or area in the sky. During meteor showers, you can often see more meteors after midnight, because the earth is rotated then so that your side is plowing into the meteors rather than having them catch up from behind.

To tentatively identify a meteor with a certain shower, you can trace its path back to see if it came from a constellation that has a known, active radiant. On the night of a shower's maximum, most meteors come from the shower, but on nights far from the shower's peak, determining whether a meteor came from a known radiant is the only way to identify meteors from spread-out showers.

Occasionally a meteor is extremely bright—as bright as or brighter than even Venus. Such an object is called a fireball. A fireball during a shower will still be a bit of comet dust, but a sporadic fireball may be a chip of a broken-up asteroid. Sometimes a fireball will leave a train, a path in the sky that remains for a few seconds. You may even be able to hear a sound, in which case there is more chance that a piece of the meteoroid is falling to earth as a meteorite.

Fireballs are so rare that you can't plan to see one, except during the peak Leonid showers every 33 years. If you should see one, you should note as much information as you can, including the time; the brightness compared to nearby stars, planets, or even the moon; and, if possible, the altitude and azimuth of the beginning and ending points of the fireball's path among the stars. This information should be reported to the Smithsonian Museum of Natural History, Washington, D.C. 20560, in the U.S.; to the Herzberg Institute of Astrophysics, Ottawa, Ontario K1 A 0R6, in Canada; or to suitable organizations in other countries. If you see a bolide and hear a sound, it is particularly important to report it quickly, for a meteorite may have landed nearby and could be picked up for scientific analysis. One may even go through a roof, as happened in Connecticut in 1982, or hit a car, as happened in upstate New York in 1992.

Meteors ionize their paths in the earth's upper atmosphere. Both professionals and some amateur astronomers use radio astronomy to detect these paths. Radio equipment needed to do so is relatively inexpensive, and advertisements for suitable equipment can be found in many magazines for amateur radio operators and occasionally in *Sky & Telescope*.

Meteorites are particularly important to astronomers because, aside from the moon rocks brought back between 1969 and 1972 by the Apollo missions and some dust (fine particles up to several millimeters across) brought back by uncrewed Soviet spacecraft, meteorites provide the only extraterrestrial material we have to study in a laboratory. NASA's Stardust spacecraft is now en route to a comet, at which it is to collect dust in 2004 and return it to earth in 2006 for analysis.

There are two basic types of meteorites (though, of course, experts make finer divisions): nickel-iron meteorites (called *irons*) and stony meteorites (called *stones*). Most meteorites found on the ground by accident are irons; they are very dense (that is, quite heavy for their size) and appear quite different from ordinary rocks. On the other hand, when someone sees a meteorite fall and then searches for it, stony meteorites are found most of the time. This indicates that most meteorites are actually "stones," but that most of them are never discovered because they resemble normal (terrestrial) stones on the ground and because they disintegrate from weathering. The largest number of meteorites have been found in recent years in Antarctica, where they have accumulated undisturbed over long periods of time.

Nobody who is interested in meteorites should miss seeing the meteorites in museums (fig. 13-3). You can also visit the Barringer Meteor Crater in southern Arizona (fig. 13-4). This crater

Fig. 13-3 *The largest meteorite ever discovered and moved was the 34-ton (31,000-kg) Ahnighito meteorite, brought back by William Peary and Matthew Henson (the first people to reach the North Pole) from the Arctic in 1892. It is now on display at the American Museum of Natural History in New York. (Jay M. Pasachoff)*

is almost a mile (1.2 km) in diameter and resulted from the most recent large meteorite to hit the earth, about 40,000 years ago. More than a dozen even larger meteorite craters are known on earth.

TO OBSERVE METEORS

You will need a comfortable place to lie back outdoors, far from lights; an accurate watch, set to the nearest second (which formerly required receiving the station WWV on a shortwave radio but now can be done with a digital watch set from the radio or television); and perhaps a friend to take notes (which helps preserve your adaptation to darkness). When a meteor moves overhead, record the time of the event to the nearest second, the duration of the meteor (usually less than 2 seconds or so), the length of the trail, the meteor's magnitude (by comparing it to nearby stars), and the color. The data usually reported are the number of meteors seen by a single observer in an hour and changes in the average hourly rate through the night; this information shows when the peak of the shower occurred. It may even be useful to measure and plot the number of meteors during 15-minute intervals. When possible, it is good to record whether the meteor seems to be coming from the radiant, or if it is a sporadic meteor. More advanced observers often do this by recording the path of each meteor and its magnitude on a star chart, using a different chart for each hour of observation.

Fig. 13-4 The Barringer meteor crater near Winslow, Arizona, is almost 1 mile (1.2 km) across and was formed about 40,000 years ago. (Meteor Crater, Northern Arizona)

Your data can be reported to the American Meteor Society, Department of Physics and Astronomy, SUNY, Geneseo, NY 14454, meisel@uno.cc.geneseo.edu; or meteorobs@charleston.net; or c/o Wanda Simmons, 3859 Woodland Heights, Challahan, FL 32011; or www.amsmeteors.org; or, on a more advanced level, to the International Meteor Organization, c/o Robert Lunsford, lunro.imo.usa@prodigy.com, www.imo.net; or to the Meteor Section of the British Astronomical Association, c/o Neil Bone, "The Harepath," Apuldram, Chichester, West Sussex, England, bafb4@central.sussex.ac.uk. The address of the BAA Web site is www.ast.cam.ac.uk/~baa. The BAA Journal also has its own Web site at www.star.ucl.ac.uk/~hwm.

The shortwave station WWV, which provides time signals, broadcasts at 2.5, 5, 10, 15, and 20 MHz.

14

OBSERVING THE SUN

Though many people think of astronomy as the study of the nighttime sky, studies of the sun are also part of astronomy. The sun has the advantage, too, that it is up in the daytime, when people are apt to be outside. The sun has the disadvantage, though, that it is too bright to see safely under most circumstances. Only during a total solar eclipse can one look at the sun without protective filters or without taking other precautions.

The photograph to the right (fig. 14-1), taken through a filter that takes out all but about 0.001 percent of the sunlight, shows the everyday solar surface. Dark sunspots cross it. The number of sunspots waxes and wanes with a cycle that lasts about 11 years.

Most of the stars are visible only at night, but one star—our sun—is visible in the daytime. The sun is much closer to us than any other star. It takes light only about eight minutes to travel from the sun to us; the next nearest star is more than four light-years away. As a result of the sun's closeness to earth, its disk covers $\frac{1}{2}°$ of the sky and we can see many details on its surface.

THE SUN'S SURFACE

The sun, like all stars, is a ball of hot gas. It has an interior and an atmosphere. The surface of the sun sends us the light and heat that provide energy for us on earth.

The solar surface is called the *photosphere,* from the Greek word *photos,* meaning "light." When we look at the sun using the protective filters and precautions described on pp. 478–481, we can see that it is largely uniform in brightness, that it becomes slightly darker toward its edges, and that there often are a few dark areas on its surface (fig. 14-1). The dark areas—*sunspots*—are relatively cool regions of the solar surface.

The energy is formed deep inside the sun's interior by the pro-

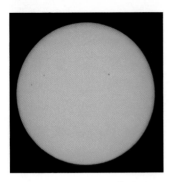

Fig. 14-1 *The everyday surface of the sun. To take your own solar images, you must photograph through a filter that absorbs all but 0.001 percent of the incoming sunlight. It is not safe to observe the sun directly without such a special filter. (Williams College, Hopkins Observatory)*

cess of nuclear fusion. In the sun's fusion process, groups of four hydrogen atoms are transformed into single helium atoms. Each helium atom that results has slightly less mass (0.7 percent) than the sum of the four hydrogens. In 1906, Albert Einstein realized that mass can change into energy according to the equation $E = mc^2$. Einstein's equation tells us that the little bit of mass that disappears is transformed into a relatively large amount of energy. (Since c—the speed of light—is such a large number, the amount of energy that results adds up to a lot.) The sun is a good example of a fusion reactor, a type that we are not yet able to build on earth. The sun is a fusion reactor located 93 million miles (150 million kilometers) from us.

Each sunspot has a dark center called its *umbra* (plural, *umbrae*) and a less dark region called its *penumbra* (fig. 14-2). Sunspots normally form in groups, and each spot can last for weeks. Since the sun rotates about once every 25 days, we can watch the sunspots form and rotate across the visible solar surface.

Sunspots are regions of the sun where the magnetic field is especially strong, perhaps 3,000 times stronger than the average field of the sun, of the earth, or of a toy magnet, which are all comparable in strength. Spots tend to occur in groups. Within each group of sunspots in one of the sun's hemispheres, the sunspots that lead in the direction the sun is rotating have north magnetic poles and the spots that trail have south magnetic poles. In the opposite hemisphere, the leading and trailing polarities are reversed.

The number of sunspots on the sun waxes and wanes over an 11-year period (fig. 14-3). The vertical axis plots the *sunspot number,* which is not actually a direct count; it is a compound value that makes allowance not only for individual spots but also for the presence of sunspot groups. The sunspot number = $k(10g + f)$,

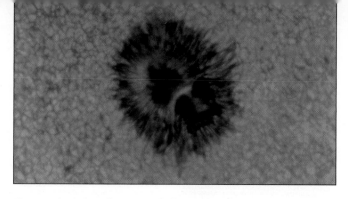

Fig. 14-2 A sunspot, showing its dark central umbra and its surrounding penumbra. The surface of the sun itself shows a salt-and-pepper effect known as granulation. Each granule is about 500 miles (700 km) across —roughly the distance from New York to Detroit. (William C. Livingston, National Solar Observatory/National Optical Astronomy Observatories)

where k is a correction factor to account for personal decisions of each individual observer (as to what constitutes an individual spot, for example), g is the number of groups, and f is the total number of spots. (Your value for k can be assigned only by a central registry, after they compare the numbers of spots you report with the established number.) An isolated spot is considered to be a group with one spot in it, so it adds 1 to each value of g and f.

After the 11-year cycle, the polarities in opposite hemispheres of the sun (northern vs. southern) reverse; that is, in a hemisphere where the leading spots (in the sun's direction of rotation) have had north magnetic poles, those spots now have south magnetic poles. Thus it takes two cycles, or 22 years, for the leading spots in each hemisphere to regain their original polarity, so the solar activity cycle is really 22 years long.

The 2000–2001 sunspot maximum will be followed about 2006 by a sunspot minimum. During sunspot maximum, there are more huge eruptions on the sun called solar flares. Flares send out particles, x-rays, and gamma rays that can hit the earth; these could cause surges in power lines, zap passengers in high-flying aircraft or astronauts on space shuttles, and contribute to creating the aurora. The flares also excite the aurora in a way that allows it to be seen by observers at lower latitudes than normal. Since the earth's magnetic field guides the particles that can cause the aurora, and since the earth's magnetic field comes out of the earth at the magnetic poles, the aurora can most often be seen close to the magnetic poles (fig. 14-4). The northern mag-

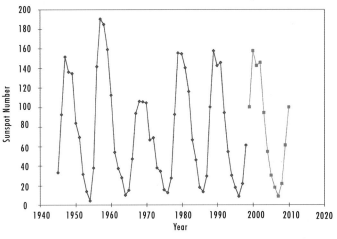

Fig. 14-3 The sunspot cycle, whose longtime period is 11.2 years. The 1999–2010 points are not observations; they simply repeat the previous cycle. (Data from Pierre Cugnon, Arielle Vigneron, and Daniel Carre, Sunspot Index Data Center, Brussels)

netic pole is near Hudson Bay in Canada. The aurora borealis, the northern lights, are less often visible at latitudes of the continental U.S.

When we look at the center of the solar disk, we are looking through gas. The effect is the same as looking into the air on a foggy day; we can see only so far. When we observe near the solar *limb*—the apparent edge of the sun—we are looking diagonally through the solar atmosphere and thus through more gas that obscures our view; we cannot see as close to the sun's center before the gas appears opaque. As a result, near the sun's edge we see higher levels of the sun's atmosphere. Since these higher levels are slightly cooler than the lower atmospheric levels we can see when we look toward the center of the sun's disk, the sun's surface looks a little darker toward its limb.

When we look at the sun through the earth's atmosphere, the atmosphere bends the different colors in sunlight by varying amounts. This effect—refraction—becomes most extreme at sunrise and sunset and leads to the beautiful, elusive phenomenon known as the *green flash* (fig. 14-5). It usually lasts only about two seconds and can be seen only when our view of the setting sun is completely unobstructed; usually you must be looking over water or from a mountain, and there must not be any haze or clouds on the horizon. (It is even harder to see the green flash at

Fig. 14-4 *The aurora borealis, the northern lights, photographed from Yellowknife, Canada. The photo shows Comet Hale-Bopp through the aurora. (Akira Fujii)*

sunrise, because you don't know where to look in advance.) The higher you are above the horizon, the more likely you are to see a green flash and the longer it may last. A green flash is a type of mirage, an effect caused by the bending of light rays by air of different temperatures. Different types of green flashes come from different but related causes, but here is a standard explanation. As the sun nears the horizon, because different colors of rays coming from the sun to us bend by different amounts, we wind up essentially with overlapping images of the sun. The colors are in the order ROY G. BIV (red, orange, yellow, green, blue, indigo, and violet), with the red image lowest. But the blue, indigo, and violet rays are scattered so much by the atmosphere that they don't reach us. The orange and yellow are absorbed by ozone and perhaps also water vapor in the earth's atmosphere. So we are left with only red and green images. The mirage effect distorts the image, multiplying the separation of the colors and expanding the size of the rim in each color. When the red image sets, we sometimes see a brief bit of green on the horizon for two seconds or, sometimes, a few seconds, since the top of the green image of the sun is all that is visible.

SAFELY OBSERVING THE SUN

Most of the time, the sun is too bright to look at safely with the naked eye. Except for certain short periods during total eclipses, the only time when we can safely look at the sun is when it is dimmed by haze, ususally close to sunset.

The safest way to study the sun is not to look at it directly at all. You can use a telescope or binoculars with its eyepiece to project

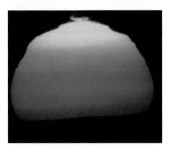

Fig 14-5 *The green flash, a green color seen at the upper edge of the sun at sunrise or sunset for about two seconds in perfect observing conditions, especially at sunset over water or from a mountain. (George V. Coyne, S.J., Vatican Observatory)*

an image of the sun onto a piece of cardboard. The process is called "eyepiece projection." Stand with your back to the sun and look at the cardboard; do not look through the telescope at the sun at all. Looking at the sun through a telescope for even a second could blind you. If a finder telescope is mounted on the telescope you are using, make certain that it is securely capped. If you are using one lens of binoculars, cap the second lens securely.

Adjust the eyepiece of your telescope or binoculars so that it is behind its normal position. Then you can vary both the position of the eyepiece and the distance of the paper behind it to put the sun's image in focus. An image the width of your hand should show enough detail to reveal sunspots clearly. You can trace the outlines of the sunspots on a daily basis and follow the way they change as the sun rotates and the sunspots evolve. Again, *never* look up through a telescope at the sun, not even to point or focus the telescope. (You can usually align the telescope fairly well by adjusting the shadows along the outside of the tube so that they are at a minimum.)

The sun is about one million times brighter than the full moon. So to observe the sun directly in order to see the sunspots, we need a filter that cuts out all but $\frac{1}{1,000,000}$ of the sun's rays. Special filters are available that go over the front end of the telescope, so that most of the sun's light never enters the telescope tube. (Filters that cut out light near the eyepiece are no longer thought to be safe for amateur astronomers; there is always the danger that they might slip or crack.)

Once you know how to observe the sun safely, you can make your own count of the sunspot number. The official Sunspot Index Data Center is in Brussels, Belgium (www.oma.be/KSB-ORB/SIDC/index.html). The AAVSO (American Association of Variable Star Observers) in Cambridge, Massachusetts, has a solar group that keeps track of sunspots and compiles the American Relative Sunspot Number each month (www.aavso.org).

THE CHROMOSPHERE, PROMINENCES, AND SPECIAL SOLAR FILTERS

Many features of the sun's atmosphere are not visible when you look in white light (all the sun's rays together); they show up only when the light of the unique color (specific wavelength) of hydrogen is isolated from visible light. This light, represented by the so-called *H-alpha line*, falls in the red part of the spectrum. Professional solar astronomers and increasing numbers of amateurs have filters that pass only this red H-alpha line of hydrogen (fig. 14-6).

When we look at the sun through an H-alpha filter, we are seeing the *chromosphere*. The chromosphere (from the Greek word for color, chromos, since it looks colorful during eclipses) is a thin layer just above the solar photosphere. The chromosphere is not really a layer; it is actually made up of spikes of gas called *spicules* that rise and fall with a period of 15 minutes each. The spicules are usually too delicate to see except with professional equipment used at exceptional sites.

H-alpha filters often come with fittings so they can be attached to the eyepiece of a telescope; in that location, they should be used with caution. When we look through an H-alpha filter, the solar surface looks mottled, and dark lines called *filaments* snake their way across the solar surface. When a filament is rotated so that it appears at the edge of the sun, it is then called a *prominence* (fig. 14-7).

On any sunny day, you can see exciting things with a telescope pointed at the sun. Eyepiece projection or a special filter over the aperture reveals the sunspots. An H-alpha filter shows the filaments and prominences. Since the sun's surface changes from day to day, looking at the sun provides a changing image.

Some technical notes about equipment: if you are getting filters or other apparatus to observe the sun, it is best to check the advertisements in *Sky & Telescope* to see what is currently available. See also sources given on p. 510 and on this book's Web page. As of this writing, Thousand Oaks Optical specializes in filters, and filters can also be bought from telescope manufacturers. The less-expensive kind of filter is made of aluminized Mylar. Higher-quality solar filters cut down the intensity of white light by depositing reflecting metal on glass, though they are not cheap. DayStar makes H-alpha filters. A filter that passes a band of wavelengths narrower than 1 angstrom (0.0001 micrometer) is necessary to see the filaments on the solar disk; a filter that passes as much as 3 or 4 angstroms is sufficient to see the prominences around the solar limb. Mylar filters cost a few dollars; alu-

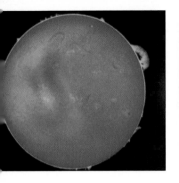 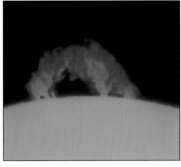

Fig. 14-6 (left) *The sun's surface photographed through an H-alpha filter that passes light in a very narrow band of spectrum. The central part of the image shows the solar disk, with dark filaments and bright active regions. Surrounding it is an image made at the same scale, but with the center of the disk occulted—hidden—by a device in the telescope. These photographs were taken by an amateur astronomer. (© DayStar Filter Corp.)*

Fig. 14-7 (right) *A solar prominence. This example was a quiescent (long-lasting) type; others may be eruptive. (R. J. Poole/DayStar Filter Corp)*

minized glass filters cost a few tens of dollars; and H-alpha filters cost more than a thousand dollars.

OLAR ECLIPSES: NATURE'S SPECTACULAR

The most awesome sight we can ever see is a total solar eclipse. The moon gradually covers the sun over a period of an hour or two in the partial phases of the eclipse (fig. 14-8), and then the crescent sun abruptly gives way to night. A few points of light on the edge of the sun—Baily's beads—are visible for a few seconds. A bright point of light on the edge of the moon, so bright that it glistens like a jewel and is called the diamond-ring effect, is the last Baily's bead (fig. 14-9). It lasts 5 or 10 seconds and then is also covered by the moon. The pinkish chromosphere is visible for another few seconds; prominences may be visible somewhat longer.

The diamond-ring effect comes from the last bit of the solar photosphere shining through a valley on the edge of the moon. When the diamond ring is over, we see a halo of light around the sun—the *corona* (from the Latin word for "crown"). The corona

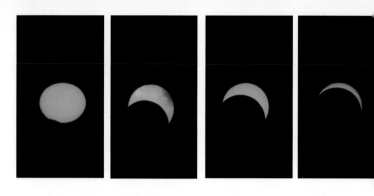

Fig. 14-8 *Partial phases of a total eclipse, photographed through a dense solar filter.* (Jay M. Pasachoff)

is the tenuous layer of gas surrounding the sun at a temperature of 4,000,000°F (2,000,000°C). It always surrounds the sun, but is a million times fainter than the surface of the sun (the photosphere) and is normally fainter than the earth's blue sky. Normally the sky looks blue because light from the solar photosphere bounces around in it. For us to be able to see the chromosphere and the corona, the sun must be up at a time when the sky is not illuminated, and that is exactly the situation we have during a total solar eclipse (fig. 14-10).

In the hour or two before the total phase of the eclipse, when the sun is only partially covered by the moon, even the remaining part of the sun is still too bright to look at safely. A few minutes before totality, the total amount of light from the photosphere is reduced enough that the eye-blink reflex that normally protects our eyes doesn't work. Even then, you can still hurt your eyes by looking at the remaining part of the sun without a special filter. Only when the diamond-ring effect is ending—and nobody ever has trouble deciding when that is—can you take your special filter away from your eyes and stare at the sun.

The corona is irregular in shape (fig. 14-11), with *streamers* extending millions of miles into space (fig. 14-12). The sun acts on the whole like a giant bar magnet, so we can sometimes see *polar tufts*—thin rays coming out of the sun's poles—as well as the streamers that appear at lower solar latitudes. The corona is expanding, forming the solar wind that extends throughout our solar system.

During the total part of the eclipse, only the corona is visible; the corona is perfectly safe to look at (fig. 14-13). It is then about

Fig. 14-9 *The diamond ring effect, the last bit of ordinary solar surface shining through a valley on the edge of the moon, is visible at the beginning and end of the total phase of a total solar eclipse. Scattering in the lens adds to the luster of the "diamond." (Jay M. Pasachoff)*

the same brightness as the full moon. When the second diamond-ring effect occurs, marking the end of totality, you must again stop looking directly at the sun.

During totality, the sky is as dark as it is at night, or at least as dark as it is during evening twilight. If you look toward the horizon in any direction, you can see beyond the darkest part of the moon's shadow. You will see a pinkish glow on the horizon that looks like a sunset extending 360° around the sky.

For a minute or two before totality begins and for a similar time

Fig. 14-10 *From the inside of the umbra of the moon's shadow, we see a total eclipse of the sun. From inside the penumbra, we see a partial eclipse. In the drawing, the point of the umbra does not quite reach the earth, which means that the eclipse is annular, like the one in the U.S. on May 10, 1994.*

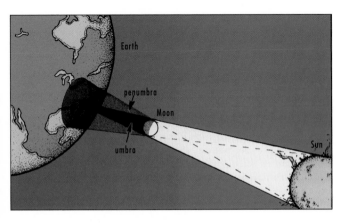

Fig. 14-11 (left) The 1998 total solar eclipse, photographed from Aruba. The spiky solar corona silhouettes the moon, which we see as a dark circle because the sun is lighting the opposite side. The sun's diameter, barely hidden by the moon, is 865,000 million miles (1.4 million km), so we can see that the coronal streamers extend millions of miles into space. (Jay M. Pasachoff)

Fig. 14-12 (right) Computer processing allows several eclipse images of longer and shorter exposures to be assembled together, bringing out detail in the solar corona that is not otherwise visible, given that the inner part of the corona is so much brighter than the middle part. (Image processing by Wendy Carlos from images of the 1998 eclipse by Jay Pasachoff. © 1998 Jay M. Pasachoff and Wendy Carlos)

after it ends, narrow bands of shadow sometimes seem to race across the landscape. These *shadow bands* occur when inhomogeneous regions of the earth's upper atmosphere bend the narrow crescent of light from the partially eclipsed sun. Shadow bands can be seen especially well if you spread out a white sheet on the ground, perhaps with distances measured across it to help you determine the spacing between adjacent bands and the distance a band appears to travel each second (which can be converted into velocity in kilometers or miles per hour). It is also interesting to note the direction in which the bands are traveling, which is often different before and after totality. They are very difficult to photograph, since they are of such low contrast.

Solar eclipses are not especially rare; partial ones, in which the central part of the moon's shadow never touches the earth, occur about three times every year. Total solar eclipses occur about every 18 months (table 21 and fig. 14-14). Since the moon and the sun appear approximately the same size in the sky—the sun is ac-

Fig. 14-13 A view during mid-totality during the 1995 total eclipse, photographed from India. No filters should be used to view the total phase of the eclipse. The sun surrounds the black dot of the moon in the sky. At the horizon, you see far enough to see out of the zone of totality, making a pinkish effect. Jupiter is also visible. (Jay M. Pasachoff)

tually 400 times larger than the moon but happens to be 400 times farther away—the shadow is not more than about 200 miles (300 km) long when it hits the earth. It traces a narrow path across the earth (fig. 14-15), thousands of miles wide. Only within that narrow path can you see the corona and experience the eclipse. Annular eclipses, when the moon's disk fits into the sun's without covering it entirely, also occur every 18 months or so (fig. 14-16).

More and more people are taking long trips to see total solar eclipses; once you see one, it is hard to resist going to see another (and perhaps taking a friend along). Among the observations that you can make at an eclipse are (1) timing the *contacts,* first contact being when the moon first begins to cover the sun, second contact when totality begins, third contact when totality ends, and fourth contact when the moon's image leaves the solar disk; (2) sketching or photographing the prominences; (3) sketching or photographing the corona; and (4) observing the shadow bands. Some people like to cover one eye with an eyepatch before totality, so the eye will be dark-adapted when totality begins. Many people like to observe the corona in binoculars during totality.

Photographing an eclipse is always interesting, and it is partic-

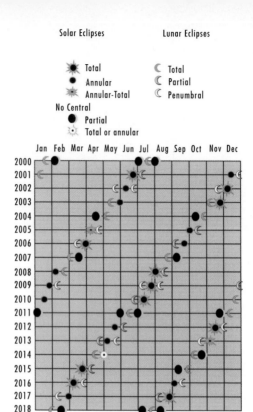

Solar Eclipses   Lunar Eclipses

Fig. 14-14 *A graphic diagram showing how eclipses are grouped in "eclipse seasons" and how solar and lunar eclipses are related.* (*Serge Koutchmy and Pierre Guillermier*)

ularly nice that the choice of exposure is not critical. Each combination of lens opening (aperture) and exposure time gives a different effect, but all are pretty. The corona falls off rapidly in brightness the farther it is from the sun, so the longer the exposure time, the more corona you can see. But you also want to keep your exposure times short enough so that the moon's edge will not blur as the sun and moon move across the sky. Exposure times of up to about 1 second with a 500 mm telephoto lens, or 10 seconds with a "normal" 50 mm lens on a 35 mm camera, will do. It is important to use as steady a tripod as possible, to release the

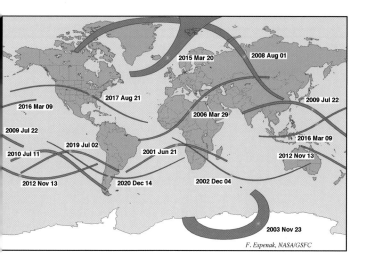

Fig. 14-15 *Total eclipses of the sun from 2001 to 2020. (Fred Espenak, NASA Goddard Space Flight Center)*

Fig. 14-16 *Annular eclipses of the sun from 2001 to 2020. Partial phases will be visible in the U.S. on December 14, 2001; June 10, 2002; and April 8, 2005. (Fred Espenak, NASA Goddard Space Flight Center)*

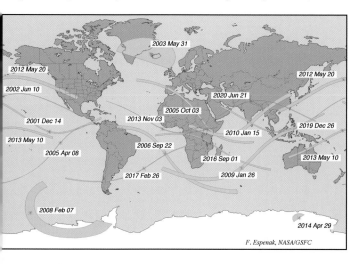

TABLE 21. SOLAR ECLIPSES

TOTAL ECLIPSES, 2000-2010

DATE/MAX DURATION	PATH
June 21, 2001/4m57s	southern Africa (Angola, Zambia (Lusaka), Zimbabwe (north of Harare), Mozambique (Chinde), Madagascar (Morombe); partial phases visible from Brazil and from most of Africa
December 4, 2002/2m04s	southern Africa (Angola, Zambia, Botswana, Zimbabwe, South Africa, Mozambique), Pacific Ocean, ending at western Australia; partial phases visible from most of Africa and from western Australia
November 23, 2003/1m57s	Antarctica (Mirnyy Station); partial phases also visible from southern tip of Chile and Argentina and from Australia and New Zealand
April 8, 2005/0m42s	total only at maximum over the Pacific Ocean
March 29, 2006/4m07s	Brazil (Natal at sunrise), Africa (Ghana (Accra), Togo, Benin, Nigeria, Niger, Chad, Libya, Egypt), Turkey (Sivas), Russia, top of Caspian Sea, Kazakhstan; partial phases visible from all of Europe, Asia to mid-Mongolia and Bangladesh, all but southern Africa from Namibia to Somalia, western tip of Brazil
August 1, 2008/4m27s	northern Canada islands, Greenland (Alert), Russia (Siberia, including Novosibirsk), western Mongolia, China (Xian); partial phases visible from Europe northeast of southern France and mid-Italy, most of Asia except Japan or southeast of mid-Malaysia or mid-Sumatra
July 22, 2009/6m39s	India, eastern tip of Nepal, Bangladesh, Sikkim, Bhutan, northernmost Myanmar, China to Shanghai, southern Japanese islands; peak over Pacific Ocean; partial phases visible from China, western Russia, southeast Asia north of mid-Sumatra, mid-Borneo, and Papua New Guinea; and northern tip of Australia's Cape York
July 11, 2010/5m20s	South Pacific Ocean including Easter Island (Chile), ending over southern Chile and Argentina (Calafate)

ANNULAR SOLAR ECLIPSES, 2000–2010

December 14, 2001/3m53s	Central America: northern Costa Rica (San José and north) and southern Nicaragua; **partial phases through all the U.S. except northern New England and northern Alaska; annularity passes 270 miles (450 km) south of Hawaii**
June 10, 2002/0m23s	Pacific Ocean (islands Sangihe and Talaud (Indonesia), north of Sulawesi; Rota and Tinian

	[Northern Marianas], near Guam); passes just south of Baja California (Cabo San Lucas) at sunset; **partial phases include western U.S. and Canada, Hawaii, and Alaska**
ay 31, 2003/3m 37s	Greenland, Iceland, northernmost Scotland; partial phases include Europe from France north and east, most of Asia including Russia but excluding Japan
ril 8, 2005/0m 42s	southernmost Costa Rica, Panama, Colombia, Venezuela (Mérida); **partial phases visible from U.S. regions south of San Diego-Denver-Chicago-Philadelphia**
tober 3, 2005/4m 32s	northern Portugal (Braga), Spain (Madrid), Algeria (Algiers), Tunisia, Libya, Sudan, Ethiopia, Kenya, southern Somalia
ptember 22, 2006/7m 09s	sunrise at Guyana, Suriname, French Guiana, adjacent Brazil, Atlantic Ocean; partial phases visible from eastern South America, western and southern Africa, part of Antarctica
bruary 7, 2008/2m 12s	Antarctica (Russkaya Station); partial phases include eastern Australia, New Zealand
auary 26, 2009/7m 54s	Indonesia (southern Sumatra, western tip of Java, most of Borneo); partial phases visible from southern Africa, southern India, southeast Asia, western Australia
auary 15, 2010/11m 07s	Africa (southern tip of Chad), Central African Republic, Congo, Uganda, Kenya (Nairobi), southern tip of Somalia), southern tip of India, northern Sri Lanka, Myanmar, China (to Qingdao); partial phases visible from eastern Europe through Asia except Japan or east and southeast of mid-Java

PARTIAL SOLAR ECLIPSES, 2000-2010

bruary 5, 2000	58 %	Antarctica
y 1, 2000	48 %	southern tips of Chile and Argentina
y 30, 2000, Pacific time	60 %	Siberia, **Alaska, western U.S. and Canada,** northern Greenland, northern Scandinavia
cember 25, 2000	72 %	**continental U.S.,** Canada, Caribbean, Central America from Nicaragua north
ril 19, 2004	74 %	southern Africa (Angola, Zambia, Mozambique and south), part of Antarctica
tober 14, 2004	93 %	Japan, northeastern Siberia, **western Alaska and Aleutians, barely to Hawaii**
arch 19, 2007	87 %	Asia north of Sumatra, Malaysia, and Philippines, including only northernmost and southernmost Japan
ptember 11, 2007	75 %	South America from mid-Brazil and mid-Peru southward, part of Antarctica

Detailed maps are available from Fred Espenak's Web site, which is linked through this Field Guide's Web site. Information drawn from Espenak's *Fifty Year Canon of Solar Eclipses: 1986–2035*. Maps also available in Philip Harrington's *Eclipse!*.

Fig. 14-17 *Looking over someone's shoulder through a solar filter made of fogged and exposed black-and-white photographic film as the observer also looks through the solar filter, we see an annular eclipse. An entire circle of solar photosphere remains visible throughout.* (Jay M. Pasachoff)

shutter carefully (preferably with a cable release), and to wait for the camera to stop vibrating after you advance the film (you may need to wait a second or two). If possible, lock up the mirror on a reflex camera to reduce vibration.

Telephotos of at least 200 mm focal length and preferably 500 mm focal length give the best images of the corona. You can use normal or wide-angle lenses to take overall views that include the corona at the top of the photo and scenery in the foreground. You will need a special solar filter to take the partial phases of an eclipse; remove the filter to photograph the diamond ring, prominences, or corona. Then you must put it on again at the end of the eclipse, perhaps after having taken a quick picture of the second diamond ring. Recommended exposures appear in table 22. A difficult tradeoff must be made between faster films, which have larger grain, and slower but less grainy films; there is no right answer to this question. Usually slide film instead of print film should be used, because it captures a wider range of intensities; prints can be made from slides later on.

Some people see a partial eclipse and wonder why others are so awestruck by a total eclipse. But the difference is like night and day. Since the photosphere is 1 million times brighter than the corona, even a 99 percent partial eclipse still leaves 1 percent of 1 million times, or 10,000 times, more light from the photosphere than there is from the corona. And so the sky is too bright to allow the corona to be seen during a partial eclipse. It is truly worth traveling to the zone of totality each time there is an eclipse. Seeing a partial eclipse and saying that you have seen an eclipse is like standing outside an opera house and saying that you have been to the opera; in both cases, you have missed the main event.

Because the orbits of the moon around the earth and the earth around the sun are elliptical, sometimes the moon appears too small to completely cover the sun during an eclipse. Then an *annulus*—a ring—of photosphere remains visible (fig. 14-17), and we have an *annular eclipse*. Since the photosphere is always visible, special filters must be used throughout an annular eclipse, both for the eye and for a camera, or pinhole cameras could be used. You should use a telephoto lens of at least 200 mm focal length to take photos through filters.

On a technical note, the same special filters described on p. 478–481 can also be used at eclipses. The less expensive materials, such as aluminized Mylar, that are often used for solar filters are available for this purpose from Tuthill, Inc., and other suppliers; look for advertisements in recent issues of *Sky & Telescope* or *Astronomy* magazines. The Mylar is inexpensive, but some of it gives a bluish cast to the sun instead of a more pleasant orange one; it is also subject to potentially dangerous pinpricks.

You can make your own solar filter for observing an eclipse if you start a few days in advance. First completely fog a roll of black-and-white film, perhaps by simply unrolling it in the sunlight. Then develop it to maximum density. Two thicknesses of this fogged and developed black-and-white film should cut down the solar intensity to a safe level; try two thicknesses first, and if they are too dense for the sun to shine through, try one thickness.

The silver in black-and-white film absorbs all sunlight across the entire spectrum, including the infrared (heat) radiation that is invisible to our eyes. Do not use any of the newer black-and-white films that don't use silver. Color film does not use silver, so it cannot be used to make a safe solar filter for eclipses or for any other observations of the sun. Also, gelatin photographic filters, such as Kodak's neutral-density Wratten filters, cannot be used for observing the sun with the naked eye, since they are effective only in the range for which photographic film is sensitive and pass infrared radiation that can be harmful to the eye.

No matter what filter you use, *never stare at the sun*. You will be able to see a great deal just by glancing through the filter at the sun for a few seconds. The partial phases do not change rapidly

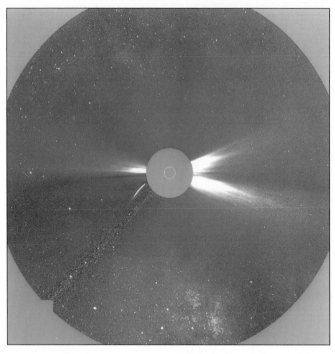

Fig. 14-18 *The solar corona, in an artificial eclipse made by one of the coronagraphs on the Solar and Heliospheric Observatory (SOHO). The inner corona has been hidden; the size of the sun is shown with a circle. A comet (curved arc at 8 o'clock) is about to go behind the occulting disk. (Naval Research Laboratory, LASCO Consortium, SOHO/ESA and NASA)*

enough to make it interesting to look at the sun longer in any case.

The corona can be observed from space on an hourly basis; images can be found through this Field Guide's Web site. The Solar and Heliospheric Observatory (SOHO) hides the moon to make images of the outer corona (fig. 14-18) and uses filters to make ultraviolet images in the light emitted by million-degree-hot gases (fig. 14-19). The Transition Region and Coronal Explorer (TRACE) makes higher-resolution observations in the ultraviolet (fig. 14-20) that show in detail how the coronal gas is held in place by magnetism, forming coronal loops.

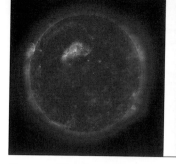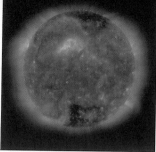

Fig. 14-19 *The solar corona seen in front of the solar disk by using special filters on the Extreme-ultraviolet Imaging Telescope on SOHO to make ultraviolet images that show only the light emitted by hot iron gas typical of temperatures of millions of degrees. Note how the corona is over the active regions, which correspond to the sunspot regions. (EIT Consortium, SOHO/ESA and NASA)*

Fig. 14-20 *Coronal loops, gas held in place by the sun's magnetic field, imaged by the Transition Region and Coronal Explorer (TRACE). (Lockheed Martin Advanced Technology Center and Smithsonian Astrophysical Observatory)*

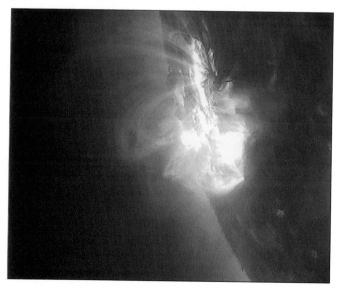

TABLE 22. SOLAR ECLIPSE PHOTOGRAPHY

Do not look through your viewfinder at the sun—except during totality—unless the lens is covered with proper (nongelatin) neutral-density filters.

Use film with the finest grain possible.

Mount your camera or telescope on as sturdy a tripod as possible. Exposure times for totality are not critical; most problems come from shaky mounts. Your camera will vibrate slightly each time you advance the film; wait a second or two to make sure the vibration has stopped before you take the next exposure. Use a cable release, and, if possible, lock up the mirror on reflex cameras.

For close-ups of totality with a 35-mm camera, use a telephoto lens of at least 300 mm and preferably 500 mm.

Bracket your exposures widely—take a variety of exposures on either side of any given lens and shutter speed.

Make sure that you don't run out of film in the middle of your exposure sequence.

A sample eclipse sequence follows, assuming that you are using ISO 64 or ISO 100 film :

PARTIAL PHASES: Every five minutes, take a bracketed series of exposures through a filter of neutral density 5, that is, a filter that passes only 1/100,000 (where the density means that there are 5 zeroes) of the incoming sunlight. Note: Do not look through any viewfinders unless they are also covered with a neutral-density 5 filter; Wratten photographic neutral-density filters are not safe for the eye. If you have made your own filter out of exposed and developed photographic film, take a wide range of exposures to make certain that you have a good one. Remember that your meter reading will probably not be accurate since it will average the light available over the entire field of view, while you are photographing a bright object surrounded by darkness. Put in a fresh roll of film five minutes before totality.

DIAMOND RING: Take off the neutral-density filters, and carefully, without shaking the camera, take one or two exposures at about $1/30$ second at $f/8$.

PROMINENCES: As soon as the diamond ring disappears, take a series of exposures at $f/8$ for $1/60$, $1/30$, $1/15$, and $1/8$ second.

CORONA: During totality, take a series of time exposures. If you are using an $f/8$ or $f/5.6$ lens, try $1/2$ second, 1 second, 2 seconds, and 4 seconds. If any time is left, repeat the sequence. Then get ready to take the diamond-ring effect at the end of totality.

If you are using a wide-angle lens, open your lens as wide as possible (to $f/2.8$, $f/2$, or $f/1.4$, for example). Use fast film, such as ISO 200 or ISO 1000. Make certain that the eclipse is in the frame, near the top, and that some objects are visible in the foreground at the bottom of the frame. Take 1, 2, 4, and 8 second exposures.

AT THE END OF THE ECLIPSE: Photograph the second diamond ring without looking through the viewfinder, then turn your camera so it is no longer facing the sun. (Otherwise, the sunlight focused through the lens could burn a hole in your shutter.) Add your neutral-density filters later for occasional photographs of the partial phases.

COORDINATES, TIME, AND CALENDARS

As the earth rotates on its axis, the sky overhead seems to turn in the opposite direction. The stars appear to rise in the east, move overhead in the course of the night, and set in the west. It is convenient for many purposes to think of the stars as being fixed on a *celestial sphere* that rotates every 24 hours.

POSITIONS IN THE SKY: RIGHT ASCENSION AND DECLINATION

On the surface of the earth, we measure the positions of cities and towns in longitude and latitude. Longitude is measured by lines that extend from north pole to south pole, usually up to 180°E and 180°W of a zero-degree (0°) line that goes through Greenwich, England.

Latitude on earth is measured by parallel lines that mark the number of degrees north or south of the equator. The north pole is +90°; the longitude scale has no meaning there, since all longitudes converge at the poles. Similarly, the south pole is −90°.

Astronomers have set up a similar set of coordinates in the sky. The celestial equator is an imaginary line around the sky, above the earth's equator. The celestial poles are the imaginary points where extensions of the earth's axis into space would meet the celestial sphere; they lie above the earth's poles.

The astronomers' coordinate that is similar to longitude is called *right ascension*. A line that extends over the sky from the celestial north pole through your zenith to the celestial south pole is your *meridian*. It crosses the celestial equator perpendicularly and at any one moment marks a line of constant right ascension. We measure right ascension along the celestial equator. Since the sky seems to turn overhead once every 24 hours, we measure right ascension in *hours*. The celestial equator is divided into 24 hours of right ascension; since the sky turns a full circle of 360°

TABLE 23. ANGULAR UNITS FOR COORDINATES IN THE SKY

RIGHT ASCENSION	UNITS OF ARC (DEGREES, MINUTES, OR SECONDS)
24^h	$360°$
1^h	$15°$
4^m	$1°$
1^m	$15'$
4^s	$1'$
1^s	$15''$

each 24 hours, each hour of right ascension equals 15°, and each minute of right ascension equals 15' (15 minutes of arc). Each *second* of right ascension is equal to ⅟₆₀ of a minute and so is equal to 15" (15 seconds of arc). When we speak of "minutes" or "seconds," we must be clear as to whether we mean units of arc or of time.

On the Atlas Charts in chapter 7, right ascension is marked on the horizontal axis. As we look up when facing north, the sky seems to rotate counterclockwise around the north celestial pole, which is marked by the North Star, Polaris. Above Polaris in the sky, we see stars rising in the east (to our right), moving overhead, and setting in the west (to our left). But some stars are *circumpolar,* that is, they always stay above the horizon. They circle under Polaris from our west (left) to our east (right). When we look at them underneath Polaris, we have a view similar to that of the Atlas Charts in chapter 7, on which north is at the top. The hours of right ascension passing overhead increase as time goes on (1^h, 2^h, 3^h, etc.), so right ascension increases from right to left on the Atlas Charts.

The vertical axis on the charts is *declination,* the number of degrees north (+) or south (–) of the celestial equator. Thus the celestial equator has declination 0°, the north celestial pole has declination +90°, and the south celestial pole has declination –90°. The regions of right ascension and declination shown on the Atlas Charts are shown in appendix 13 on the drawings on pp. 224–227.

In order to fix the scale of right ascension, we must arbitrarily assume some particular line to be zero. We choose this zero point, 0^h of right ascension, to be the *vernal equinox.* The equinoxes are the two points where the celestial equator crosses the *ecliptic,* the path the sun follows across the sky in the course of the year. The vernal equinox is the one of these points that the sun crosses on its way north each year; the other is the *autumnal equinox.*

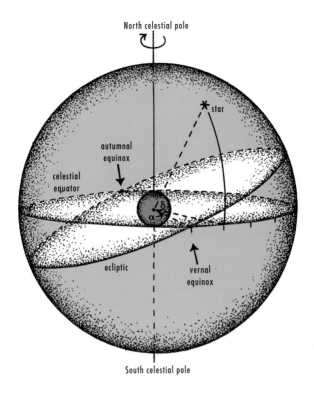

North celestial pole

star

autumnal
equinox

celestial
equator

δ
α

ecliptic

vernal
equinox

South celestial pole

Fig. 15-1 *Right ascension, noted by* α *(alpha), is measured in hours east-ward around the celestial equator from the vernal equinox. Declination, noted by* δ *(delta), is measured in degrees north (+) or south (−) of the celestial equator.*

Although day and night are theoretically equal in length on the days of the equinoxes, that would be true only if the sun were a point, not a disk, and if the earth's atmosphere did not bend sunlight. However, the top of the sun actually rises a few minutes before the center of the sun's disk—the point used in calculations. Also, the earth's atmosphere bends sunlight, so we can see the sun for several minutes before the time sunrise would occur and after the time sunset would occur if the earth had no atmosphere.

To find a star's celestial coordinates (fig. 15-1), we measure the number of hours around the celestial equator to its hour circle, and the number of degrees north or south of the celestial equator to its declination. Right ascension is often abbreviated either *r.a.* or with the Greek letter α (alpha). Declination is often abbreviated either *dec.* or with the Greek letter δ (delta). A star's right ascension is equal to the length of time that elapses after the vernal equinox crosses your meridian until the star crosses your meridian. The interval is measured in *sidereal time;* one rotation of the sky takes 24 hours of sidereal time.

The stars are essentially fixed in the sky, and so their right ascension and declination do not change measurably over short periods of time. The sun, moon, and planets, though, wander through the sky with respect to the stars; their right ascension and declination change during the course of a year.

PRECESSION

The axis on which the earth spins is not perpendicular to the plane of the earth's orbit. It is, rather, tipped by 23½°. The celestial coordinate system is complicated somewhat by the fact that the earth's axis wobbles as the earth spins, just as a spinning top wobbles. For the earth, gravity from the sun and moon cause the wobbling, the same way that the earth's gravity makes a top wobble. The wobbling is called *precession.*

The earth's axis actually traces out a huge curve in the sky over a 26,000-year period. Polaris is only temporarily the North Star; in about 13,000 years, Vega will be within a few degrees of the north pole. As a result of the precession of the earth's axis, the celestial coordinates of stars "precess" slightly with time; that is, they change their values from year to year. The vernal equinox moves slowly westward along the ecliptic at about 50" per year. The effect is not large, but must be taken into account to plot the positions of astronomical objects accurately.

Stellar positions are usually given in one of a few "standard epochs," such as the year 2000, and the standard epoch may not change until 2050. Dates within a year are given as decimals, such as 2004.1 for February 5, 2004, or 2010.0 for the beginning of the year 2010. Changing from the epoch in which positions are given to the current date is now easily done with a pocket calculator; the computers that operate many telescopes do it automatically. Useful formulae for precession are

new r.a. = r.a. + [3.074 sec. + 1.336 sin (r.a) tan (dec.)] × N,
and new dec. = dec. + 20.04" cos (r.a.) × N,

where N is the number of years since a standard epoch. In this Field Guide, the year 2000.0 is the standard epoch. In any case, casual sky observers do not need to worry about calculating precession.

Since the constellation boundaries fixed by the International Astronomical Union in 1930 were defined to lie along right ascension and declination lines for epoch 1875.0, the constellation boundaries no longer coincide with round numbers in epoch 2000.0 coordinates, as is clear from the Atlas Charts in chapter 7.

A smaller effect than precession, called *nutation,* is a "nodding" of the earth's axis caused chiefly by changes in the location of the moon's orbit. The coordinates thus change in a small ellipse, with axes of only 18.5" and 13.7", over a 19-year period. No amateur telescope would have to be set to that degree of accuracy.

Another small effect, the *aberration of starlight,* is a shift of up to 20.5" resulting from the fact that the earth is moving through space. Just as you have to tilt your umbrella slightly forward to keep dry when moving rapidly through a rainstorm, since your motion makes the raindrops appear to slant toward you, aberration makes astronomers tilt their telescopes slightly forward as the earth orbits the sun.

Positions in most star catalogs, including the tables in this guide, take precession into account, but not nutation or aberration.

TIME BY THE STARS

The stars return to the same point overhead where they were on the preceding night, but in that time, the earth has gone part (about $\frac{1}{365}$) of the way around the sun. So the earth has to rotate a little farther for the sun to return to the same place in our sky (one *solar day*) compared to the length of time it takes the stars to return to the same place in our sky (one *sidereal day*). One day divided by 365 is about 4 minutes, so the solar day is longer than the sidereal day by about 4 minutes, actually $3^m 56^s$.

At about the vernal equinox each year, sidereal midnight (the beginning of a sidereal day) and solar "noon" coincide: 0 hours sidereal time = 12^h solar time. Then they drift apart by 4 minutes per day for another year. At the autumnal equinox, 12 hours sidereal time = 12^h solar time. The conversions between sidereal time and solar time are tabulated in appendix 12.

Your sidereal time is the number of hours since the vernal equinox has crossed your meridian. Your sidereal time is also equal to the right ascension of stars now crossing your meridian. Thus if you know your sidereal time, you know which stars are most favorably placed for observing, i.e., those within a few hours of the current sidereal time, if they are in the right range of declination.

The Big Dipper and the North Star actually make a convenient sidereal clock in the sky. A line from Polaris through the Pointers sweeps counterclockwise around the sky once a day. When that line is pointed straight upward, the sidereal time is 11^h, since the right ascension of the Pointers is almost exactly 11^h. When the line from Polaris through the Pointers goes due left, toward due west on the horizon, the sidereal time is 6 hours later, or 17^h. When the Pointers are below Polaris so the line from Polaris through the Pointers goes straight downward, the sidereal time is 23^h.

CALENDARS AND TIME BY THE SUN

One solar day for any observer is the length of time that the sun takes to return to our meridian. It takes about 365¼ of those solar days to make a year. Instead of changing our clocks by ¼ day each year, we have three 365-day years followed by one 366-day *leap year,* a scheme worked out by the astronomer Sosigenes for Julius Caesar in 46 B.C.E.

But a year is 365.2422 days long, which is not exactly 365¼ days. In 1582, the time had slipped substantially from the proper season because of this difference; the Gregorian calendar skipped 10 days to compensate and made some new rules about leap year: the "century years" (1900, 2000, etc.) would not be leap years unless they are evenly divisible by 400. That is, 1800 and 1900 were not leap years, but 2000 was a leap year.

Actual *apparent solar time* does not advance at a constant rate through the year, because the earth travels at varying speeds as it traverses its elliptical orbit. As a result, it is more convenient for us to keep *mean solar time,* the average rate of solar time. Watches keep time at the same rate as mean solar time.

Your *local apparent solar time* is when the sun crosses your meridian. But your neighbors a few miles to the east or west have different meridians than yours. To compensate for these slight differences, we use a system of *standard time,* in which the solar time is the same in a band of earth longitudes 15° wide. The meridian of Greenwich is centered in one of these bands. The actual boundaries of the time zones often bend to follow political divisions between states, provinces, or countries.

The local apparent solar time minus the local mean solar time is called the *equation of time.* The difference can be as large as 16 minutes; we see its effect in figure 15-2.

The equation of time shows how far the real sun is ahead or behind the mean sun. As this value changes through the year, the sun also goes higher and lower in the sky. Thus the position of the sun at a given standard (mean) time each day appears to follow a figure-8 called the *analemma* (figs. 15-2 and 15-3).

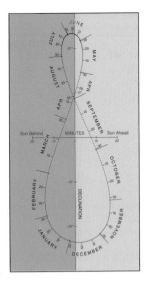

Fig. 15-2 (left) *The analemma, the figure-8 that shows how far mean solar time is ahead of or behind true solar time for a given day of the year.* (Wil Tirion, after Sky & Telescope)

Fig. 15-3 (right) *The analemma in a single multiple-exposure photo, with an exposure of the sun taken at 8:30 A.M. EST every two weeks. The summer solstice is the highest point and the winter solstice is the lowest point; the lens was left open from dawn until shortly before the time for the standard exposure on those days and on the day of the crossover. The exposure of the foreground building was added on still another day, without the dense filter that normally allowed only the sun to be visible.* (Dennis di Cicco)

Astronomers keep time in *Universal Time* (U.T.), basically the time for the meridian of Greenwich, England. (The name "Greenwich mean time," G.M.T., is no longer commonly used in astronomy.) Universal Time is set by the rotation of the earth, which is slightly irregular. Predictions are based on steady time unaffected by the earth's rotation, which is kept by atomic clocks. The average of predictions of steady time calculated by time services around the world is released as Coordinated Universal Time (U.T.C.). U.T.C. is kept aligned with U.T. by the occasional addition of leap seconds to UTC at the end of December or June. These additions are publicly announced. The exact difference between the U.T. and U.T.C. can be calculated only after the fact, but is kept to less than 1 second.

Time gets earlier as you move westward on earth. At a given instant, you subtract the following number of hours to go from U.T. to time zones in the U.S. and Canada.

TABLE 24. TIME ZONES

TO CONVERT FROM UNIVERSAL TIME (U.T.) TO	TO GET STANDARD TIME, SUBTRACT:	TO GET DAYLIGHT-SAVING TIME, SUBTRACT:
Atlantic Time Zone	4 hours	3 hours
Eastern Time Zone	5 hours	4 hours
Central Time Zone	6 hours	5 hours
Mountain Time Zone	7 hours	6 hours
Pacific Time Zone	8 hours	7 hours
Alaska Time Zone	9 hours	8 hours
Hawaii Time Zone	1 0 hours	—

Most states employ daylight-saving time, in which you add one hour to the standard time for the summer months; Arizona, Hawaii, and parts of Indiana have standard time all year. To remember to add one hour in the spring and subtract it again in the fall, think of "spring ahead, fall back." That is, move your clocks ahead one hour in the spring and move them back one hour in the fall.

Halfway around the world from Greenwich, in the middle of the Pacific, the date changes by one at the *International Date Line,* so that the same hours on the same day don't keep endlessly circling the world. The time gets earlier by one hour each time you cross a time zone as you go from east to west; as you cross the International Date Line from east to west, one day is added.

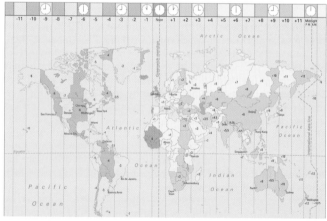

Fig. 1 5-4 Time zones, not including daylight-saving time. (Map Creation Ltd.)

TELESCOPES AND BINOCULARS

The human eye collects light very efficiently, but all the light it collects must pass through its tiny pupil, the black spot in the center of the eye. A telescope or pair of binoculars gathers much more light in a given time and so allows you to see fainter objects. Further, the larger the lens or mirror the telescope has, the finer the detail that you can see. Telescopes and binoculars can also magnify, but this ability is usually less important than their light-gathering power or resolution. It is not useful to magnify a faint or blurry image.

INOCULARS

Binoculars have the advantage of being less expensive, lighter, and more portable than telescopes. A good pair of binoculars is better for observing the sky than Galileo's telescopes, which Galileo used to discover the moons of Jupiter, the phases of Venus, and other important things. So it is often the eye and brain of the observer that is as important as the quality of the observing instrument.

A common type of binoculars used for astronomy is called 7 × 50 (fig. 16-1). The 7 means that it magnifies 7 times, and the 50 means that its front lenses are 50 mm across. Since the pupil of a dark-adapted eye can be 8 mm across, a 50-mm lens funneling light into an 8-mm pupil means that $(50/8)^2$ = about 40 times more light enters the eye. This factor of 40 corresponds to seeing stars 4 magnitudes fainter. It also means that faint nebulosity or star clouds become visible.

Not all 7 × 50 binoculars fill the eye's full pupil and don't over-fill it, so some binoculars are better for astronomy than others. One should test and/or ask before purchasing them. Other common sizes, like 8 × 35, may provide slightly more magnification

Fig. 16-1 *Binoculars of different sizes. The 7 × 50 binoculars are noticeably larger than the 8 × 35 binoculars; each 50-mm lens collects twice as much light as each 35-mm lens. (Jay M. Pasachoff)*

but collect less light and so do not allow objects to be seen as faint as those seen with 7 × 50s.

Handholding binoculars often does not allow objects to be seen, because the hand shakes. Propping the binoculars up at the very least, and getting a binocular holder for a tripod, greatly aids in astronomy. Binoculars that have moving elements in them to keep the image steady are available, though they are more expensive than other binoculars.

TELESCOPES

Telescopes of adequate quality are not inexpensive. If you do not want to pay more than $200 for a telescope, a good pair of binoculars is the better piece of equipment to purchase.

If you do plan to buy a telescope, pay special attention to the sturdiness of the mount, for a telescope on a flimsy mount is useless. A mount not only supports the telescope but also may track astronomical objects across the sky as the earth rotates. With a mount that isn't tracking, an object drifts out of sight from the center of a 1° field of view in only two minutes. The effect of the earth's motion shows up in photographs of stars taken without a tracking mount. When the shutter is left open for several minutes to capture the images of fainter stars, the stars' images show up as curved trails rather than as points of light.

Many telescopes come on *equatorial mounts,* in which one axis —the polar axis—points up at an angle so that it is directed at a celestial pole. Another, perpendicular, axis allows the telescope to point anywhere in the sky. A single motor on the polar axis allows the telescope to track stars, planets, or galaxies so they won't drift out of view. When this type of telescope is portable, it must be aligned with an axis toward the celestial pole each time it is set up.

The alternative—an *altazimuth mount*—has one axis that points straight up, around which rotation around the compass (azimuth) takes place, and another axis that points from side to side, around which up-and-down rotation (altitude) takes place. Astronomical objects move across the sky at a slant, so varying motions on both axes are necessary for an altazimuth mount.

The wonders of modern electronics have allowed great simplifications in setting up telescopes. With suitable electronics, you need merely point your telescope at two bright, identifiable stars and punch in which stars they are. From that time on, the telescope can point automatically at objects you request by name or catalog number.

pes of Telescopes

A *reflecting telescope* uses a mirror to collect and focus light, while a *refracting telescope* uses a lens. The larger the telescope, the better the resolution it can provide, so the more magnification images can tolerate. Large telescopes allow you to see fainter objects and more detail, but you can enjoy studying the planets and other objects with a small telescope. After all, Galileo discovered the phases of Venus and the moons of Jupiter with a refracting telescope only about 1 ¼ inches (4 cm) across.

The focal ratio of a telescope or camera, its f-number, is the focal length of the main lens or mirror divided by its diameter. Thus a telescope whose main mirror has a focal length of 80 inches and is 8 inches across is $f/10$ (or, in the metric system, 200 cm focal length and 20 cm across, again making $f/10$). The magnification of a telescope, a concept useful only for extended objects like planets and nebulae rather than for stars, can be calculated as the focal length of the telescope divided by the focal length of the eyepiece. You can increase the magnification of a small telescope by using an eyepiece with a shorter focal length. For example, the magnification of a telescope with a focal length of 2 m used with an eyepiece of 20 mm is calculated by converting both lengths to a common unit, such as mm, and calculating 2,000 mm divided by 20 mm = 100, or a magnification of 100 times. With a 10-mm-focal-length eyepiece, the magnification is 2,000/10 = 200 times. The quality of a telescope or of the earth's atmosphere often does not allow useful magnifications beyond about 200 times, and the claims in advertisements for magnifications much higher are often not meaningful and, indeed, make professionals suspicious of the quality of the product. A rule of thumb is x60 of magnification for each inch of aperture; thus a 2-inch (50-mm) telescope can usefully magnify 120 times; beyond that, you are merely making a blurry image larger, and it often looks worse. A 6-inch (15 cm)

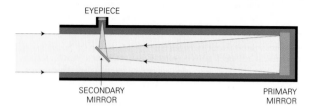

EYEPIECE

SECONDARY
MIRROR

PRIMARY
MIRROR

Fig. 16-2 *A reflecting telescope of the Newtonian type. Isaac Newton in-vented the method of using a small diagonal flat mirror to reflect the fo-cused light out to the side, where it can be viewed without blocking the incoming light. (Telescope drawings by Wil Tirion)*

telescope could be used to magnify up to about 360 times.

The traditional Newtonian telescope is still very popular and typically uses a mirror 4 to 8 inches across (10 to 20 cm) and a tube 48 inches (120 cm) long, although it is available in other sizes as well. Another type of telescope that is very popular with amateurs is neither a pure reflector nor a pure refractor; in these *compound telescopes,* the light passes through a lens and is then focused by mirrors inside the telescope's tube. Since the light bounces back and forth between the mirrors, the tube can be short and therefore more portable. Most telescopes of this type are Schmidt-Cassegrains: Schmidt telescopes have wide fields; Cassegrain telescopes are those in which the image is directed through a hole in the main mirror to your pupil or to a camera. Most of these telescopes come on equatorial mounts. The main competitors are Meade Instruments Corp. and Celestron International.

A unique kind of small, portable, inexpensive, wide-field (3°) telescope is marked as an *Astroscan 2001*. Its round base allows it to be pointed anywhere in the sky. It comes in only one small size, however.

Fig. 16-3 *A refracting telescope, in which the light is focused by a lens.*

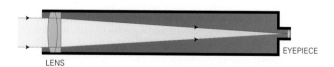

LENS

EYEPIECE

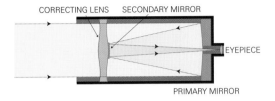

CORRECTING LENS   SECONDARY MIRROR

EYEPIECE

PRIMARY MIRROR

Fig. 16-4 A compound telescope of the Schmidt-Cassegrain type can show a wide field in good focus because of the combination of lens and mirrors used.

Dobsonian telescopes use thin mirrors, wooden cases with plastic bearings, and altazimuth mounts to obtain large apertures (mirror sizes) relatively inexpensively. These reflectors are designed for visual observation rather than for photography, since they do not track the stars.

An eyepiece or two comes with the telescope, but you would probably want more. It is helpful if your telescope takes eyepieces in the standard 1.25" size, and not in the smaller size typical of the cheapest telescopes.

Anyone evaluating telescopes should look at advertisements in recent issues of *Sky & Telescope* and *Astronomy* magazines to see what types and models are currently available. An article called "Selecting Your First Telescope" is available from the Astronomical Society of the Pacific, 390 Ashton Ave., San Francisco, CA

Fig. 16-5 (left) *A Newtonian reflecting telescope.* (Jay M. Pasachoff)
Fig. 16-6 (right) *A refracting telescope.* (Jay M. Pasachoff)

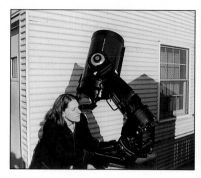
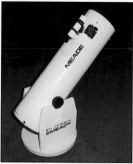

Fig. 16-7 An 8-inch (20-cm) Schmidt-Cassegrain telescope. (Jay M. Pasachoff)

Fig. 16-8 A Dobsonian telescope, with its simple construction and altazimuth mount. (Jay M. Pasachoff)

94112. The ASP Web site address is www.aspsky.org. Telescope information also appears on the *Sky & Telescope* Web site www.skypub.com and on the *Astronomy* magazine Web site www.kalmbach.com. I maintain current lists of manufacturers on this book's Web site. The Astronomical League (www.astroleague.org) is a federation of amateur astronomy societies.

ASTROPHOTOGRAPHY

A telescope is equivalent to a giant camera lens. All one has to do is hold film properly in place at the focus. The normal way to do that is to attach a camera, with its own lens removed, to the back of the telescope. A telescope with a 1000-mm focal length is then a 1000-mm telephoto. Accessories to make direct connections between telescopes and cameras are common. Point-and-shoot cameras, however, do not have removable lenses, and cannot be used in this way.

To obtain high magnifications, people sometimes use "eyepiece projection," in which the camera, with its lens in place, is put beyond the focal point. I leave further description of this technique to more technical books.

Films for astrophotography keep getting better. They get faster and faster, which means that images can be taken in less time. But faster films usually have coarser grain, which makes grainy images. Deciding on the appropriate tradeoff between film speed and grain remains the choice of each individual astrophotographer.

More and more people are using fast print films, with speeds of ISO 800 or ISO 1600, to photograph the sky. The characteristics of these films, such as their sensitivity to light of different colors, is changed by the manufacturers from time to time, so you should keep current with discussions in astronomy magazines if you want seriously to photograph faint objects. Further, films are optimized for the durations of exposures used for most outdoor photos, which are often $\frac{1}{100}$ second or faster. For long exposures, of seconds or minutes, the film does not gain in sensitivity at the same rate as the exposure is lengthened, and the rates can be different for different colors. So films that are seemingly similar for snapshots may be very different for astrophotography.

CD CAMERAS AND VIDEO CAMERAS

Ordinary video cameras have become so sensitive that they can be used for astrophotography. Most video cameras do not have removable lenses and must be affixed with special brackets to the backs of telescopes. For some kinds of astrophotography, such as imaging the moon or an eclipse or a meteor shower, the camera's own lens is all you need. Many video cameras include excellent zoom lenses that can provide a lot of magnification for viewing eclipses and the moon, for example.

The imaging device in many ordinary video cameras is a kind of light-sensitive chip, like a computer chip, known as a charge-coupled device or CCD. "CCD" has become such a common term among astronomers, both amateurs and professionals, that I wish a more user-friendly term, like "cilm" (for CCD used as film) had gained currency.

For most amateur astrophotography, special CCD cameras capable of storing images for many seconds are used. These are available from a variety of suppliers, and are attached to telescopes without separate lenses. The better CCD cameras are cooled to substantially below zero to make them more sensitive. Many of the photographs in this book were taken by amateur astronomers using CCD cameras.

Adding and manipulating astronomical images on computers has become a popular pastime. Adobe Photoshop has been the most used program for the purpose, though there are others. For astronomical purposes, people often take individual monochromatic images in individual colors and then assemble them on the computer to make color displays.

The following manufacturers make telescopes, mounts, filters, and other accessories that are popular with amateurs. This list is far from complete, and you can look at the latest sets of ads in *Sky & Telescope* (www.skypub.com) or *Astronomy* (www. astronomy.com) magazines.

CELESTRON INTERNATIONAL, 2835 Columbia Street, Torrance, CA 90503. (310) 328-9560, fax: (310) 212-5835; www.celestron. com. A wide variety of telescopes and binoculars.

DAYSTAR FILTER CORP., P.O. Box 5110, Diamond Bar, CA 91765-5110. (909) 591-4673. Solar hydrogen-alpha filters.

EDMUND SCIENTIFIC, 101 East Gloucester Pike, Barrington, NJ 08007-1380. (800) 728-6999, (609) 573-6250; e-mail: scientifics@edsci.com. Newtonian telescopes, refractors, and the relatively inexpensive ($329 as of this writing) wide-field *Astroscan 2001* telescope, plus miscellaneous equipment.

LUMICON ACCESSORIES, 2111 Research Drive, Livermore, CA 94550. (800) 767-9576, (925) 447-9570, fax: (925) 447-9589; www.lumicon.com. Accessories, as well as binoculars and telescopes.

MEADE INSTRUMENTS, 6001 Oak Canyon, Irvine, CA 92620. (949) 451-1450; fax: (949) 451-1460; www.meade.com. A wide variety of telescopes and binoculars.

ORION, P.O. Box 1715, Santa Cruz, CA 95061. (800) 447-1001. www.oriontel.com. A variety of binoculars, telescopes, and accessories.

QUESTAR CORP., 6204 Ingham Road, New Hope, PA 18938. (215) 862 5277, (800) 247-9607; fax: (215) 862-0512; e-mail: questar@erols.com; www.questar-corp.com. Expensive but high-quality 3.5-in. compact, Maksutov-type folded telescopes.

SBIG (Santa Barbara Instrument Group), P.O. Box 50437, Santa Barbara, CA 93150. (805) 969-1851; fax: (805) 969-4069; e-mail: sbig@sbig.com; www.sbig.com. Relatively inexpensive CCD cameras.

TELE VUE, 100 Route 59, Suffern, NY 10901. (914) 357-9522; www.televue.com. Eyepieces and refractors.

THOUSAND OAKS OPTICAL, Box 4813, Thousand Oaks, CA 91359. (800) 996-9111, (805) 491-3642; thousandoaksoptical.com; e-mail: info@thousandoaksoptical.com. Solar filters.

TUTHILL, 11 Tanglewood Lane, P.O. Box 1086, Mountainside, NJ 07092. (800) 223-6999; www.tuthillscope.com. Tested telescopes and various accessories.

APPENDIXES
GLOSSARY
BIBLIOGRAPHY
INDEX

ABBRE-VIATION	LATIN NAME	PRONOUN-CIATION	GENITIVE	TRANSLATION
And	Andromeda	an-drom'ə-də	Andromedae	Andromeda
Ant	Antlia	ant'le-ə	Antliae	Pump
Aps	Apus	a'pəs	Apodis	Bird of Paradise, Bee
Aqr	Aquarius	ə-kwar'e-əs	Aquarii	Water Bearer
Aql	Aquila	ak'wə'lə	Aquilae	Eagle
Ara	Ara	a'rə	Arae	Altar
Ari	Aries	ar'ez, ar'e-ez'	Arietis	Ram
Aur	Auriga	o-ri'gə	Aurigae	Charioteer
Boo	Boötes	bo-o'tez	Boötis	Herdsman
Cae	Caelum	se'ləm	Caeli	Chisel
Cam	Camelopardalis	kə-mel'o-pär-də-lis	Camelopardalis	Giraffe
Cnc	Cancer	kan'sər	Cancri	Crab
CVn	Canes Venatici	ka'nez vi-nat'ə-si	Canum Venaticorum	Hunting Dogs
CMa	Canis Major	ka'nis ma'jər	Canis Majoris	Big Dog
CMi	Canis Minor	ka'nis mi'nər	Canis Minoris	Little Dog
Cap	Capricornus	kap'ri-kor'nəs	Capricorni	Goat
Car	Carina	kə'ri'nə	Carinae	Ship's Keel
Cas	Cassiopeia	kas'e-ə-pe'ə	Cassiopeiae	Cassiopeia
Cen	Centaurus	sen-tor'əs	Centauri	Centaur
Cep	Cepheus	se'fyoos'	Cephei	Cepheus
Cet	Cetus	se'təs	Ceti	Whale
Cha	Chamaeleon	kə-mel'yən	Chamaeleonis	Chameleon
Cir	Circinus	sur'sə-nəs	Circini	Compass
Col	Columba	kə-lum'bə	Columbae	Dove
Com	Coma Berenices	ko'mə ber'ə-ni'sez	Comae Berenices	Berenice's Hair
CrA	Corona Australis	kə-ro'nə ostra'lis	Coronae Australis	Southern Crown
CrB	Corona Borealis	kə-ro'nə bor'e-al'is	Coronae Borealis	Northern Crown
Crv	Corvus	kor'vəs	Corvi	Crow
Crt	Crater	kra'tər	Crateris	Cup
Cru	Crux	kruks	Crucis	Southern Cross
Cyg	Cygnus	sig'nəs	Cygni	Swan
Del	Delphinus	del-fi'nəs	Delphini	Dolphin
Dor	Dorado	də-rä'do	Doradus	Swordfish
Dra	Draco	dra'ko	Draconis	Dragon
Equ	Equuleus	i-kwooæle-əs	Equulei	Little Horse
Eri	Eridanus	i-rid'ən-əs	Eridani	River Eridanu
For	Fornax	for'naks'	Fornacis	Furnace
Gem	Gemini	jem'ə-ni'	Geminorum	Twins
Gru	Grus	grus	Gruis	Crane
Her	Hercules	hur'kyə-lez'	Herculis	Hercules
Hor	Horologium	hor'ə-lo'je-əm	Horologii	Clock
Hya	Hydra	hi'drə	Hydrae	Water Snake

BRE-ATION	LATIN NAME	PRONOUN-CIATION	GENITIVE	TRANSLATION
	Hydrus	hi'drəs	Hydri	Water Snake
	Indus	in'dəs	Indi	Indian
	Lacerta	lə-sur'tə	Lacertae	Lizard
	Leo	le'o	Leonis	Lion
li	Leo Minor	le'o mi'nər	Leonis Minoris	Little Lion
	Lepus	le'pəs	Leporis	Hare
	Libra	li'brə	Librae	Scales
p	Lupus	loo'pəs	Lupi	Wolf
	Lynx	lingks	Lyncis	Lynx
	Lyra	li'rə	Lyrae	Harp
n	Mensa	men'sə	Mensae	Table
c	Microsco-pium	mi'krə-sko'pe-əm	Microscopii	Microscope
n	Monoceros	mə-nos'ər-əs	Monocerotis	Unicorn
s	Musca	mus'kə	Muscae	Fly
r	Norma	nor'mə	Normae	Level
t	Octans	ok'tanz'	Octantis	Octant
h	Ophiuchus	of'e-yoo'kəs	Ophiuchi	Ophiuchus
	Orion	o'ri'ən	Orionis	Orion
v	Pavo	pä'vo	Pavonis	Peacock
g	Pegasus	peg'ə-səs	Pegasi	Pegasus
	Perseus	pur'se-əs	Persei	Perseus
e	Phoenix	fe'niks	Phoenicis	Phoenix
	Pictor	pik'tər	Pictoris	Easel
c	Pisces	pi'sez	Piscium	Fish
A	Piscis Austrinus	pi'sis os-tri'nəs	Piscis Austrini	Southern Fish
p	Puppis	pup'is	puppis	Ship's Stern
t	Pyxis	pik'sis	Pyxidis	Ship's Compass
	Reticulum	ri-tik'yə-ləm	Reticuli	Net
e	Sagitta	sə-jit'ə	Sagittae	Arrow
r	Sagittarius	saj'ə-tar'e-əs	Sagittarii	Archer
o	Scorpius	skor'pe-əs	Scorpii	Scorpion
l	Sculptor	skulp'tər	Sculptoris	Sculptor
t	Scutum	skyoo'təm	Scuti	Shield
	Serpens	sur'pənz	Serpentis	Serpent
x	Sextans	seks'təns	Sextantis	Sextant
u	Taurus	tor'əs	Tauri	Bull
	Telescopium	tel'ə-sko'pe-əm	Telescopii	Telescope
	Triangulum	tri-ang'gyə-ləm	Trianguli	Triangle
A	Triangulum Australe	tri-ang'gyə-ləm o-stra'le	Trianguli Australis	Southern Triangle
c	Tucana	too-ka'nə	Tucanae	Toucan
la	Ursa Major	ur'sə ma'jər	Ursae Majoris	Big Bear
li	Ursa Minor	ur'sə mi'nər	Ursae Minoris	Little Bear
	Vela	ve'lə	Velorum	Ship's Sails
	Virgo	vur'go	Virginis	Virgin
	Volans	vo'länz	Volantis	Flying Fish
	Vulpecula	vul-pek'yə-lə	Vulpeculae	Little Fox

513

This list includes 314 stars, all those brighter than apparent magnitude 3.55. For each star, the first column gives the star's Greek-letter designation from Bayer's atlas (1603) and the abbreviation of its constellation name. For visual double stars, the letter A indicates that the data given are for the brighter component; the data for the fainter, B, component and the separation of the components appear in the Remarks column. AB indicates that a merged brightness is given.

Star positions were calculated for 1999.5, giving values close to epoch 2000.0 positions. V is the "visual magnitude," magnitude measured through a standard yellowish filter that simulates the sensitivity of the human eye. (For variable stars, a small letter v is appended, and the range is given in the remarks.) B-V is a "color index," the magnitude seen through a standard blue filter (B) filter minus the visual (V) magnitude; the color index indicates the star's temperature, though some stars are reddened by the passage of their light through interstellar dust. The star's spectral type is followed with the Roman numeral indicating the star's luminosity class: V indicates normal dwarf stars like the sun; IV indicates subgiants; III, giants; II, bright giants; and I, supergiants. The absolute magnitude is calculated from the distances measured with the Hipparcos spacecraft, and those distances follow, translated from angular measurement on the sky to light years. A few exceptions, marked with a colon, have absolute magnitudes calculated from the spectral classification. The remarks include notations of eclipsing binaries (ecl.).

Names for navigational stars are in bold.

STAR NAME	RA 2000.0	DEC 2000.0	MAGNITUDES V	MAGNITUDES B-V	SPECTRAL TYPE	M_V	D (LY)	REMARKS (ECLIPSING VARIABLE, SEPARATIONS, AND MAGNITUDES OF DOUBLES; NAMES)
Sun	...	...	−26.75	0.63	G2 V	+4.82	8 light minutes	**Sun**
α And	00 08.4	+29 05	2.1v	−0.11	B9 IV	−0.5	97	**Alpheratz**
β Cas	00 09.2	+59 09	2.3v	0.34	F2 III	1.2	54	var: 2.25−2.31, 0.10d **Caph**
γ Peg	00 13.3	+15 11	2.8v	−0.23	B2 IV	−2.4	333	var: 2.80−2.87, 0.15d **Algenib**
β Hyi	00 25.6	−77 16	2.80	0.62	G1 IV	3.4	24	
α Phe	00 26.3	−42 19	2.39	1.09	K0 III	−0.3	77	**Ankaa**

STAR NAME	RA 2000.0	DEC	MAGNITUDES V	B−V	SPECTRAL TYPE	Mv	D (LY)	REMARKS (ECLIPSING VARIABLE, SEPARATIONS, AND MAGNITUDES OF DOUBLES; NAMES)	
α Cas	00 40.5	+56 32	2.23	1.17	K0 III	-2.5	228		Schedar
β Cet	00 43.6	-18 00	2.04	1.02	K0 III	-1.0	96		Diphda
η Cas A	00 49.1	+57 49	3.44	0.57	G0 V	4.6	19	B: 7.51, K4 Ve, 12"	
γ Cas	00 56.7	+60 43	2.5v	-0.15	B0 IV	-5.0	613	var: 1.6-3.0; B: 8.8, 2"	
β Phe AB	01 06.0	-46 43	3.31	0.89	G8 III	0.3:	147	AB similar in light, spectrum, 1"	
η Cet	01 08.7	-10 11	3.45	1.16	K1.5 III	-0.1	118		
β And	01 09.7	+35 37	2.06	1.58	M0 III	-1.9	199		Mirach
δ Cas	01 25.8	+60 14	2.7v	0.13	A5 IV	0.2	99		Ruchbah
γ Phe	01 28.4	-43 19	3.4v	1.57	K7 III	-1.6	234	var: 3.39-3.49	
α Eri	01 37.6	-57 14	0.46	-0.16	B3 V	-2.9	144		**Achernar**
τ Cet	01 44.1	-15 57	3.50	0.72	G8 V	5.7	12		
α Tri	01 53.1	+29 35	3.41	0.49	F6 IV	1.8	64		Metallah
ε Cas	01 54.4	+63 40	3.38	-0.15	B3 IV	-2.5	442		Segin
β Ari	01 54.7	+20 48	2.64	0.13	A4 V	1.3	60		Sharatan
α Hyi	01 58.7	-61 35	2.86	0.28	F0 III–IV	1.1	71		
γ And A	02 04.0	+42 20	2.26	1.37	K3 II	-3.0	355	B: 5.4,B9V,10";C:6.2,A0V;BC: 1"	Almaak
α Ari	02 07.2	+23 28	2.00	1.15	K2 IIIa	0.5	66		Hamal
β Tri	02 09.5	+34 59	3.00	0.14	A5 IV	0.1	124		
ο Cet A	02 19.4	-3 00	2-10v	1.42	M5 1 0 III	3.0	418	LPV, 2-10; B: VZ Cet,9.5v,Bpe,1"	Mira
α UMi A	02 31.1	+89 15	2.0v	0.60	F5 8 I	-4.1	431	low amp Cep,4d; B: 8.2,F3 V,18"	**Polaris**
γ Cet AB	02 43.4	+3 14	3.47	0.09	A2 V	1.3	82	A: 3.57; B: 6.23, 3"	Kaalidhma
θ Eri A	02 58.2	-40 19	3.42	0.14	A5 V	-0.3	161	B: 4.35, A1 V, 8"	Acamar
α Cet	03 02.3	+4 05	2.53	1.64	M2 III	-1.7	220	composite spectrum	Menkar
γ Per	03 04.7	+53 30	2.93	0.70	G8 III + A2 V	-0.8	196		
ρ Per	03 05.2	+38 50	3.4v	1.65	M4 II	-1.3	325	semi-regular var: 3.3-4.0	

APPENDIX 2: THE BRIGHTEST STARS (CONTINUED)

STAR NAME	RA 2000.0	DEC	MAGNITUDES V	B-V	SPECTRAL TYPE	M_V	D (LY)	REMARKS (ECLIPSING VARIABLE, SEPARATIONS, AND MAGNITUDES OF DOUBLES; NAMES)
β Per	03 08.2	+40 57	2.1v	−0.05	B8 V + F	−0.5	93	ecl: 2.12−3.4−2.87d; composite — Algol
α Per	03 24.3	+49 51	1.79	0.48	F5 I	−4.9	592	in cluster — Mirphak
δ Per	03 42.9	+47 47	3.01	−0.13	B5 III	−3.1	528	
δ Eri	03 43.2	−9 46	3.54	0.92	K0 IV	3.4	29	
η Tau	03 47.5	+24 06	2.87	−0.09	B7 III	−1.6	212	in Pleiades — Alcyone
γ Hyi	03 47.4	−74 14	3.24	1.62	M2 III	−1.0	214	
ζ Per A	03 54.2	+31 53	2.85	0.12	B1 I	−6.0	982	B: 9.16, B8 V, 13"
ε Per A	03 57.8	+40 01	2.89	−0.18	B0.5 IV	−3.4	538	B: 7.39, B9.5 V, 9"
γ Eri	03 58.0	−13 30	2.95	1.59	M1 III	−1.6	221	— Zaurak
λ Tau A	04 00.7	+12 29	3.5v	−0.12	B3 V	−2.3	370	ecl: 3.3−3.8, 3.95d; B: A4 IV
α Ret A	04 14.4	−62 29	3.35	0.91	G8 II−III	−0.5	163	
θ² Tau	04 28.6	+15 52	3.40	0.18	A7 III	0.2	149	in Hyades
ε Tau	04 28.7	+19 10	3.53	1.01	K0 III	0.1	155	in Hyades — Ain
α Dor AB	04 34.0	−55 03	3.27	−0.10	A0 V	0.2	176	A: 3.8; B: 4.3, B9 IV, 0.2"
α Tau A	04 35.9	+16 30	0.85v	1.54	K5 III	−0.8	65	var: 0.75−0.95 — Aldebaran
π³ Ori	04 49.9	+6 58	3.19	0.45	F6 V	3.7	26	— Hassaleh
ι Aur	04 57.0	+33 10	2.69	1.53	K3 II	−3.6	512	var?
ε Aur A	05 01.9	+43 50	3.04v	0.54	A9 I + B	−8.0	7824	ecl: 2.94−3.83, 9892d — Al Anz
ε Lep	05 05.4	−22 22	3.19	1.46	K4 III	−2.0	227	
η Aur	05 06.5	+41 14	3.17	−0.18	B3 V	−1.2	219	— Hoedus II
β Eri	05 07.8	−5 05	2.79	0.13	A3 IV	0.4	89	— Kursa
μ Lep	05 12.9	−16 12	3.3v	−0.11	B9 IV	−0.4	184	var: 2.97−3.36, 2d
β Ori A	05 14.5	−8 12	0.12	−0.03	B8 I	−6.6	773	B: 7.6, B5 V.9"; C: 7.6; BC: 0.1" — Rigel
α Aur AB	05 16.7	+46 00	0.08	0.80	G6:III + G2:III	−0.8	42	composite: A: 0.6; B: 1.1, 0.04" — Capella
η Ori AB	05 24.5	−2 24	3.4v	−0.17	B0.5 V	−.9	901	ecl: 3.14−3.35, 8d; A: 3.6; B: 5.0, 1.6"
γ Ori	05 25.2	+6 20	1.64	−0.22	B2 III	−2.8	243	— Bellatrix

STAR NAME	RA 2000.0	DEC	MAGNITUDES V	B–V	SPECTRAL TYPE	Mᵥ	D (LY)	REMARKS (ECLIPSING VARIABLE, SEPARATIONS, AND MAGNITUDES OF DOUBLES; NAMES)	
β Tau	05 26.4	+28 36	1.65	−0.13	B7 III	−1.3	131		Alnath
β Lep A	05 28.3	−20 46	2.84	0.82	G5 II	−0.7	159	B: 7·4, 2.6″	
δ Ori A	05 32.1	−0 18	2.23v	−0.22	O9.5 II	−5.4	916	ecl: 1.94–2.13·5·7d	Mintaka
α Lep	05 32.8	−17 50	2.58	0.21	F0 I	−5.5	1283		Arneb
β Dor	05 33.6	−62 30	3.8v	0.84	F7–G2 I	−4.2	1038	Cepheid: 3.43–4.06, 9.8d	
λ Ori A	05 35.2	+9 56	3.54	−0.18	O8 III	−4.6	1055	B: 5.61, B0 V, 4″	Meissa
ι Ori A	05 35.5	−5 55	2.77	−0.24	O9 III	−5.6	1325	B: 7·3, B7 IIIp(He weak), 1″	Nair al Saif
ε Ori	05 36.3	−1 12	1.70	−0.19	B0 I	−6.6	1342		Alnilam
ζ Tau	05 37.6	+21 08	3.0v	−0.19	B2 III	−2.8	417	var: 2.90–3.03; B: 5.0, 0.007″	
α Col A	05 39.6	−34 05	2.64	−0.12	B7 IV	−1.9	268		Phaet
ζ Ori A	05 40.8	−1 57	2.05	−0.21	O9.5 I	−5.5	817	B: 4.2, B0 III, 2.4″	Alnitak
ζ Lep	05 47.0	−14 50	3.55	0.10	A2 V	1.7	70		
κ Ori	05 47.8	−9 40	2.06	−0.17	B0.5 I	−5.0	815		Saiph
β Col	05 50.9	−35 46	3.12	1.16	K1.5 III	0.2	86		Wezn
α Ori	05 55.2	+7 24	0.5v	1.85	M2 I	−5.0	522	var: 0.4–1.3	**Betelgeuse**
β Aur	05 59.6	+44 57	1.90v	0.03	A1 IV	−0.2	82	ecl: 1.93–2.02, 4d	Menkalinan
θ Aur AB	05 59.7	+37 13	2.62	−0.08	A0 II	−1.0	173	B: 7.2, G2 V, 4″	
η Gem	06 14.8	+22 31	3.3v	1.60	M3 III	−1.8	349	var: 3.3–3.9; B: 8.8, 1.6″	Propus
ζ CMa	06 20.2	−30 03	3.02	−0.19	B2.5 V	−2.2	336		Phurud
β CMa	06 22.7	−17 57	2.0v	−0.23	B1 II–III	−4.0	499	var: 1.93–2.00, 0.25d	Murzim
μ Gem	06 22.9	+22 31	2.9v	1.64	M3 III	−1.5	232	var: 2.76–3.02	Tejat Posterior
α Car	06 23.9	−52 41	−0.72	0.15	A9 I	−5.4	313		**Canopus**
γ Gem	06 37.7	+16 25	1.93	0.00	A1 IV	−0.6	105		Alhena
ν Pup	06 37.7	−43 11	3.17	−0.11	B8 III	−2.4	423		
ε Gem	06 44.0	+25 08	2.98	1.40	G8 I	−5.0	903		Mebsuta
ξ Gem	06 45.3	+12 53	3.36	0.43	F5 IV	2.2	57		Alzirr

APPENDIX 2: THE BRIGHTEST STARS (CONTINUED)

STAR NAME	RA 2000.0	DEC	MAGNITUDES V	B–V	SPECTRAL TYPE	Mv	D (LY)	REMARKS (ECLIPSING VARIABLE, SEPARATIONS, AND MAGNITUDES OF DOUBLES; NAMES)
α CMa A	06 45.2	–16 43	–1.46	0.00	A1 V	1.5	9	B: 8.5, white dwarf; 50y, 10" — Sirius
α Pic	06 48.2	–61 57	3.27	0.21	A6 V	0.7	99	
τ Pup	06 49.9	–50 37	2.93	1.20	K1 III	–1.9	183	
ε CMa A	06 58.6	–28 58	1.50	–0.21	B2 II	–4.1	431	Adara
σ CMa	07 01.7	–27 56	3.5v	1.73	K7 I	–4.7	1216	var: 3.43–3.49
o² CMa	07 03.1	–23 50	3.02	–0.08	B3 I	–6.6	2567	
δ CMa	07 08.4	–26 24	1.84	0.68	F8 I	–7.2	1791	Wezen
L₂ Pup	07 13.5	–44 38	2.6v	1.56	M5 III	1.5	198	Long Period Var: 2.6–6.2 — HR2748
π Pup	07 17.1	–37 06	2.70	1.62	K3 I	–5.1	1094	
δ Gem AB	07 20.1	+21 59	3.53	0.34	F0 IV	2.2	59	B: 8.2, K3 V, 0.2" — Wasat
η CMa	07 24.1	–29 18	2.45	–0.08	B5 I	–7.5	3196	Aludra
β CMi	07 27.2	+8 18	2.90	–0.09	B8 V	–0.8	170	Gomeisa
σ Pup A	07 29.2	–43 18	3.25	1.51	K5 III	–1.6	184	B: 8.6, G5: V, 22"
α Gem A	07 34.6	+31 54	1.94	0.03	A2 V	0.6	52	AB: 3" separation — Castor
α Gem B	07 34.7	+31 54	2.92	0.04	A5 V	1.0	52	BA: 3" separation — Castor
α CMi A	07 39.4	+5 14	0.38	0.42	F5 IV–V	2.8	11	B: 10.3, 4" — Procyon
β Gem	07 45.4	+28 02	1.14	1.00	K0 III	1.1	34	Pollux
ξ Pup	07 49.4	–24 51	3.34	1.24	G6 I	–7.5	3260	
χ Car o 7 56,7		–52 58	3.47	–0.18	B3 IV	–2.0	387	
ζ Pup	08 03.5	–40 00	2.25	–0.26	O5 I	–6.1	1399	Naos
ρ Pup	08 07.6	–24 18	2.8v	0.43	F5 II	1.4	63	var: 2.68–2.78, 0.14d
γ² Vel	08 09.5	–47 20	1.8v	–0.22	WC8 + O9 I:	–5.8	840	var: 1.6–1.8, 154s — Suhail al Muhlif
β Cnc	08 16.5	+9 11	3.52	1.48	K4 III	–1.2	290	Altarf
ε Car	08 22.5	–59 31	1.86v	1.28	K3:III + B2:V	–4.8	632	ecl?: 3.1–3.4p, 785d — Avior
o UMa A	08 30.2	+60 43	3.4v	0.84	G5 III	–0.3	184	var: 3.3–3.8, 358d
δ Vel AB	08 44.6	–54 42	1.96	0.04	A1 V	0.0	80	B: 5.0, 2"

STAR NAME	RA	DEC 2000.0	MAGNITUDES V	B–V	SPECTRAL TYPE	M_V	D (LY)	REMARKS (ECLIPSING VARIABLE, SEPARATIONS, AND MAGNITUDES OF DOUBLES; NAMES)
ε Hya ABC	08 46.8	+6 25	3.38	0.68	G5:III + A:	0.0	135	composite A: 3.8; B: 4.7, 0.2"; C: 7.8,3"
ζ Hya	08 55.4	+5 57	3.11	1.00	G9 II III	−0.2	151	
ι UMa A	08 59.2	+48 03	3.14	−0.19	A7 IV	2.2	48	BC: 10.8, M1 V, 4" Talitha
λ Vel	09 07.9	−43 25	2.21v	1.66	K4 Ib II	−4.8	573	var: 2.14−2.22 Suhail
a Car	09 10.9	−58 57	3.44	0.19	B2 IV–V	−2.2	418	ecl: 3.2−3.6, 6.7d HR3659
β Car	09 13.2	−69 42	1.68	0.00	A1 III	−1.1	111	Miaplacidus
ι Car	09 17.0	−59 16	2.2v	−0.18	A7 I	−4.4	694	var: 2.2−2.5 Turais
α Lyn	09 21.0	+34 24	3.13	1.55	K7 III	−1.3	222	
κ Vel	09 22.1	−55 00	2.50	0.18	B2 IV–V	−3.9	539	
α Hya	09 27.6	−8 39	1.98	1.44	K3 II–III	−2.1	177	Alphard
N Vel	09 31.1	−57 01	3.13	1.55	K5 III	−1.3	238	HR3803
θ UMa	09 32.9	+51 42	3.17	0.46	F6 IV	2.6	44	
o Leo AB	09 41.2	+9 54	3.52	0.49	F5 II + A5	0.1	135	A: occ.bin. Subra
l Car	09 45.2	−62 31	3.8v	0.14	F9 G5 I	−5.8	1509	Cepheid var: 3.39−4.12, 35.5d HR3884
ε Leo	09 45.9	+23 46	2.98	0.80	G1 II	−1.6	251	Ras Elased Australis
υ Car AB	09 47.0	−65 04	3.01	0.28	A6 II	−2.5	326	B: 6.26, B7 III, 5"
φ Vel	09 56.8	−54 34	3.54	−0.08	B5 I	−5.5	1929	
η Leo	10 07.4	+16 46	3.52	−0.03	A0 I	−5.5	2131	B: 4.5, 0.1"
α Leo A	10 08.4	+11 58	1.35	−0.11	B7 V	−0.6	77	Regulus
ω Car	10 13.7	−70 02	3.32	−0.08	B8 III	−2.1	370	
ζ Leo	10 16.7	+23 25	3.44	0.31	F0 III	−1.1	260	Adhafera
λ UMa	10 17.1	+42 55	3.45	0.03	A1 IV	0.4	134	Tania Borealis
q Car	10 17.0	−61 20	3.4v	1.54	K3 II	−4.2	736	var: 3.36−3.42 HR4050
γ Leo A	10 20.0	+19 51	2.61	1.15	K1 III	−0.5	0.7 126	AB: 5" separation Algieba
γ Leo B	10 20.0	+19 51	3.47	1.10	G7 III	−1.9	126	BA: 5" separation
μ UMa	10 22.3	+41 30	3.05	1.59	M0 III	−1.5	249	Tania Australis

STAR NAME	RA 2000.0	DEC	MAGNITUDES V	B–V	SPECTRAL TYPE	M_V	D (LY)	REMARKS (ECLIPSING VARIABLE, SEPARATIONS, AND MAGNITUDES OF DOUBLES; NAMES)
ρ Car	10 31.9	–61 41	3.3v	–0.09	B4 V	–2.0	326	var: 3.27–3.37 HR4140
θ Car	10 42.9	–64 23	2.76	–0.22	B0.5 V	–3.1	439	
μ Vel AB	10 46.8	–49 25	2.69	0.90	G5 III + F8:V	–0.9	116	B: 6.4, 2"
ν Hya	10 49.6	–16 11	3.11	1.25	K2 III	–0.3	138	
β UMa	11 01.8	+56 23	2.37	–0.02	A1 IV–V	0.4	79	Merak
α UMa AB	11 03.8	+61 45	1.79	1.07	K0 III	–1.3	124	Dubhe A: 1.86, B: 4.8, A8 V, <1"
ψ UMa	11 09.7	+44 30	3.01	1.14	K1 III	–0.5	147	
δ Leo	11 14.2	+20 32	2.56	0.12	A4 IV	1.3	58	Zosma
θ Leo	11 14.2	+15 26	3.34	–0.01	A2 IV	–0.2	178	Chort
ν UMa	11 18.5	+33 06	3.48	1.40	K3 III	–3.2	421	Alula Borealis B: 9.5, 7"
ξ Hya	11 33.0	–31 51	3.54	0.94	G7 III	0.0	129	
λ Cen	11 35.8	–63 01	3.13	–0.04	B9.5 II	–2.5	410	
β Leo	11 49.1	+14 35	2.14	0.09	A3 V	1.9	36	Denebola
γ UMa	11 53.9	+53 42	2.44	0.00	A0 V	0.2	84	Phad
δ Cen	12 08.3	–50 43	2.6v	–0.12	B2 IV	–3.1	395	var: 2.51–2.65
ε Crv	12 10.2	–22 37	3.00	1.33	K2 III	–2.3	303	
δ Cru	12 15.2	–58 44	2.80v	–0.23	B2 IV	–2.6	364	var: 2.25–2.31 p. 3.7h
δ UMa	12 15.5	+57 02	3.31	0.08	A2 V	1.4	81	Megrez
γ Crv	12 15.9	–17 32	2.59	–0.11	B8 III	–0.9	165	Gienah Ghurab
α Cru A	12 26.6	–63 05	1.33	–0.24	B0.5 IV	–4.0	321	Acrux AB: 5"
α Cru B	12 26.6	–63 05	1.73	–0.26	B1 V	–3.6	321	BA: 5"
δ Crv A	12 29.9	–16 30	2.95	–0.05	B9.5 IV	–2.0	88	Algorab B: 8.26, K2 V, 24"
γ Cru	12 31.2	–57 06	1.63v	1.59	M3.5 III	–0.7	88	Gacrux var: 1.6–1.9
β Crv	12 34.4	–23 23	2.65	0.89	G5 II	–0.5	140	Kraz
α Mus	12 37.1	–69 08	2.69v	–0.20	B2 IV–V	–2.3	306	var: 2.17–2.24 p. 2h
γ Cen A	12 41.5	–48 57	2.87	–0.03	A1 IV	–0.1	130	AB: 5"

STAR NAME	RA	DEC 2000.0	MAGNITUDES V	B–V	SPECTRAL TYPE	Mv	D (LY)	REMARKS (ECLIPSING VARIABLE, SEPARATIONS, AND MAGNITUDES OF DOUBLES; NAMES)
γ Cen B	12 41.5	−48 57	2.96	0.01	A0 IV	−0.1	130	BA: 5″
γ Vir AB	12 41.6	−1 27	2.76	0.36	F1 V + F2 V	2.2	39	A: 3.48, B: 3.50, 4″ Porrima
β Mus AB	12 46.3	−68 06	3.05	−0.18	B2 V + B2.5 V	−2.2	424	A: 3.58, B: 4.10, 1″
β Cru	12 47.7	−59 41	1.2v	−0.23	B0.5 III	−4.0	352	var: 1.23–1.31, 0.7d? Becrux
ε UMa	12 54.1	+55 58	1.8v	−0.02	A0 IV	−0.2	81	var: 1.76–1.79, 5.1d Alioth
δ Vir	12 55.6	+3 24	3.38	1.58	M3 III	−0.5	202	Auva
α² CVn A	12 56.0	+38 19	2.9v	−0.12	A0	0.4	110	B: 5.6, F0 V, 20″ Cor Caroli
ε Vir	13 02.3	+10 58	2.83	0.94	G9 III	0.4	102	Vindamiatrix
γ Hya	13 18.9	−23 10	3.00	0.92	G8 III	−0.6	132	
ι Cen	13 20.6	−36 42	2.75	0.04	A2 V	1.4	59	
ζ UMa A	13 23.9	+54 58	2.27	0.02	A1 V	0.3	78	B: 3.94, A1 IV–V, 14″ Mizar
α Vir	13 25.2	−11 09	1.0v	−0.23	B1 V	−3.6	262	var: 0.97–1.04; mult. 3.1:4.5:7.5 Spica
ζ Vir	13 34.8	−0 36	3.37	0.11	A2 IV	1.6	73	Heze
ε Cen	13 39.9	−53 28	2.3v	−0.22	B1 III	−3.3	376	
η UMa	13 47.5	+49 19	1.86	−0.19	B3 V	−1.8	101	Alkaid
ν Cen	13 49.5	−41 41	3.41	−0.22	B2 IV	−2.4	475	
μ Cen	13 49.6	−42 28	3.0v	−0.17	B2 IV–V	−2.8	527	variable shell: 2.92–3.43
η Boo	13 54.7	+18 24	2.68	0.58	G0 IV	2.4	37	Mufrid
ζ Cen	13 55.6	−47 17	2.55	−0.22	B2.5 IV	−2.9	384	
β Cen AB	14 03.8	−60 22	0.6v	−0.23	B1 III	−5.5	526	var: 0.61–0.68; B: 3.9, 1″ Hadar
π Hya	14 06.4	−26 41	3.27	1.12	K2 III	0.2	101	
θ Cen	14 06.7	−36 22	2.06	1.01	K0 III	0.1	61	Menkent
α Boo	14 15.7	+19 12	−0.04	1.23	K1.5 III	−0.6	37	high space velocity Arcturus
ι Lup	14 19.4	−46 03	3.55	−0.18	B2.5 IV	−1.7	352	
γ Boo	14 32.1	+38 19	3.03	0.19	A7 IV	1.0	85	Seginus
η Cen	14 35.5	−42 09	2.3v	−0.19	B1.5 IV	−2.8	308	variable shell

APPENDIX 2: THE BRIGHTEST STARS (CONTINUED)

STAR NAME	RA 2000.0	DEC	MAGNITUDES V	B-V	SPECTRAL TYPE	M_V	D (LY)	REMARKS (ECLIPSING VARIABLE, SEPARATIONS, AND MAGNITUDES OF DOUBLES; NAMES)
α Cen B	14 39.7	−60 50	1.33	0.88	K1 V	6.2	4	BA: 21"; C: Proxima, 12.4, M5e, 2°
α Cen A	14 39.7	−60 49	−0.01	0.71	G2 V	4.2	4	AB: 21" Rigil Kentaurus
α Lup	14 41.9	−47 23	2.3v	−0.20	B1.5 III	−4.1	548	var: 2.28−2.31, 0.26d
α Cir	14 42.4	−64 58	3.19	0.24	A7	1.9	53	B: 8.6, K5 V, 16"
ε Boo AB	14 44.9	+27 05	2.37	0.97	K0 II-III+A0 V	−2.6	210	A: 2.70; B: 5.12, 3" Izar
α Lib A	14 50.9	−16 02	2.75	0.15	A3 III-IV	0.7	77	Zuben Elgenubi
β UMi	14 50.7	+74 10	2.08	1.47	K4 III	−1.1	126	Kocab
β Lup	14 58.5	−43 08	2.68	−0.22	B2 IV	−3.5	523	
κ Cen	14 59.2	−42 06	3.13	−0.20	B2 V	−2.8	539	
β Boo	15 01.8	+40 24	3.50	0.97	G8 III	−0.7	219	Nekkar
σ Lib	15 03.9	−25 16	3.3v	1.70	M2.5 III	−1.9	292	var: 3.20−3.36 Brachium
ζ Lup	15 12.3	−52 06	3.41	0.92	G8 III	0.1	116	
δ Boo	15 15.5	+33 19	3.47	0.95	G8 III	0.6	117	
β Lib	15 17.0	−9 23	2.61	−0.11	B8 III	−1.0	160	Zuben Elschemali
γ TrA	15 18.9	−68 40	2.89	0.00	A1 III	−0.8	183	
γ UMi	15 20.7	+71 50	3.05	0.05	A3 III	−0.1	147	Pherkad
δ Lup	15 21.4	−40 39	3.2v	−0.22	B1.5 IV	−2.8	510	
ε Lup AB	15 22.7	−44 41	3.37	−0.18	B2 IV-V	−2.5	504	A: 3.5; B: 5.0, <1"
ι Dra	15 24.9	+58 58	3.29	1.16	K2 III	0.8	102	Ed Asich
α CrB	15 34.7	+26 43	2.2v	−0.02	A0 IV	0.3	75	ecl: 2.21−2.32, 17.4d Alphekka
γ Lup AB	15 35.1	−41 10	2.78	−0.20	B2 IV	−2.8	567	A: 3.5; B: 3.6, <1", similar spectra
α Ser	15 44.2	+6 26	2.65	1.17	K2 III CN	0.9	73	var? Unukalhai
μ Ser	15 49.7	−3 26	3.53	−0.04	A0 III	0.3	156	
β TrA	15 55.1	−63 26	2.85	0.29	F0 IV	2.3	40	
π Sco A	15 58.9	−26 07	2.89	−0.19	B1 V + B V	−3.0	459	A: occ.bin: 3.4 + 4.5, 0.0003" sep.

STAR NAME	RA 2000.0	DEC 2000.0	MAGNITUDES V	B–V	SPECTRAL TYPE	M_V	D (LY)	REMARKS (ECLIPSING VARIABLE, SEPARATIONS, AND MAGNITUDES OF DOUBLES; NAMES)	
T CrB	15 59.4	+25 55	2.0v	0.10	gM3: + B	–9.3	...	recurrent nova 1866, 1946; now V=11	
η Lup A	16 00.1	–38 24	3.41	–0.22	B2.5 IV	–2.5	493	A: 3:47; B: 7.70, 15"	
δ Sco AB	16 00.3	–22 38	2.32	–0.12	B0.3 IV	–4.4	522	AB:mult<1";C:4.9,B2IV V,8"	Dschubba
β Sco AB	16 05.4	–19 48	2.62	–0.07	B0.5 V	–4.2	530	A: 2.78; B: 5.04, 1"; C: 4.93, 14"	Graas
δ Oph	16 14.4	–3 41	2.74	1.58	M1 III	–0.8	170		Yed Prior
ε Oph	16 18.4	–4 41	3.24	0.96	G9.5 III	0.8	107		Yed Posterior
σ Sco A	16 21.1	–25 35	2.9v	0.13	B1 III	–4.8	522	var: 2.94–3.06,0.25d; B: 8.3; B9 V, 20"	
η Dra A	16 23.9	+61 31	2.74	0.91	G8 III	0.7	88	B: 8.7, 6"	
α Sco A	16 29.5	–26 26	0.9v	1.83	M1.5 I	–5.8	604	B: 5.37; B2.5 V, 3"	Antares
β Her	16 30.3	+21 30	2.77	0.94	G7 III	–0.5	148		Kornephoros
τ Sco	16 35.9	–28 13	2.82	–0.25	B0 V	–3.1	430		
ζ Oph	16 37.2	–10 34	2.56	0.02	O9.5 V	–4.3	458		
ζ Her AB	16 41.2	+31 36	2.81	0.65	G1 IV	2.5	35	A: 2.90: B: 5.53; G7 V, 1.1"	
η Her	16 42.8	+38 56	3.53	0.92	G7.5 III	0.1	112		
α TrA	16 48.6	–69 02	1.92	1.44	K2 II–III	–5.0	415		Atria
ε Sco	16 50.2	–34 17	2.29	1.15	K2 III	0.1	65		
μ¹ Sco	16 51.9	–38 03	3.1v	–0.20	B1.5 IV	–4.1	821	ecl: 2.80–3.08, 1.4d	
κ Oph	16 57.7	+9 22	3.20	1.15	K2 III	1.1	86		
ζ Ara	16 58.5	–55 59	3.13	1.60	K4 III	–4.5	574		
ζ Dra	17 08.8	+65 43	3.17	–0.12	B6 III	–2.0	340		Aldhibah
η Oph AB	17 10.4	–15 44	2.43	0.06	A2.5 V	0.8	84	A: 3.0; B: 3.5. A3 V, 1"	Sabik
η Sco	17 12.1	–43 15	3.33	0.41	F2pC	1.4	72		
α Her AB	17 14.7	+14 23	3.1v	1.44	M5 I–II	–3.5	1369	var: 3.0–4.0; B: 5.4. 5"	Ras Algethi
δ Her	17 15.1	+24 50	3.14	0.08	A1 V	1.2	78	B: 8.8, 9"	Sarin
π Her	17 15.0	+36 48	3.16	1.44	K3 II	–2.2	367		
θ Oph	17 22.1	–24 59	3.3v	–0.22	B2 IV	–3.1	563	occ.bin: 3.4, 5.4; var: 3.25–3.29, 0.14d	

STAR NAME	RA	DEC 2000.0	MAGNITUDES V	B-V	SPECTRAL TYPE	Mv	D (LY)	REMARKS (ECLIPSING VARIABLE, SEPARATIONS, AND MAGNITUDES OF DOUBLES; NAMES)
β Ara	17 25.2	-55 31	2.85	1.46	K3 I-II	-4.1	603	
γ Ara A	17 25.3	-56 22	3.34	-0.13	B1 I	-5.7	652	B: 10.0, 1800
υ Sco	17 30.8	-37 17	2.69	-0.22	B2 IV	-3.5	518	
β Dra A	17 30.4	+52 19	2.79	0.98	G2 I-II	-2.9	361	B: 11.5, 4" Restaban
α Ara	17 31.8	-49 52	2.95	-0.17	B2 V	-1.8	242	
λ Sco	17 33.6	-37 06	1.6v	-0.22	B1.5 IV	-3.6	359	var: 1.59-1.65; 0.21d Shaula
α Oph	17 35.0	+12 34	2.08	0.15	A5 V	1.3	47	Rasalhague
θ Sco	17 37.3	-43 00	1.87	0.40	F1 III	-3.0	272	Sargas
ξ Ser	17 37.6	-15 24	3.54	0.26	F0 III	0.9	105	
κ Sco	17 42.5	-39 02	2.4v	-0.22	B1.5 III	-3.6	464	var: 2.39-2.42; 0.2d Cebalrai
β Oph	17 43.4	+4 34	2.77	1.16	K2 III	0.7	82	
μ Her A	17 46.3	+27 44	3.42	0.75	G5 IV	3.6	27	BC: 9.78, 33"
τ¹ Sco	17 47.6	-40 07	3.03	0.51	F2 I	-8.0	3619	
G Sco	17 49.7	-37 02	3.21	1.17	K2 III	-0.6	127	HR6630
γ Dra	17 56.6	+51 29	2.23	1.52	K5 III	-1.1	148	Etamin
ν Oph	17 59.0	-9 46	3.34	0.99	G9.5 III	0.0	153	
γ² Sgr	18 05.8	-30 25	2.99	1.00	K0 III	0.1	96	Nash
η Sgr A	18 17.6	-36 46	3.11v	1.56	M3.5 III	-0.1	149	var: 3.08-3.12; B: 8.33; G8: IV/; 4"
δ Sgr	18 21.0	-29 50	2.70	1.38	K2.5 III	-3.3	306	Kaus Meridionalis
η Ser	18 21.4	-2 54	3.26	0.94	K0 III-IV	1.4	62	
ε Sgr	18 24.2	-34 23	1.85	-0.03	A0 II	-1.4	145	Kaus Australis
α Tel	18 27.0	-45 58	3.51	-0.17	B3 IV	-1.0	249	
λ Sgr	18 28.0	-25 26	2.81	1.04	K1 III	0.4	77	Kaus Borealis
α Lyr	18 36.9	+38 46	0.03	0.00	A0 V	0.6	25	Vega
φ Sgr	18 45.7	-26 59	3.17	-0.11	B8 III	-1.0	231	similar companion, 0.1"
β Lyr	18 50.0	+33 22	3.4v	0.00	B7 V	-4.1	881	ecl: 3.34-4.34; 12.9d Sheliak

STAR NAME	RA 2000.0	DEC	MAGNITUDES V	B−V	SPECTRAL TYPE	M_V	D (LY)	REMARKS (ECLIPSING VARIABLE, SEPARATIONS, AND MAGNITUDES OF DOUBLES; NAMES)
σ Sgr	18 55.3	−26 18	2.02	−0.22	B3 IV	−2.4	224	Nunki
ξ² Sgr	18 57.7	−21 06	3.51	1.18	K1 III	0.1	913	
γ Lyr	18 58.8	+32 41	3.24	−0.05	B9 II	−3.3	634	Sulaphat
ζ Sgr AB	19 02.6	−29 53	2.60	0.08	A2 IV:V + A4:V	1.1	89	A: 3.2; B: 3.5, <1" Ascella
ζ Aql A	19 05.4	+13 52	2.99	0.01	A0 V	0.9	83	
λ Aql	19 06.3	−4 53	3.44	−0.09	B9 V	0.6	125	
τ Sgr	19 07.0	−27 40	3.32	1.19	K1.5 III	−0.4	120	
π Sgr ABC	19 09.7	−21 01	2.89	0.35	F2 II–III	−3.1	440	A: 3.7; B: 3.8; C: 6.0, <1" Albaldah
δ Dra	19 12.6	+67 40	3.07	1.00	G9 III	0.6	100	Nodus Secundus
δ Aql	19 25.6	+3 07	3.36	0.32	F2 IV	2.6	50	
β Cyg A	19 30.7	+27 57	3.08	1.13	K3 II + B9.5 V	−1.5	385	B: 5.11, 35"; C: Δm=1.5, 0.4" Albireo
δ Cyg AB	19 44.9	+45 07	2.87	−0.03	B9.5 III	−0.7	171	B: 6.4, F1 V, 2"
λ Aql	19 46.2	+10 36	2.72	1.52	K3 II	−2.6	326	Tarazed
α Aql	19 50.8	+8 51	0.77	0.22	A7 V	2.1	17	Altair
η Aql	19 52.5	+1 00	3.9v	0.83	F6 G1 I	−4.3	1173	Cepheid var: 3.53–4.33, 7.2d
γ Sge	19 58.8	+19 29	3.47	1.57	M0 III	−0.9	274	
θ Aql	20 11.4	−0 50	3.23	−0.07	B9.5 III	−1.4	287	
β Cap A	20 21.0	−14 47	3.08	0.79	K0: II: + A5: V	−1.4	344	A: mult: 4.0+4.3+4.8+6.7, <1" Dabih
γ Cyg	20 22.1	+40 15	2.20	0.68	F8 I	−4.1	522	Sadr
α Pav	20 25.5	−56 44	1.94	−0.20	B2.5 V	−2.1	183	Peacock
α Ind	20 37.5	−47 18	3.11	1.00	K0 III CN	0.1	101	
α Cyg	20 41.4	+45 17	1.25	0.09	A2 I	−7.5	1467	Deneb
β Pav	20 44.9	−66 12	3.42	0.16	A6 IV	0.6	141	
η Cep	20 45.3	+61 50	3.43	0.92	K0 IV	2.7	47	
ε Cyg	20 46.2	+33 58	2.46	1.03	K0 III	0.7	72	
ζ Cyg	21 13.0	+30 13	3.20	0.99	G8 III	0.2	151	

STAR NAME	RA	DEC	MAGNITUDES		SPECTRAL TYPE	M_V	D (LY)	REMARKS (ECLIPSING VARIABLE, SEPARATIONS, AND MAGNITUDES OF DOUBLES; NAMES)
	2000.0		V	B–V				
α Cep	21 18.5	+62 34	2.44	0.22	A7 V	1.4	49	Alderamin
β Cep	21 28.6	+70 33	3.2v	−0.22	B1 III	−4.0	815	Alphirk; var: 3.16−3.27, 0.2d; B: 7.8, 13"
β Aqr	21 31.6	−5 35	2.91	0.83	G0 I	−3.5	612	Sadalsuud
ε Peg	21 44.1	+9 51	2.4v	1.53	K2 I	−5.2	672	Enif; var: 0.7−3.5 (flare in 1972)
δ Cap	21 47.1	−16 09	2.9v	0.29	A3 IV	2.2	39	var: 2.83−3.05, 1d; occ.bin: 3.2 + 5.2
γ Gru	21 53.9	−37 22	3.01	−0.12	B8 IV–V	−1.1	203	
α Aqr	22 05.8	−0 20	2.96	0.98	G2 I	−4.3	756	Sadalmelik
α Gru	22 08.2	−46 58	1.74	−0.13	B7 V	−0.9	101	Al Nair
θ Peg	22 10.3	+6 12	3.53	0.08	A2mA1 IV–V	1.0	97	Baham
ζ Cep	22 10.8	+58 12	3.35	1.57	K1.5 I	−4.2	726	
α Tuc	22 18.5	−60 17	2.86	1.39	K3 III	−2.2	199	
δ Cep A	22 29.1	+58 25	4.0v	0.71	F5 G2 I	−4.4	950	Cepheid variable: 3.55−4.41, 5.4d
ζ Peg	22 41.5	+10 50	3.40	−0.09	B8.5 III	−0.7	208	Homam
β Gru	22 42.6	−46 53	2.1v	1.60	M5 III	−1.4	170	var: 2.0−2.3
η Peg	22 43.0	+30 13	2.94	0.86	G8 II + F0 V	−1.2	1113	Matar
ε Gru	22 48.5	−51 19	3.49	0.08	A2 V	0.4	130	
ι Cep	22 49.6	+66 12	3.52	1.05	K0 III	0.6	115	
μ Peg	22 50.0	+24 36	3.48	0.93	G8 III	0.8	117	
δ Aqr	22 54.8	−15 50	3.27	0.05	A3	−0.1	159	Skat
α PsA	22 57.7	−29 38	1.16	0.09	A3 V	1.6	25	**Fomalhaut**
β Peg	23 03.8	+28 05	2.4v	1.67	M2 II–III	−1.7	199	Scheat
α Peg	23 04.7	+15 12	2.49	−0.04	A0 III–IV	−0.9	140	Markab
γ Cep	23 39.3	+77 37	3.21	1.03	K1 III–IV	2.1	45	Alrai

This table is by Robert Garrison and Brian Beattie, University of Toronto, based on Hipparcos data and their own spectral types, from the *Observers Handbook of the Royal Astronomical Society of Canada*, with permission.

Star names come from *The Bright Star Catalogue*, compiled by Dorrit Hoffleit (Yale University Observatory).

APPENDIX 3: PROPERTIES OF THE PRINCIPAL SPECTRAL TYPES

SPECTRAL TYPE	APPARENT COLOR	COLOR INDEX (B–V)	SURFACE TEMPERATURE	(K) PRIMARY ABSORPTION LINES IN SPECTRUM	EXAMPLES
O	blue	less than –0.2	25,000–40,000	Strong lines of ionized helium and highly ionized metals; hydrogen lines weak	ζ Orionis (O9.5)
B	blue	–0.2 to 0.0	11,000–25,000	Lines of neutral helium prominent; hydrogen lines stronger than in type O	Spica (B1) Rigel (B8)
A	blue to white	0.0 to 0.3	7,500–11,000	Strong lines of hydrogen, ionized calcium, and other ionized metals; weak helium lines	Vega (A0) Sirius (A1)
F	white	0.3 to 0.6	6,000–7,500	Hydrogen lines weaker than in type A; ionized calcium strong; lines of neutral metals becoming prominent	Canopus (F0) Procyon (F5) Polaris (F8)
G	white to yellow	0.6 to 1.1	5,000–6,000	Numerous strong lines of ionized calcium and other ionized and neutral metals; hydrogen lines weaker than in type F	Sun (G2) Capella (G5)
K	orange to red	1.1 to 1.5	3,500–5,000	Numerous strong lines of neutral metals	Arcturus (K2) Aldebaran (K5)
M	red	greather than 1.5	3,000–3,500	Numerous strong lines of neutral metals; strong molecular bands (primarily titanium oxide	Antares (M1) Betelgeuse (M2)

NOTE: The number after the letter in each spectral type (see last column, above) indicated a further subdivision within each type. For example, Sirius (type A1) is hotter than Deneb (type A2).

From *Sky Catalogue 2000.0* by Alan Hirshfeld and Roger W. Sinnott, courtesy of Sky Publishing Corp.

NO. (RANK)	NAME	POSITION (2000.0) R.A. H M	DEC. ° ′	DISTANCE (LIGHT-YEARS)	SPECTRAL TYPE	MAGNITUDE V	M_v	LUMINOSITY ($L_{SUN} = 1$)
1	Sun				G2 V	−26.75	4.82	1.0
2	Proxima Cen	14 29.7	−62 41	4.21	M5.5 V	11.05	15.49	0.00005
	a (alpha) Cen A	14 39.6	−60 50	4.40	G2 V	0.02	4.37	1.51
	a (alpha) Cen B				K0 V	1.36	5.71	0.44
3	Barnard's star	17 57.8	+04 42	5.94	M4 V	9.54	13.24	0.0004
4	Wolf 359 (CN Leo)	10 56.5	+07 01	7.80	M6 V	13.45	16.56	0.00002
5	BD +36°2147 HD 95735 (Lalande 21185)	11 03.4	+35 58	8.32	M2 V	7.49	10.46	0.006
6	Sirius A	06 45.1	−16 43	8.61	A1 V	−1.45	1.44	22.49
	Sirius B				DA2	8.44	11.33	0.0025
7	L 726-8, BL Cet = A	01 39.0	−17 57	8.74	M5.5 V	12.41	15.27	0.00007
	UV Cet = B				M6 V	13.25	16.11	0.00003
8	Ross 154 (V1216 Sgr)	18 49.8	−23 50	9.69	M3.5 V	10.45	13.08	0.0005
9	Ross 248 (HH And)	23 41.9	+44 10	10.31	M5.5 V	12.29	14.79	0.0001
10	e (epsilon) Eri	03 32.9	−09 27	10.50	K2 V	3.72	6.18	0.286
11	CD −36°15693 HD217987 (Lacaille 9352)	23 05.9	−35 51	10.73	M2 V	7.35	9.76	0.01
12	Ross 128 (FI Vir)	11 47.7	+00 48	10.89	M4 V	11.12	13.50	0.00034
13	L 789-6 (EZ Aqr) = A = B	22 38.5	−15 18	11.25	M5 V	12.69 13.6	15.00 15.9	0.00008 0.00004
14	61 Cyg A	21 06.9	+38 45	11.38	K5 V	5.22	7.51	0.084
	61 Cyg B				K7 V	6.04	8.32	0.0398
15	Procyon A	07 39.3	+05 14	11.42	F5 IV-V	0.36	2.64	7.45
	Procyon B				DA	10.75	13.03	0.0005

NOTES: V = visual magnitude; Mv = absolute magnitude. Under Spectral Type, Roman numeral V = normal dwarfs like the sun and DA is a class of

APPENDIX 5. SELECTED BRIGHT PLANETARY NEBULAE

PLANETARY NEBULA	NAME AND CONSTELLATION	R.A. (2000.0) H M	DEC. ° '	MAGNITUDE (VISUAL)	DIAMETER (ARC SEC)
NGC 7293	Helix in Aquarius	22 30	−20 48	6.5	>77°
NGC 6853	Dumbbell in Vulpecula, M27	19 60	+22 43	7.6	350/910
NGC 246	in Cetus	00 47	−11 53	8.0	225
NGC 3132	Eight-burst in Vela	10 08	−40 26	8.2	>47
NGC 7009	Saturn in Aquarius	21 04	−11 23	8.3	25/100
NGC 3242	Ghost of Jupiter in Hydra	10 25	−18 38	8.6	16/1250
NGC 6543	Cat's Eye in Draco	17 59	+66 38	8.8	18/350
NGC 6572	in Ophiuchus	18 12	+06 51	9.0	8
NGC 7662	in Andromeda	23 26	+42 33	9.2	20/130
NGC 6210	in Hercules	16 45	+23 49	9.3	>14
NGC 1535	in Eridanus	04 14	−12 44	9.6	18/44
NGC 6720	Ring Nebula in Lyra, M57	18 54	+33 02	9.7	70/150
NGC 6826	Blinking Planetary in Cygnus	19 45	+50 32	9.8	30/140
NGC 6818	Little Gem in Sagittarius	19 44	−14 09	9.9	>17
NGC 2392	Eskimo in Gemini	07 29	+20 55	9.9	13/44

NOTES: Under diameter, two values separated by a slash indicates a bright inner core surrounded by a faint, extended halo; the symbol ">" indicates that, while only the size of the core is listed, a larger halo exists.

Courtesy of Yervant Terzian, Cornell University, with data from *Sky Catalogue 2000.0*, Vol. 2.

ADS	STAR	R.A. H M	DEC. ° '	MAGNITUDES	2000 PA1 °	SEP1 "	2010 PA °	SEP2 "	PERIOD (YEARS)
558	55 Psc	00 39.9	+21 26	5.6 8.5	194	6.6	194	6.6	–
561	α Cas	00 40.5	+56 32	2.2 8.9	283	70.3	283	70.9	–
671	η Cas	00 49.1	+57 49	3.5 7.5	318	12.9	322	13.2	480
940	42 φ And	01 09.5	+47 15	4.5 5.8	126	0.5	120	0.5	372
996	ζ Psc	01 13.7	+07 35	5.2 6.3	63	23.0	63	23.3	–
1129	ψ Cas	01 25.9	+68 08	4.7 9.6	127	19.4	127	18.9	–
1507	γ Ari	01 53.6	+19 18	4.6 4.6	1	7.5	1	7.5	–
1563	λ Ari	01 57.9	+23 36	4.8 7.3	47	37.5	47	37.5	–
1615	α Psc	02 02.0	+02 46	4.1 4.9	271	1.8	265	1.8	933
1630	γ¹ And	02 03.9	+42 20	2.3 5.0	63	9.7	63	9.7	–
1630	γ² And	02 03.9	+42 20	5.5 6.3	103	0.4	96	0.2	61
1697	ι Tri	02 12.4	+30 18	5.2 6.7	70	3.9	70	3.9	–
1860	ι Cas	02 29.2	+67 25	4.6 6.6	231	2.6	229	2.6	620
1477	α UMi	02 31.8	+89 16	2.0 8.9	222	18.3	222	18.3	–
2080	γ Cet	02 43.3	+03 14	3.6 5.9	298	2.6	298	2.6	–
2157	η Per	02 50.7	+55 54	3.8 8.5	301	28.4	301	28.4	–
–	θ¹,² Eri	02 58.3	–40 18	3.2 4.3	91	8.3	91	8.3	–
2850	32 Eri	03 54.3	–02 57	4.8 6.1	348	6.9	348	6.9	–
2888	ε Per	03 57.9	+40 01	2.9 8.0	9	9.0	9	9.0	–
3137	φ Tau	04 20.4	+27 21	5.1 8.7	255	48.2	254	47.8	–
3321	α Tau	04 35.9	+16 31	0.9 10.7	31	133.5	31	134.9	–
3823	β Ori	05 14.5	–08 12	0.2 6.7	202	9.4	202	9.4	–
4179	λ Ori	05 35.1	+09 56	3.5 5.5	44	4.3	44	4.3	–
4186	θ¹ Ori	05 35.3	–05 23	5.1 6.7	31	8.8	31	8.8	–
4241	σ Ori	05 38.8	+02 36	3.8 6.9	84	13.0	84	13.0	–
4566	θ Aur	05 59.7	+37 13	2.6 7.1	308	3.8	304	3.9	–
5107	β Mon	06 28.8	–07 02	4.6 5.0	133	7.2	133	7.2	–
5400	12 Lyn	06 46.2	+59 27	5.5 6.1	69	1.7	65	1.7	699
5423	α CMa	06 45.2	–16 43	–1.5 8.5	140	5.0	92	8.8	50
5654	ε CMa	06 58.6	–28 58	1.5 7.8	161	7.5	161	7.5	–
5983	δ Gem	07 20.1	+21 59	3.6 8.2	225	5.8	227	5.6	1200
6175	α Gem	07 34.6	+31 53	1.9 3.0	64	3.9	57	4.6	445
6321	κ Gem	07 44.5	+24 24	3.6 9.4	241	6.9	240	7.0	–
6988	ι¹ Cnc	08 46.7	+28 46	5.3 6.6	307	30.4	307	30.4	–
7203	σ² UMa	09 10.5	+67 07	4.9 8.9	353	3.8	350	4.1	1141
7402	23 UMa	09 31.5	+63 04	3.7 9.2	268	22.7	268	22.7	–
7724	γ Leo	10 19.9	+19 51	2.5 3.6	125	4.4	126	4.5	619
7979	54 Leo	10 55.6	+24 45	4.5 6.4	110	6.5	110	6.5	–
8119	ξ UMa	11 18.2	+31 32	4.3 4.8	269	1.8	211	1.6	60
8148	ι Leo	11 23.9	+10 32	4.1 6.9	115	1.7	105	2.0	192
8489	2 CVn	12 16.1	+40 40	5.9 8.7	260	11.5	260	11.5	–
8531	17 Vir	12 22.5	+05 18	6.5 9.3	337	20.8	337	20.8	–
–	α Cru	12 26.6	–63 06	1.2 1.5	113	4.0	113	4.0	–

DS	STAR	R.A. H M	DEC. ° '	MAGNITUDES		2000 PA1 °	SEP1 "	2010 PA °	SEP2 "	PERIOD (YEARS)
630	γ Vir	12 41.7	−01 27	3.6	3.6	253	1.3	21	1.5	169
706	α CVn	12 56.0	+38 19	2.9	5.5	228	19.3	228	19.3	—
891	ζ UMa	13 24.0	+54 56	2.2	3.9	152	14.4	152	14.4	—
	α Cen	14 39.6	−60 50	0.1	1.2	223	13.5	245	6.8	80
338	π Boo	14 40.7	+16 25	4.8	5.8	109	5.5	109	5.5	—
343	ζ Boo	14 41.1	+13 44	4.5	4.6	299	0.8	294	0.6	123
372	ε Boo	14 44.9	+27 04	2.6	4.7	344	2.6	345	2.6	—
375	54 Hya	14 46.0	−25 27	5.2	7.2	123	8.4	123	8.4	—
413	ξ Boo	14 51.3	+19 06	4.7	6.9	317	6.6	308	6.0	152
494	44 Boo	15 03.8	+47 39	5.3	6.1	54	2.2	58	2.1	220
617	η CrB	15 23.3	+30 18	5.6	5.9	69	0.7	167	0.6	42
701	δ Ser	15 34.8	+10 32	4.2	5.2	174	4.1	173	4.2	—
737	ζ CrB	15 39.4	+36 38	5.0	5.9	305	6.3	305	6.3	—
909	ξ Sco AB	16 04.4	−11 23	4.9	4.9	320	0.5	356	1.0	46
913	β Sco	16 05.4	−19 48	2.6	4.8	20	13.7	19	13.8	—
0074	α Sco	16 29.4	−26 26	0.9v	5.5	276	2.9	276	2.9	—
0087	λ Oph	16 30.9	+01 59	4.2	5.2	31	1.5	37	1.6	130
0157	ζ Her	16 41.3	+31 36	3.0	5.7	342	0.5	176	1.2	34
0345	μ Dra AB	17 05.3	+54 28	5.6	5.7	6	1.9	352	2.1	482
0417	36 Oph	17 15.5	−26 35	5.2	5.2	146	4.9	143	5.1	549
0418	α Her	17 14.6	+14 24	3.5	5.4	104	4.6	103	4.6	3600
0424	δ Her	17 15.0	+24 50	3.1	8.7	280	11.2	286	12.4	—
0526	ρ Her	17 23.6	+37 08	4.5	5.4	318	4.1	318	4.1	—
0993	95 Her	18 01.5	+21 36	4.9	5.2	257	6.4	257	6.4	—
1005	τ Oph	18 03.1	−08 11	5.2	5.9	284	1.7	287	1.6	280
1046	70 Oph	18 05.5	+02 30	4.2	6.0	146	3.9	131	5.7	88
1336	39 Dra	18 24.0	+58 48	5.1	8.1	350	3.8	350	3.8	—
1635	ε¹ Lyr	18 44.3	+39 40	5.0	6.0	350	2.6	347	2.5	1165
1635	ε² Lyr	18 44.4	+39 37	5.2	5.4	82	2.3	78	2.4	724
2540	β Cyg	19 30.7	+27 58	3.2	5.4	54	34.2	54	34.2	—
2880	δ Cyg	19 45.0	+45 08	2.9	6.6	221	2.5	216	2.6	828
3007	ε Dra	19 48.2	+70 16	4.0	6.7	20	3.2	23	3.2	—
3632	α¹ Cap	20 17.6	−12 31	4.3	9.0	222	45.2	222	45.2	—
3645	α² Cap	20 18.1	−12 33	3.6	10.0	182	6.6	185	6.6	—
4279	γ Del	20 46.7	+16 7.0	4.4	5.1	266	9.2	266	9.0	—
4296	λ Cyg	20 47.5	+36 29	4.8	6.2	6	0.9	2	0.9	391
4636	61 Cyg	21 06.9	+38 45	5.2	6.0	150	30.6	151	31.2	722
5032	β Cep	21 28.6	+70 34	3.2	7.8	249	13.3	249	13.3	—
5270	μ Cyg	21 44.2	+28 45	4.7	6.0	308	1.8	314	1.6	713
5971	ζ Aqr	22 28.8	+00 01	4.4	4.6	184	2.0	171	2.2	760
7140	σ Cas	23 51.9	+55 46	5.0	7.0	326	3.1	326	3.1	—

Courtesy of R. W. Argyle, Institute of Astronomy, University of Cambridge, Cambridge, U.K.

NAME	TYPE	R.A. (2000.0) H	M	DEC °	′	MAG-NITUDE (RANGE)	PERIOD (DAYS)
T Cas	M	00	23.2	+55	48	[6.9–11.9]	445
o Cet	M	02	19.3	−02	59	[3.4–9.2]	332
R Tri	M	02	37.0	+34	16	[6.2–11.7]	266
R Hor	M	02	53.9	−49	53	[6.0–13.0]	404
X Cam	M	04	45.7	+75	06	[8.1–12.6]	144
R Pic	SR	04	46.2	−49	15	6.7–10.0	164
L² Pup	SR	07	13.5	−44	39	2.6–6.2	140
S CMi	M	07	32.7	+08	19	[7.5–12.6]	333
U Gem	UG	07	55.1	+22	00	8.2–14.9	103
R Car	M	09	32.2	−62	47	[4.6–9.6]	309
ZZ Car	C	09	45.2	−62	30	3.3–4.2	36
R Leo	M	09	47.6	+11	26	[5.8–10.0]	312
S Car	M	10	09.4	−61	33	[5.7–8.5]	150
R UMa	M	10	44.6	+68	47	[7.5–13.0]	302
T UMa	M	12	36.4	+59	29	[7.7–12.9]	257
S UMa	M	12	43.9	+61	06	[7.8–11.7]	226
T Cen	SR	13	41.8	−33	36	5.5–9.0	90
R CVn	M	13	49.0	+39	33	[7.7–11.9]	329
R Cen	M	14	16.6	−59	55	5.3–11.8	546
S Boo	M	14	22.9	+53	49	[8.4–13.3]	271
V Boo	SR	14	29.8	+38	52	7.0–12.0	258
R CrB	RCB	15	48.6	+28	09	5.7–14.8	†
R Ser	M	15	50.7	+15	08	[6.9–13.4]	356
U Her	M	16	25.8	+18	54	[7.5–12.5]	406
R Dra	M	16	32.7	+66	45	[7.6–12.4]	245
R Oph	M	17	07.8	−16	06	[7.6–13.3]	303
T Her	M	18	09.1	+31	01	[8.0–12.8]	165
R Sct	RV	18	47.5	−05	42	4.5–8.2	140
R Cyg	M	19	36.8	+50	12	[7.5–13.9]	426
R Vul	M	21	04.4	+23	49	[8.1–12.6]	136
T Cep	M	21	09.5	+68	29	[6.0–10.3]	388
SS Cyg	UG	21	42.7	+43	35	8.2–12.4	50
R Peg	M	23	06.6	+10	33	[7.8–13.2]	378
V Cas	M	23	11.7	+59	42	6.9–13.4	229

NOTES: C = Cepheid variable; M = Mira variable; RCB = R Coronae Borealis variable; RV = RV Tauri variable; SR = semiregular variable; UG = U Geminorum variable. Brackets indicate an average range of magnitude; † = irregular, no period.
Drawn from *Sky Catalogue* 2000.0, vol. 2, by Alan Hirshfeld and Roger W. Sinnott (Sky Publishing Corp., 1985).

NAME	TYPE	R.A.(2000.0) H	M	DEC ° '	MAGNITUDE(RANGE)	PERIOD(DAYS)
Algol (β Persei)	E	03	08.2	+40 57	2.1–3.4	2.87
λ Tauri	E	04	00.7	+12 29	3.4–3.9	3.95
RT Aurigae	C	06	28.6	+30 30	5.1–5.8	3.73
ζ Geminorum	C	07	04.1	+20 34	3.7–4.2	10.15
δ Librae	E	15	01.0	−08 31	4.9–5.9	2.33
u Herculis	E	17	17.3	+33 06	4.8–5.4	2.05
W Sagittarii	C	18	05.0	−29 35	4.3–5.1	7.59
β Lyrae	E	18	50.1	+33 22	3.3–4.3	12.94
RR Lyrae	RR	19	25.5	+42 47	7.0–8.1	0.57
η Aquilae	C	19	52.5	+01 00	3.5–4.5	7.18
δ Cephei	C	22	29.2	+58 25	3.5–4.4	5.37

NOTES: C = Cepheid variable; E = eclipsing binary; RR = RR Lyrae variable (cluster variable—see pp. 198–202). Finding charts for beta Lyrae and beta Persei (Algol) appear at the end of chapter 6.

Drawn from *Sky Catalogue* 2000.0, vol. 2, by Alan Hirshfeld and Roger W. Sinnott (Sky Publishing Corp., 1985).

INTRINSIC AND ROTATIONAL PROPERTIES OF THE PLANETS

NAME	EQUATORIAL RADIUS (KM)	EQUATORIAL RADIUS ÷ EARTH'S	MASS ÷ EARTH'S	MEAN DENSITY (G/CM³)	OBLATENESS	SURFACE GRAVITY (EARTH = 1)	SIDEREAL ROTATION PERIOD	INCLINATION OF EQUATOR TO ORBIT (DEGREES)	RANGE IN APPARENT MAGNITUDE	APPARENT EQUATORIAL DIAMETER (ARC SEC)	ALBEDO
Mercury	2,439	0.3824	0.0553	5.43	0	0.378	58.646^d	0.0°	−2.3 to +5.6	4 to 12	0.1
Venus	6,052	0.9489	0.8150	5.24	0	0.894	$243.01^{d}R$	177.3	−4.6 to −3.9	9 to 58	0.7
Earth	6,378.140	1	1	5.515	0.0034	1	$23^h56^m04.1^s$	23.45	—	—	0.3
Mars	3,397	0.5326	0.1074	3.94	0.005	0.379	$24^h37^m22.662^s$	25.19	−1.3 to +1.8	3 to 25	0.2
Jupiter	71,492	11.194	317.896	1.33	0.064	2.54	9^h50^m to $>9^h55^m$	3.12	−2.6 to −1.9	31 to 49	0.5
Saturn	60,268	9.41	95.185	0.70	0.108	1.07	$10^h39.9^m$	26.73	+0.1 to +1.0	15 to 19	0.5
Uranus	25,559	4.0	14.537	1.30	0.03	0.8	17^h14^m	97.86	+5.6 to +5.9	3.4 to 3.9	0.5
Neptune	24,764	3.9	17.151	1.76	0.017	1.2	16^h7^m	29.56	+7.9 to +8.0	2.1 to 2.3	0.4
Pluto	1,151	0.2	0.0025	2.03	?	0.01	$6^d9^h17^m$	120	+13.7 to +13.8	0.1	0.5

R = retrograde; d = days; h = hours; m = minutes; s = seconds.

ORBITAL PROPERTIES OF THE PLANETS

NAME	SEMIMAJOR AXIS (A.U.)	SEMIMAJOR AXIS (10⁶ KM)	SIDEREAL PERIOD (YEARS)	SIDEREAL PERIOD (DAYS)	SYNODIC PERIOD (DAYS)	ECCENTRICITY OF ORBIT	INCLINATION TO ECLIPTIC
Mercury	0.387 099	57.909	0.240 84	87.96	115.9	0.205 63	7.004 87°
Venus	0.723 332	108.209	0.615 18	224.68	583.9	0.006 77	3.394 71°
Earth	1	149.598	0.999 98	365.25		0.016 71	0.000 05°
Mars	1.523 662	227.939	1.8807	686.95	779.96	0.093 41	1.850 61°
Jupiter	5.203 363	778.298	11.857	4337	398.9	0.048 39	1.305 30°
Saturn	9.537 070	1429.394	29.424	10,760	378.1	0.054 15	2.484 46°
Uranus	19.191 264	2875.039	83.75	30,700	369.7	0.047 168	0.769 86°
Neptune	30.068 963	4504.450	163.72	60,200	367.5	0.008 59	0.769 17°
Pluto	39.481 687	5915.799	248.02	90,780	366.7	0.248 81	17.141 75°

Mean elements of planetary orbits for 2000 (E. M. Standish, X. X. Newhall, J. G. Williams, and D. K. Yeomans), Explanatory Supplement to the Astronomical Almanac.

SATELLITE	SEMIMAJOR AXIS OF ORBIT (KM)	SIDEREAL REVOLUTION PERIOD (D H M)			ORBITAL ECCENTRICITY	ORBITAL INCLINATION (°)	RADIUS (KM)	VISIBLE MAGNITUDE AT MEAN OPPOSITION DISTANCE
SATELLITE OF THE EARTH								
The Moon	384,400	27	07	43	0.055	18–29	1738	−12.7
SATELLITES OF MARS								
1 Phobos	9,378	0	07	39	0.015	1.0	13 × 11 × 9	11.3
2 Deimos	23,459	1	06	18	0.0005	0.9–2.7	8 × 6 × 5	12.4
SATELLITES OF JUPITER								
XVI Metis	128,000	0	07	04			20	17.5
XV Adrastea	129,000	0	07	06			13 × 10 × 8	19.1
V Amalthea	181,000	0	11	57	0.003	0.4	131 × 73 × 75	14.1
XIV Thebe	222,0000	0	16	11	0.015	0.8	55 × 45	15.7
I Io	422,000	1	18	28	0.004	0.04	1830 × 1819 × 1815	5.0
II Europa	671,000	3	13	14	0.009	0.5	1565	5.3
III Ganymede	1,070,000	7	03	43	0.002	0.2	2634	4.6
IV Callisto	1,883,000	16	16	32	0.007	0.5	2403	5.7
XIII Leda	11,094,000	238			0.148	26.1	5	20
VI Himalia	11,480,000	251			0.158	27.6	85	14.8
X Lysithea	11,720,000	259			0.107	29.0	12	18.4
VII Elara	11,737,000	260			0.207	24.8	40	16.8
XII Ananke	21,200,000	671 R			0.169	147	10	18.9
XI Carme	22,600,000	692 R			0.207	164	15	18.0
VIII Pasiphae	23,500,000	735 R			0.378	145	18	17.0
IX Sinope	23,700,000	758 R			0.275	153	14	18.3
								20

SATELLITE	SEMIMAJOR AXIS OF ORBIT (KM)	SIDEREAL REVOLUTION PERIOD (D H M)			ORBITAL ECCENTRICITY	ORBITAL INCLINATION (°)	RADIUS (KM)	VISIBLE MAGNITUDE AT MEAN OPPOSITION DISTANCE
SATELLITES OF SATURN								
18 Pan	133,583		13	48			10	18
15 Atlas	137,670		14	27	0.000	0.3	19 × 17 × 14	18
16 Prometheus	139,353		14	43	0.003	0.0	74 × 50 × 34	16
17 Pandora	141,700		15	05	0.004	0.0	55 × 44 × 31	16
11 Epimetheus	151,422		16	40	0.009	0.3	69 × 55 × 55	15
10 Janus	151,472		16	40	0.007	0.1	97 × 95 × 77	14
1 Mimas	185,520		22	37	0.0202	1.5	209 × 196 × 191	12.9
2 Enceladus	238,020	1	09		0.00452	0.0	256 × 247 × 245	11.7
3 Tethys	294,660	1	22		0.00000	1.9	536 × 528 × 526	10.2
13 Telesto	294,660	1	22				15 × 13 × 8	19
14 Calypso	294,660	1	22				15 × 8 × 8	19
4 Dione	377,400	2	18		0.00223	0.02	560	10.4
12 Helene	377,400	2	18		0.005	0.0	18 × 16 × 15	18
5 Rhea	527,040	4	12		0.00100	0.4	764	9.7
6 Titan	1,221,830	15	22		0.029192	0.3	2575	8.3
7 Hyperion	1,481,100	21	07		0.104	0.4	180 × 140 × 113	14.2
8 Iapetus	3,561,300	79			0.02828	14.7	718	11.1
9 Phoebe	12,952,000	550 R			0.16326	177†	110	16.5

APPENDIX 10: PLANETARY SATELLITES (CONTINUED)

SATELLITE	SEMIMAJOR AXIS OF ORBIT (KM)	SIDEREAL REVOLUTION PERIOD (D H M)	ORBITAL ECCENTRICITY	ORBITAL INCLINATION (°)	RADIUS (KM)	VISIBLE MAGNITUDE AT MEAN OPPOSITION DISTANCE
SATELLITES OF URANUS						
6 Cordelia	49,770	08 02	0.00026	0.1	13	24
7 Ophelia	53,790	09 02	0.0099	0.1	15	24
8 Bianca	59,170	10 25	0.0009	0.2	21	23
9 Cressida	61,780	11 07	0.0004	0	31	22
10 Desdemona	62,680	11 22	0.00013	0.1	27	23
11 Juliet	64,350	11 50	0.00066	0.1	42	22
12 Portia	66,090	12 19	0.0000	0.1	54	21
13 Rosalind	69,940	11 54	0.0001	0.3	27	23
14 Belinda	75,260	14 57	0.00007	0	33	22
18 S1986 U10	76,400	15 18			20	
15 Puck	86,010	18 17	0.00012	0.3	77	20
5 Miranda	129,390	1 09 56	0.0027	4.2	240 × 234 × 233	16.3
1 Ariel	191,020	2 12 29	0.0034	0.3	581 × 578 × 578	14.2
2 Umbriel	266,300	4 03 27	0.005	0.4	585	14.8
3 Titania	435,910	8 16 56	0.0022	0.1	789	13.7
4 Oberon	583,520	13 11 07	0.0008	0.1	761	13.9
16 Caliban	7,169,000	1.6 years R	0.082	140†	30	22.4
17 Sycorax	12,214,000	3.5 years R	0.509	153†	60	20.9

APPENDIX 10: PLANETARY SATELLITES (CONTINUED)

SATELLITE	SEMIMAJOR AXIS OF ORBIT (KM)	SIDEREAL REVOLUTION PERIOD (D H M)			ORBITAL ECCENTRICITY	ORBITAL INCLINATION (°)	RADIUS (KM)	VISIBLE MAGNITUDE AT MEAN OPPOSITION DISTANCE
SATELLITES OF NEPTUNE								
Naiad	48,230	7			0.000	4.74	29	24.7
Thalassa	50,070	3		1	0.000	0.21	40	23.8
Despina	52,530	8			0.000	0.07	74	22.6
Galatea	61,950	10			0.000	0.05	79	22.3
Larissa	73,550	13			0.00139	0.20	104 × 89	22.0
Proteus	117,650	27			0.0004	0.55	218 × 208 × 201	20.3
Triton	354,760	22			0.000016	157.345	1353	13.5
Nereid	5,513,400	6			0.7512	27.6	170	18.7
RINGS AND RING ARCS OF NEPTUNE								
Galle	42,000							
Leverrier	53,200							
Lassell	53,200–57,500							
Arago	57,500							
Adams	62,900							
Courage	62,900							
Liberté	62,900							
Egalité	62,900							
Fraternité	62,900							
SATELLITE OF PLUTO								
Charon	19,682	9 19 17R			<0.001	99	593	16.8

R indicates retrograde motion

† measured relative to the ecliptic plane

Based on *The Astronomical Almanac for the year 2000.*

YEAR	DATE (NOON)	JULIAN DAY 2450000+	SUN	MERCURY	VENUS	MARS	JUPITER	SATURN	URANUS
2000	Jan 1	1545	279	271	240	327	25	40	314
2000	Jan 11	1555	290	286	253	335	25	40	315
2000	Jan 21	1565	300	303	265	343	26	40	315
2000	Jan 31	1575	310	320	277	350	27	40	316
2000	Feb 10	1585	320	337	289	358	29	41	317
2000	Feb 20	1595	330	346	302	6	30	41	317
2000	Mar 1	1605	340	341	314	13	32	42	318
2000	Mar 11	1615	350	333	326	21	34	43	318
2000	Mar 21	1625	0	334	339	28	36	44	319
2000	Mar 31	1635	10	342	351	35	38	45	319
2000	Apr 10	1645	20	355	3	43	41	46	320
2000	Apr 20	1655	30	11	16	50	43	47	320
2000	Apr 30	1665	39	29	28	57	45	49	320
2000	May 10	1675	49	50	40	64	48	50	320
2000	May 20	1685	59	71	53	71	50	51	320
2000	May 30	1695	68	89	65	78	53	52	320
2000	Jun 9	1705	78	102	77	84	55	54	320
2000	Jun 19	1715	88	109	90	91	57	55	320
2000	Jun 29	1725	97	108	102	98	59	56	320
2000	Jul 9	1735	107	103	114	104	61	57	320
2000	Jul 19	1745	116	100	126	111	63	58	319
2000	Jul 29	1755	126	106	139	118	65	59	319
2000	Aug 8	1765	135	121	151	124	67	59	318
2000	Aug 18	1775	145	141	163	130	68	60	318
2000	Aug 28	1785	154	160	176	137	69	60	318
2000	Sep 7	1795	164	178	188	143	70	60	317
2000	Sep 17	1805	174	194	200	149	70	60	317
2000	Sep 27	1815	184	208	212	156	71	60	317
2000	Oct 10	1825	194	219	225	162	71	60	317
2000	Oct 17	1835	203	225	237	168	70	59	316
2000	Oct 27	1845	213	220	249	175	69	59	316
2000	Nov 6	1855	223	210	261	181	68	58	316
2000	Nov 16	1865	233	214	273	187	67	57	317
2000	Nov 26	1875	244	228	285	193	66	56	317
2000	Dec 6	1885	254	243	297	199	65	56	317
2000	Dec 16	1895	264	258	308	205	63	55	317
2000	Dec 26	1905	274	274	320	211	62	54	318
2001	Jan 5	1915	284	290	331	217	61	54	318
2001	Jan 15	1925	294	307	341	223	61	54	319
2001	Jan 25	1935	305	322	351	228	61	54	319
2001	Feb 4	1945	315	330	1	234	61	54	320
2001	Feb 14	1955	325	323	8	239	61	54	321
2001	Feb 24	1965	335	315	14	244	62	54	321
2001	Mar 6	1975	345	318	17	249	63	55	322
2001	Mar 16	1985	355	328	16	254	65	56	322
2001	Mar 26	1995	5	341	12	258	66	57	323
2001	Apr 5	2005	15	357	6	262	68	58	323
2001	Apr 15	2015	25	16	2	265	70	59	324

APPENDIX 11: JULIAN DAYS AND PLANETARY LONGITUDES (CONTINUED)

YEAR	DATE (NOON)	JULIAN DAY 2450000+	SUN	MERCURY	VENUS	MARS	JUPITER	SATURN	URANUS
2001	Apr 25	2025	34	36	1	267	72	60	324
2001	May 5	2035	44	57	5	268	74	61	324
2001	May 15	2045	54	74	11	268	76	63	324
2001	May 25	2055	63	85	18	267	78	64	324
2001	Jun 4	2065	73	89	27	265	81	65	324
2001	Jun 14	2075	83	86	37	262	83	66	324
2001	Jun 24	2085	92	81	47	259	85	68	324
2001	Jul 4	2095	102	82	58	256	87	69	324
2001	Jul 14	2105	111	91	69	255	90	70	324
2001	Jul 24	2115	121	107	80	255	92	71	323
2001	Aug 3	2125	130	127	91	256	94	72	323
2001	Aug 13	2135	140	147	103	258	96	73	322
2001	Aug 23	2145	149	165	115	262	98	73	322
2001	Sep 2	2155	159	181	126	266	100	74	322
2001	Sep 12	2165	169	194	138	271	101	74	321
2001	Sep 22	2175	179	205	151	277	102	74	321
2001	Oct 2	2185	188	209	163	283	104	74	321
2001	Oct 12	2195	198	203	175	289	104	74	320
2001	Oct 22	2205	208	194	188	296	105	74	320
2001	Nov 1	2215	218	200	200	302	105	73	320
2001	Nov 11	2225	228	214	213	309	105	73	320
2001	Nov 21	2235	238	230	225	317	105	72	321
2001	Dec 1	2245	248	246	238	324	104	71	321
2001	Dec 11	2255	259	262	250	331	103	70	321
2001	Dec 21	2265	269	278	263	338	102	70	321
2001	Dec 31	2275	279	294	275	346	100	69	322
2002	Jan 10	2285	289	308	288	353	99	68	322
2002	Jan 20	2295	299	314	301	0	98	68	323
2002	Jan 30	2305	309	304	313	8	97	68	323
2002	Feb 9	2315	320	304	326	15	96	68	324
2002	Feb 19	2325	330	303	338	22	96	68	324
2002	Mar 1	2335	340	314	351	29	95	68	325
2002	Mar 11	2345	350	328	3	36	95	68	326
2002	Mar 21	2355	0	344	16	43	96	69	326
2002	Mar 31	2365	10	2	28	50	96	70	327
2002	Apr 10	2375	19	22	40	57	98	71	327
2002	Apr 20	2385	29	43	52	64	99	72	328
2002	Apr 30	2395	39	59	65	71	100	73	328
2002	May 10	2405	49	68	77	77	102	74	328
2002	May 20	2415	58	69	89	84	104	75	328
2002	May 30	2425	68	64	101	90	106	77	328
2002	Jun 9	2435	78	61	113	97	108	78	328
2002	Jun 19	2445	87	65	124	104	110	79	328
2002	Jun 29	2455	97	76	136	110	112	80	328
2002	Jul 9	2465	106	92	147	117	114	82	328
2002	Jul 19	2475	116	113	159	123	116	83	328
2002	Jul 29	2485	125	134	170	129	119	84	327
2002	Aug 8	2495	135	152	180	136	121	85	327

YEAR	DATE (NOON)	JULIAN DAY 2450000+	SUN	MERCURY	VENUS	MARS	JUPITER	SATURN	URANUS
2003	Apr 15	2745	24	44	351	295	128	84	331
2003	Apr 25	2755	34	50	4	301	128	85	332
2003	May 5	2765	44	47	16	307	129	86	332
2003	May 15	2775	53	42	28	313	130	87	332
2003	May 25	2785	63	41	40	318	131	88	332
2003	June 4	2795	72	48	52	323	133	89	332
2003	Jun 14	2805	82	61	64	328	134	91	332
2003	Jun 24	2815	92	78	75	332	136	92	332
2003	Jul 4	2825	101	99	89	336	138	93	332
2003	Jul 14	2835	111	120	101	338	140	95	332
2003	Jul 24	2845	120	139	113	339	142	96	332
2003	Aug 3	2855	130	154	125	339	144	97	331
2003	Aug 13	2865	139	167	138	338	146	98	331
2003	Aug 23	2875	149	174	150	336	149	99	330
2003	Sep 2	2885	159	175	163	333	151	100	330
2003	Sep 12	2895	168	167	175	331	153	101	330
2003	Sep 22	2905	178	162	187	330	155	102	329
2003	Oct 2	2915	188	171	200	330	157	102	329
2003	Oct 12	2925	198	188	212	331	159	103	329
2003	Oct 22	2935	208	205	225	333	161	103	329
2003	Nov 1	2945	218	222	237	337	163	103	328
2003	Nov 11	2955	228	237	250	341	164	103	328
2003	Nov 21	2965	238	253	262	346	165	102	328
2003	Dec 1	2975	248	267	274	351	167	102	329

YEAR	DATE (NOON)	JULIAN DAY 2450000+	SUN	MERCURY	VENUS	MARS	JUPITER	SATURN	URANUS
2002	Aug 18	2505	144	168	190	142	123	86	326
2002	Aug 28	2515	154	181	200	148	125	87	326
2002	Sep 7	2525	164	190	209	155	127	88	326
2002	Sep 17	2535	173	192	216	161	129	88	325
2002	Sep 27	2545	183	185	222	168	131	88	325
2002	Oct 7	2555	193	178	225	174	133	89	325
2002	Oct 17	2565	203	186	224	180	134	89	325
2002	Oct 27	2575	213	201	220	187	135	88	324
2002	Nov 6	2585	223	218	214	193	136	88	324
2002	Nov 16	2595	233	234	210	199	137	87	324
2002	Nov 26	2605	243	250	210	206	137	87	325
2002	Dec 6	2615	253	265	214	212	138	86	325
2002	Dec 16	2625	263	280	220	219	137	85	326
2002	Dec 26	2635	274	293	228	225	137	84	326
2003	Jan 5	2645	284	297	237	232	136	84	326
2003	Jan 15	2655	294	287	247	238	135	83	326
2003	Jan 25	2665	304	282	258	245	134	82	327
2003	Feb 4	2675	314	289	269	251	132	82	328
2003	Feb 14	2685	324	301	280	257	131	82	328
2003	Feb 24	2695	334	315	292	264	130	82	329
2003	Mar 6	2705	344	331	304	270	129	82	329
2003	Mar 16	2715	354	349	315	277	128	82	330
2003	Mar 26	2725	4	9	327	283	128	83	330
2003	Apr 5	2735	14	28	339	289	128	83	331

APPENDIX 11: JULIAN DAYS AND PLANETARY LONGITUDES (CONTINUED)

YEAR	DATE (NOON)	JULIAN DAY 2450000+	SUN	MERCURY	VENUS	MARS	JUPITER	SATURN	URANUS
2003	Dec 11	2985	258	279	287	356	167	101	329
2003	Dec 21	2995	268	281	299	2	168	100	329
2003	Dec 31	3005	278	269	311	8	168	99	330
2004	Jan 10	3015	289	267	324	14	168	99	330
2004	Jan 20	3025	299	275	336	20	168	98	330
2004	Jan 30	3035	309	288	348	27	167	97	331
2004	Feb 9	3045	319	302	0	33	166	96	332
2004	Feb 19	3055	329	318	12	39	165	96	332
2004	Feb 29	3065	339	336	23	46	164	96	333
2004	Mar 10	3075	349	355	34	52	163	96	333
2004	Mar 20	3085	359	14	45	59	161	96	334
2004	Mar 30	3095	9	28	55	65	160	96	334
2004	Apr 9	3105	19	31	64	71	159	97	335
2004	Apr 19	3115	29	25	73	78	159	97	335
2004	Apr 29	3125	39	21	80	84	158	98	336
2004	May 9	3135	48	23	84	91	158	99	336
2004	May 19	3145	58	32	86	97	159	100	336
2004	May 29	3155	67	46	83	103	159	101	336
2004	Jun 8	3165	77	64	78	109	160	102	336
2004	Jun 18	3175	87	85	72	116	161	104	336
2004	Jun 28	3185	96	107	69	122	162	105	336
2004	Jul 8	3195	106	125	70	128	164	106	336
2004	Jul 18	3205	115	141	75	135	166	108	336
2004	Jul 28	3215	125	152	81	141	167	109	335
2004	Aug 7	3225	134	158	89	147	169	110	335
2004	Aug 17	3235	144	156	98	154	171	111	335
2004	Aug 27	3245	154	148	108	160	173	112	334
2004	Sep 6	3255	163	146	119	166	175	113	334
2004	Sep 16	3265	173	157	129	173	178	114	334
2004	Sep 26	3275	183	175	141	179	180	115	333
2004	Oct 6	3285	193	193	152	186	182	116	333
2004	Oct 16	3295	202	210	164	192	184	116	333
2004	Oct 26	3305	212	225	176	199	186	117	332
2004	Nov 5	3315	222	240	188	205	188	117	332
2004	Nov 15	3325	232	254	200	212	190	117	332
2004	Nov 25	3335	243	264	213	219	192	117	332
2004	Dec 5	3345	253	264	225	225	193	116	333
2004	Dec 15	3355	263	252	237	232	195	116	333
2004	Dec 25	3365	273	252	250	239	196	115	333
2005	Jan 4	3375	283	261	262	246	197	114	334
2005	Jan 14	3385	293	275	275	253	198	113	334
2005	Jan 24	3395	304	290	287	260	198	113	334
2005	Feb 3	3405	314	306	300	267	198	112	335
2005	Feb 13	3415	324	322	312	274	198	111	336

YEAR	DATE (NOON)	JULIAN DAY 2450000+	SUN	MERCURY	VENUS	MARS	JUPITER	SATURN	URANUS
2005	Feb 23	3425	334	341	325	281	198	111	336
2005	Mar 5	3435	344	359	337	288	197	110	337
2005	Mar 15	3445	354	12	350	295	196	110	337
2005	Mar 25	3455	4	12	2	303	195	110	338
2005	Apr 4	3465	14	4	15	310	193	110	338
2005	Apr 14	3475	24	1	27	317	192	110	339
2005	Apr 24	3485	33	7	40	324	191	111	339
2005	May 4	3495	43	17	52	332	190	112	340
2005	May 14	3505	53	32	64	339	189	112	340
2005	May 24	3515	62	50	76	346	189	113	340
2005	Jun 3	3525	72	72	89	353	188	114	340
2005	Jun 13	3535	82	93	101	0	189	115	340
2005	Jun 23	3545	91	111	113	7	189	117	340
2005	Jul 3	3555	101	126	125	14	189	118	340
2005	Jul 13	3565	110	136	137	20	190	119	340
2005	Jul 23	3575	120	140	149	26	192	120	340
2005	Aug 2	3585	129	136	161	32	193	122	339
2005	Aug 12	3595	139	129	173	38	194	123	339
2005	Aug 22	3605	149	130	185	42	196	124	339
2005	Sep 1	3615	158	143	197	47	198	125	338
2005	Sep 11	3625	168	161	209	50	200	126	338
2005	Sep 21	3635	178	180	220	52	202	127	338
2005	Oct 1	3645	187	197	232	53	204	128	337
2005	Oct 11	3655	197	213	243	52	206	129	337
2005	Oct 21	3665	207	227	254	50	208	130	337
2005	Oct 31	3675	217	240	264	47	211	130	336
2005	Nov 10	3685	227	249	274	44	213	131	336
2005	Nov 20	3695	237	248	283	40	215	131	336
2005	Nov 30	3705	247	236	291	38	217	131	336
2005	Dec 10	3715	258	237	297	38	219	131	337
2005	Dec 20	3725	268	248	301	38	221	130	337
2005	Dec 30	3735	278	262	300	40	222	130	337
2006	Jan 9	3745	288	277	296	43	224	129	338
2006	Jan 19	3755	298	293	290	46	225	128	338
2006	Jan 29	3765	308	310	286	50	227	127	338

APPENDIX 12: LOCAL SIDEREAL TIME AT 00:00 LOCAL STANDARD TIME

(COMPUTED FOR THE YEAR 2001)

DATE	JAN	FEB	MAR	APR	MAY	JUN	JUL	AUG	SEP	OCT	NOV	DEC
1	6:42	8:45	10:35	12:37	14:35	16:38	18:36	20:38	22:40	0:39	2:41	4:39
2	6:46	8:49	10:39	12:41	14:39	16:42	18:40	20:42	22:44	0:43	2:45	4:43
3	6:50	8:52	10:43	12:45	14:43	16:46	18:44	20:46	22:48	0:47	2:49	4:47
4	6:54	8:56	10:47	12:49	14:47	16:50	18:48	20:50	22:52	0:51	2:53	4:51
5	6:58	9:00	10:51	12:53	14:51	16:53	18:52	20:54	22:56	0:54	2:57	4:55
6	7:02	9:04	10:55	12:57	14:55	16:57	18:56	20:58	23:00	0:58	3:01	4:59
7	7:06	9:08	10:59	13:01	14:59	17:01	19:00	21:02	23:04	1:02	3:05	5:03
8	7:10	9:12	11:03	13:05	15:03	17:05	19:04	21:06	23:08	1:06	3:09	5:07
9	7:14	9:16	11:07	13:09	15:07	17:09	19:08	21:10	23:12	1:10	3:12	5:11
10	7:18	9:20	11:10	13:13	15:11	17:13	19:11	21:14	23:16	1:14	3:16	5:15
11	7:22	9:24	11:14	13:17	15:15	17:17	19:15	21:18	23:20	1:18	3:20	5:19
12	7:26	9:28	11:18	13:21	15:19	17:21	19:19	21:22	23:24	1:22	3:24	5:23
13	7:30	9:32	11:22	13:25	15:23	17:25	19:23	21:26	23:28	1:26	3:28	5:27
14	7:34	9:36	11:26	13:28	15:27	17:29	19:27	21:29	23:32	1:30	3:32	5:30
15	7:38	9:40	11:30	13:32	15:31	17:33	19:31	21:33	23:36	1:34	3:36	5:34
16	7:42	9:44	11:34	13:36	15:35	17:37	19:35	21:37	23:40	1:38	3:40	5:38
17	7:45	9:48	11:38	13:40	15:39	17:41	19:39	21:41	23:43	1:42	3:44	5:42
18	7:49	9:52	11:42	13:44	15:43	17:45	19:43	21:45	23:47	1:46	3:48	5:46
19	7:53	9:56	11:46	13:48	15:46	17:49	19:47	21:49	23:51	1:50	3:52	5:50
20	7:57	9:59	11:50	13:52	15:50	17:53	19:51	21:53	23:55	1:54	3:56	5:54
21	8:01	10:03	11:54	13:56	15:54	17:57	19:55	21:57	23:59	1:58	4:00	5:58
22	8:05	10:07	11:58	14:00	15:58	18:00	19:59	22:01	0:03	2:01	4:04	6:02
23	8:09	10:11	12:02	14:04	16:02	18:04	20:03	22:05	0:07	2:05	4:08	6:06

(COMPUTED FOR THE YEAR 2001)

DATE	JAN	FEB	MAR	APR	MAY	JUN	JUL	AUG	SEP	OCT	NOV	DEC
24	8:13	10:15	12:06	14:08	16:06	18:08	20:07	22:09	0:11	2:09	4:12	6:10
25	8:17	10:19	12:10	14:12	16:10	18:12	20:11	22:13	0:15	2:13	4:16	6:14
26	8:21	10:23	12:14	14:16	16:14	18:16	20:15	22:17	0:19	2:17	4:19	6:18
27	8:25	10:27	12:17	14:20	16:18	18:20	20:18	22:21	0:23	2:21	4:23	6:22
28	8:29	10:31	12:21	14:24	16:22	18:24	20:22	22:25	0:27	2:25	4:27	6:26
29	8:33		12:25	14:28	16:26	18:28	20:26	22:29	0:31	2:29	4:31	6:30
30	8:37		12:29	14:32	16:30	18:32	20:30	22:33	0:35	2:33	4:35	6:34
31	8:41		12:33		16:34		20:34	22:36		2:37		6:37

* For leap years (2000, 2004, 2008), the local sidereal time on February 29 is 10:33.

COMPARED WITH THE SIDEREAL TIME FOR DATES IN 2001:

From January 1, 2000, through February 28, 2000, subtract 3 minutes.
From February 29, 2000, through December 31, 2000, add 1 minute.
Times for 2001 are in the table.
From January 1, 2002, through December 31, 2002 (also 2006), subtract 1 minute.
From January 1, 2003, through December 31, 2003 (also 2007), subtract 2 minutes.
From January 1, 2004, through February 28, 2004 (also 2008), subtract 3 minutes.
From February 29, 2004, through December 31, 2004 (also 2008), add 1 minute.

2005, 2009: Same as 2001; see table.
2006: Same as 2002.
2007: Same as 2003.
2008: Same as 2004.

Each chart covers roughly the area between the coordinates given below. The right ascension column in the middle can be used with either the left-hand or the right-hand columns of declination:

CONSTELLATION FOR + DECLINATION ZONE	ATLAS CHART FOR DECLINATION +50° TO +90°	RIGHT ASCENSION	ATLAS CHART FOR DECLINATION −50° TO −90°
Cas, Cep	1	$22\frac{1}{2}^h–1\frac{1}{2}^h$	45
Cam, Cas, Cep, Per	2	$1\frac{1}{2}^h–4\frac{1}{2}^h$	46
Cam, Lyn	3	$4\frac{1}{2}^h–7\frac{1}{2}^h$	47
Cam, Dra, UMa	4	$7\frac{1}{2}^h–10\frac{1}{2}^h$	48
Dra, UMa, UMi	5	$10\frac{1}{2}^h–13\frac{1}{2}^h$	49
Dra, UMa, UMi	6	$13\frac{1}{2}^h–16\frac{1}{2}^h$	50
Dra, UMi	7	$16\frac{1}{2}^h–19\frac{1}{2}^h$	51
Cep, Cyg, Dra, Lac	8	$19\frac{1}{2}^h–22\frac{1}{2}^h$	52
FOR DECLINATION +20° TO +50°			**FOR DECLINATION −20° TO −50°**
And, Psc, Tri	9	$0^h–2^h$	33
And, Air, Per, Tau, Tri	10	$2^h–4^h$	34
Aur, Per, Tau	11	$4^h–6^h$	35
Aur, Gem, Lyn, Tau	12	$6^h–8^h$	36
Can, LMi, Lyn, UMa	13	$8^h–10^h$	37
Leo, LMi, UMa	14	$10^h–12^h$	38
Com, CVn	15	$12^h–14^h$	39
Boo, CrB	16	$14^h–16^h$	40
Her	17	$16^h–18^h$	41
Cyg, Lyr, Vul	18	$18^h–20^h$	42
Cyg, Vul	19	$20^h–22^h$	43
And, Lac, Peg	20	$22^h–24^h$	44
FOR DECLINATION −20° TO +20°			
Aqr, Cet, Peg, Psc		$23^h–1^h$	21
Ari, Cet, Psc		$1^h–3^h$	22
Eri, Tau		$3^h–5^h$	23
CMa, Lep, Mon, Ori		$5^h–7^h$	24
CMi, Cnc, Hya, Mon, Pup		$7^h–9^h$	25
Leo, Hya, Sex		$9^h–11^h$	26
Com, Crt, Crv, Leo, Vir		$11^h–13^h$	27
Boo, Lib, Vir		$13^h–15^h$	28
Her, Lib, Oph, Sco, Ser		$15^h–17^h$	29
Her, Oph, Sct, Ser, Sgr		$17^h–19^h$	30
Aql, Cap, Del, Sge, Sgr		$19^h–21^h$	31
Aqr, Cap, Equ, Peg		$21^h–23^h$	32

NOTE: Constellation abbreviations are shown in appendix 1. The Atlas Charts where each constellation is found is also listed in chapter 4.

DATE	DAY (REGULAR YEAR)	DAY (LEAP YEAR)
January 1	1	1
January 10	10	10
January 20	20	20
February 1	32	32
February 10	41	41
February 20	51	51
March 1	60	61
March 10	69	70
March 20	79	80
April 1	91	92
April 10	100	101
April 20	110	111
May 1	121	122
May 10	130	131
May 20	140	141
June 1	152	153
June 10	161	162
June 20	201	202
July 1	182	183
July 10	191	192
July 20	201	202
August 1	213	214
August 10	222	223
August 20	232	233
September 1	244	245
September 10	253	254
September 20	263	264
October 1	274	275
October 10	283	284
October 20	293	294
November 1	305	306
November 10	314	315
November 20	324	325
December 1	335	336
December 10	344	345
December 20	354	355
December 31	365	366

NOTE: 2004 and 2008 are leap years.

GLOSSARY

ABERRATION OF STARLIGHT: The tiny apparent displacement of stars resulting from the motion of the earth through space.

ABSOLUTE MAGNITUDE (M): The magnitude a celestial object would appear to have if it were at a distance of 10 parsecs.

ABSOLUTE VISUAL MAGNITUDE (M_V): The absolute magnitude of an object measured through a special yellowish filter that approximates the visual range of the human eye.

ABSORPTION NEBULA: A nebula seen in silhouette as it absorbs light from behind; also called a dark nebula.

ALTITUDE: Angular distance (usually measured in degrees) above the horizon.

ANALEMMA: The figure-8 representing the equation of time and the variation of the sun's altitude in the sky during the course of a year.

ANGSTROM: A unit of wavelength or distance, equivalent to 1/10,000 micrometer or 1/10,000,000,000 meter.

ANNULAR ECLIPSE: A solar eclipse in which a ring—an annulus—of solar photosphere remains visible.

APHELION: The farthest point from the sun in an object's orbit around it.

APPARENT MAGNITUDE (M): Magnitude as seen by an observer.

APPARENT SOLAR TIME: Time determined by the actual position of the sun in the sky; corresponds to time on most sundials.

ASTERISM: A noticeable pattern of stars that makes up part of one or more constellations; not a constellation itself.

ASTEROID: A minor planet, smaller than any major planet in our solar system; not one of the satellites (moons) of a major planet such as the earth or Jupiter.

ASTRONOMICAL UNIT (A.U.): The average distance from the earth to the sun, which equals 149,598,770 kilometers.

AUTUMNAL EQUINOX: The intersection of the ecliptic and the celestial equator that the sun passes each year on its way to southern (negative) declinations.

LY'S BEADS: A chain of several bright "beads" of white light, visible just before or after totality at a solar eclipse. The effect occurs when bits of photosphere shine through valleys at the moon's edge. See also diamond-ring effect.

ER DESIGNATIONS: The Greek letters assigned to the stars in a constellation, usually in order of brightness, by Johann Bayer in his sky atlas (1603).

TS: Dark bands in the clouds on giant planets such as Jupiter; compare with zones.

ARY STAR: A double star; a system containing two or more stars. In an eclipsing binary, one star goes behind the other periodically, changing the total amount of light we see.

CK HOLE: A region of space in which mass is packed so densely that (according to Einstein's general theory of relativity) nothing, not even light, can escape.

SINI'S DIVISION: The major division in Saturn's rings, which separates the A-ring from the B-ring.

ESTIAL EQUATOR: The imaginary great circle that lies above the earth's equator on the celestial sphere.

ESTIAL LONGITUDE: Longitude measured (in degrees) along the ecliptic to the east from the vernal equinox.

ESTIAL POLES: The points in the sky where the earth's axis, extended into space, intersects with the celestial sphere.

ESTIAL SPHERE: The imaginary sphere surrounding the earth, with the stars and other astronomical objects attached to it.

HEID VARIABLE: A star that varies in the manner of delta Cephei. The absolute magnitudes of these variable stars can be calculated from their periods of variation; by comparing the absolute and apparent magnitudes, the distances to these stars and the galaxies they are in can be determined.

ROMOSPHERE: A layer in the sun and many other stars just above the photosphere. During eclipses, the solar chromosphere glows reddish from hydrogen emission.

CUMPOLAR: Refers to a star, asterism, or constellation that is close enough to the celestial pole that, from the latitude at which you are observing, it never appears to set.

MET: A body, probably resembling a "dirty snowball," between 0.1 and 100 miles across — that travels through the solar system in an elliptical orbit of random inclination to the ecliptic. A comet grows a tail if it comes close enough to the sun.

NJUNCTION: The alignment of two celestial bodies that occurs when they reach the same celestial longitude. The bodies then appear approximately closest to each other in the sky. See also inferior conjunction, superior conjunction.

CONSTELLATION: One of the 88 parts into which the sky is divided; also refers to the historical, mythological, or other figures that represented earlier divisions of the sky.

CONTACT(S): The stage(s) of an eclipse, occultation, or transit when the edges of the apparent disks of astronomical bodies seem to touch. At a solar eclipse, first contact is when the advancing edge of the sun first touches the moon; second contact is when the advancing edge of the sun touches the other side of the moon, beginning totality; the third contact is when the trailing edge of the sun touches the trailing edge of the moon, ending totality; and fourth contact marks the end of the eclipse.

CORONA: The outermost layer of the sun and many other stars; a faint halo of extremely hot (million-degree) gas.

CREPE RING: Saturn's inner ring, also known as the C-ring, which extends inward to the planet from the brightest ring (the B-ring).

CRESCENT: One of the phases of the moon or the inner planets (Venus and Mercury) as seen from earth, caused by the relative angles of sunlight and the observer's viewpoint. From spacecraft, crescent phases of the earth, Mars, Jupiter, Saturn, Uranus, and Neptune have also been seen.

DECLINATION: The celestial coordinate analogous to latitude, usually measured in degrees, minutes, and seconds of arc north (+) or south (−) of the celestial equator.

DIAMOND-RING EFFECT: An effect created as the total phase of a solar eclipse is about to begin, when the last Baily's bead—a remaining bit of photosphere—glows so intensely by contrast with the sun's faint corona that it looks like the jewel on a ring. Also refers to the equivalent phase at the end of totality.

DOUBLE STAR: A system containing two or more stars. In a true double, the stars are physically close to each other; in an optical double, they lie in approximately the same direction from the earth and appear close to each other, but are actually far apart. See also binary star.

EARTHSHINE: Sunlight reflected off the earth, which dimly lights the side of the moon that does not receive direct sunlight.

ECLIPSE, LUNAR: The passage of the moon into the earth's shadow.

ECLIPSE, SOLAR: The passage of the moon's shadow across the earth. See also annular eclipse, contact(s), penumbra, umbra.

ECLIPTIC: The apparent path the sun follows across the sky during the year; the same path is also followed approximately by the moon and planets.

EJECTA BLANKET: Chunks of rock, usually extending from one side of a crater, that were ejected during the crater's formation.

ELONGATION: Angular distance in celestial longitude from the sun in the sky.

EMISSION LINES: Extra radiation at certain specific wavelengths in a spectrum, compared with neighboring wavelengths (colors).

EMISSION NEBULA: A gas cloud that receives energy from a hot star, allowing it to give off radiation in emission lines such as those of hydrogen. The characteristic reddish radiation of many emission nebulae is mostly from the hydrogen-alpha line.

ENCKE'S DIVISION: A thin division in the A-ring of Saturn.

EQUATION OF TIME: The variation of local apparent solar time minus local mean solar time over the year.

EQUINOX: One of the two intersections of the ecliptic and the celestial equator: see autumnal equinox, vernal equinox.

FILAMENT: A dark region snaking across the sun; a prominence seen in projection against the solar disk.

FIREBALL: An extremely bright meteor, usually with an apparent magnitude brighter than −5; some fireballs are as bright as magnitude −20.

FLAMSTEED NUMBER: The number assigned to a star in a given constellation, in order of right ascension, in the 1725 catalogue of John Flamsteed.

GALACTIC: Pertaining to our galaxy, the Milky Way Galaxy.

GALACTIC CLUSTER: An irregular grouping of stars of a common and possibly recent origin. Also called an open cluster.

GALACTIC EQUATOR, GALACTIC POLES: The equator and poles in a coordinate system in which the equator is placed along the plane of our galaxy, the Milky Way Galaxy.

GALAXY: A giant collection of stars, gas, and dust. Our galaxy, the Milky Way Galaxy, contains 1 trillion times the mass of our sun.

GIANT: A star brighter and larger than most stars of its color and temperature. Stars become giants (normally red giants) when they use up all the hydrogen in their cores and leave the "main sequence" part of their life cycle. See also supergiant.

GIBBOUS: A phase of a moon or a planet in which more than half of the side we see is illuminated. Remember: it "gibb us" ("gives us") more light.

GLOBULAR CLUSTER: A spherical grouping of stars of a common origin; globular clusters and the stars in them are very old.

GRABEN: On the surface of the earth, the moon, or other planets or moons, a long and narrow region between two faults that has subsided.

HALF MOON: The first-quarter or third-quarter phase, when half the visible side of the moon is illuminated.

HOUR ANGLE: The sidereal time elapsed since an object was on the meridian, or, if the hour angle is negative, before the object reaches the meridian. (The hour angle equals the difference between the right ascension of an object and of your meridian.)

HOUR CIRCLE: A line along which right ascension is constant, lying on a great circle that passes through the celestial poles and the object.

HUBBLE'S LAW: The relationship between the velocity and distance of galaxies and other distant objects; it shows that the universe is expanding.

HYDROGEN-ALPHA LINE: The strongest spectral line of hydrogen in the visible part of the spectrum. It falls in the red, so that an emission hydrogen-alpha line is red; an absorption hydrogen-alpha line is the absence of that wavelength of red.

IC: The prefix used before numbers assigned to nonstellar objects in the Index Catalogues published as supplements to the New General Catalogue of J. L. E. Dryer.

INFERIOR CONJUNCTION: The conjunction in which a planet whose orbit is inside that of the earth passes between the earth and the sun.

INTRINSIC BRIGHTNESS: The amount of energy (usually light) an object gives off; its true brightness, independent of the effects of distance or dimming by intervening material.

IONIZED HYDROGEN: Hydrogen that has lost its electron; ionized hydrogen gas, commonly found in stars and nebulae, has free protons and free electrons.

JULIAN DAY: The number of days since noon on 1 January 4713 B.C.E. Variable star observers and other astronomers commonly calculate the interval between dates of events by subtracting Julian days, eliminating the necessity to keep track of leap years and other calendar details.

LIBRATION: The turning of the visible face of the moon, which allows us to see different amounts of the lunar surface around the limb (edge).

LIGHT-YEAR: The distance that light travels in a year, which equals 9,460,000,000,000 km or 63,240 A.U. (astronomical units).

LIMB: The edge of the apparent disk of an astronomical body, such as the sun, moon, or a planet.

MAGNITUDE: A logarithmic scale of brightness, in which each change of five magnitudes is equivalent to a change by a factor of 100. Adding one magnitude corresponds to a decrease in brightness by a factor of 2.512.... See also absolute magnitude, apparent magnitude.

AIN-SEQUENCE STAR: A star in the prime of its life, when hydrogen inside it is undergoing nuclear fusion; such stars form a band—the main sequence—across a graph of stellar temperatures vs. stellar brightness.

AXIMA: The times when a variable star reaches its maximum brightness.

EAN SOLAR TIME: Time as kept by a fictitious "mean" sun that travels at a steady rate across the sky throughout the year.

ERIDIAN: The great circle passing through the celestial poles and your zenith.

ESSIER CATALOGUE: The list of 103 nonstellar, deep-sky objects compiled by Charles Messier in the 1770s, and subsequently expanded to 109 or 110 objects.

ETEOR: A meteoroid streaking across the sky; a shooting star.

ETEORITE: The part of a meteoroid that survives its passage through the earth's atmosphere.

ETEOR SHOWER: The appearance of many meteors during a short period of time, as the earth passes through a comet's orbit.

INIMA: The times of a variable star's minimum brightness.

IRA VARIABLE: A long-period variable star, like the star omicron Ceti (called "Mira").

EBULA: A region of gas or dust in a galaxy that can be observed optically. See also emission nebula, absorption nebula, and reflection nebula.

GC: The prefix used before numbers assigned to nonstellar objects in the New General Catalogue, published by J. L. E. Dryer in 1888.

EUTRON STAR: A small (20-km diameter), dense (a billion tons per cubic cm) star, resulting from the collapse of a dying star to the point where only the fact that its neutrons resist being pushed still closer together prevents further collapse.

OVA: A newly visible star, or one that suddenly increases dramatically in brightness.

UTATION: A small nodding motion of the earth's axis of rotation with a period of 19 years; this motion is superimposed on precession.

BLATE: A nonspherical shape formed by rotating an ellipse around its narrower axis; the equatorial diameter of an oblate body (such as Jupiter) is greater than its polar diameter.

CCULTATION: The hiding of one celestial body by another.

PEN CLUSTER: An irregular grouping of stars of a common and possibly recent origin. Also called a galactic cluster.

PPOSITION: The point in a planet's orbit at which its celestial longitude is 180° from that of the sun. A planet at opposition is visible all night long.

PARSEC: The distance from which 1 A.U. appears to subtend (cover) 1 second of arc; 1 parsec equals 3.261 633... light-years.

PENUMBRA: At an eclipse, the part of the earth or moon's shadow from which part of the solar disk is visible. Also refers to the outer, less dark portion of a sunspot.

PERIHELION: The nearest point to the sun in an object's orbit around it.

PHOTOSPHERE: The visible surface of the sun or of another star.

PLANETARY NEBULA: A shell of gas ejected by a dying star that contains about as much mass as the sun.

POLAR TUFTS: Small spikes visible in the solar corona near the sun's poles, formed by gas following the sun's magnetic field.

POSITION ANGLE: The angle, centered at the brighter component of a double star, that an observer follows counterclockwise from north around to the fainter component.

PRECESSION: The slow drifting of the orientation of the earth's axis over a period of 26,000 years. Also refers to its effect on the location of the equinoxes, and thus on the coordinate system of right ascension and declination used to plot positions of stars and other objects.

PROMINENCE: Gas suspended above the solar photosphere by the sun's magnetic field; ordinarily visible at the solar limb (edge). A prominence glows reddish during eclipses because of its characteristic hydrogen-alpha radiation.

PROPER MOTION: Apparent angular motion across the sky, shown as a change in an object's position with respect to the background stars.

PULSAR: A rotating neutron star that gives off sharp pulses of radio waves with a period ranging from about 0.001 to 4 seconds.

QUASAR: a "quasi-stellar object" with an extremely large redshift: presumably a powerful event going on in the central region of a galaxy. Using Hubble's Law tells us that quasars must be among the most distant objects in the universe.

RADIANT: The location on the celestial sphere from which meteors in a given shower appear to radiate, because of perspective.

RED GIANT: A swollen star; a stage occurring at the end of a star's main-sequence period of life. See also giant.

REFLECTING TELESCOPE: A telescope that uses a mirror in the principal stage of forming an image.

REFLECTION NEBULA: A dust cloud that reflects a star's light to us.

REFRACTING TELESCOPE: A telescope that uses a lens in the principal stage of forming an image.

RETROGRADE MOTION: The apparent backward (westward) loop in a planet's motion across the sky over a lengthy period of time. Copernicus explained it as the projection effect when the earth overtakes another planet as they both orbit the sun.

REVOLUTION: The orbiting of a planet or other object around the sun or another central body. (Compare with rotation.)

RIGHT ASCENSION: The angle of an object around the celestial equator, measured in hours, minutes, and seconds eastward from the vernal equinox.

ROTATION: The spinning of a planet or other object on its axis. (Compare with revolution.)

SEPARATION: The angular distance (measured in degrees, minutes, and seconds of arc) between components of a double star.

SHADOW BANDS: Light and dark bands that appear to sweep across the ground in the minutes before and after totality at a solar eclipse; caused by irregularities in the earth's upper atmosphere.

SIDEREAL TIME: Time by the stars; technically, the hour angle of the vernal equinox, which is equal to the right ascension of objects on your meridian.

SOLAR FLARE: An explosive eruption on the sun reaching temperatures of millions of degrees. Note: a flare is not a prominence.

SOLSTICE(S): The positions of the sun when it reaches its northernmost declination (in northern-hemisphere summer) or southernmost declination (in northern-hemisphere winter).

SPECTRAL LINE: A wavelength of the spectrum at which the intensity is greater than (an emission line) or less than (an absorption line) neighboring values.

SPECTRUM (PLURAL: SPECTRA): The radiation from an object, spread out into its component colors, wavelengths, or frequencies.

STAR CLOUD: One of several regions of the Milky Way where great numbers of stars appear.

STREAMERS: Large-scale structures in the sun's corona, usually near the solar equator, shaped by the sun's magnetic field.

SUNSPOTS: Relatively dark regions on the solar photosphere, corresponding to areas with exceptionally high magnetic fields.

SUPERGIANT: A star brighter and larger than even giants of the same color and temperature. Only the most massive stars become supergiants, after passing through the giant stage.

SUPERIOR CONJUNCTION: The conjunction in which a planet whose orbit is inside that of the earth passes on the far side of the sun with respect to the earth.

SUPERNOVA: The explosion and devastation of a very massive star.

SUPERNOVA REMNANT: Gas left over from a supernova that can be seen in the sky or detected from its radio or x-ray emission. (The Crab Nebula, for example, can be detected all three ways.)

SURFACE BRIGHTNESS: The brightness of a unit area of an object's surface.

For spread-out objects such as nebulae, the surface brightness determines the amount of contrast the object has against the background sky, and whether the object's surface is bright enough to make an image on your retina. Even though the object's total brightness may be high, it still may be hard to see if it is spread out enough so that its surface brightness is low.

SYNODIC: Related to the alignment of three bodies, often the earth, the sun, and a third body, such as the moon or a planet.

TERMINATOR: The edge of the lighted region of a moon or planet; the line between day and night.

TRAIN: A path left in the sky by a meteor.

TRANSIENT LUNAR PHENOMENA (TLP'S): Changes, such as emissions of gas, observed on the moon.

TRANSIT: The passage of an inner planet (Mercury or Venus) across the sun's disk as seen from earth, or of a moon (such as one of Jupiter's Galilean satellites) across its planet's disk. Also, the passage of an object across an observer's meridian.

UMBRA: At an eclipse, the part of the moon or earth's shadow from which the solar disk is entirely hidden. Also refers to the inner, darker portion of a sunspot. See also penumbra.

UNIVERSAL TIME (U.T.): Solar time based on time at the meridian of Greenwich, England.

VARIABLE STAR: A star whose apparent brightness changes over time.

VERNAL EQUINOX: The intersection of the ecliptic and the celestial equator that the sun passes on its way to northern (positive) declinations.

ZENITH: The point directly overhead (wherever an observer is), 90° above the horizon.

ZODIAC: Traditionally, a set of 12 constellations through which the sun, moon, and planets pass in the course of a year. Actually, that band of sky contains many more parts of constellations, and because of precession, the sun is no longer in the constellations associated with its "traditional" dates at those times.

ZONES: Bright bands in the cloud layers of the giant planets (Jupiter, Saturn, Uranus, and Neptune).

BIBLIOGRAPHY

*See **www.williams.edu/astronomy/fieldguide** for updated lists and ordering links.*

GENERAL ASTRONOMY

Pasachoff, Jay M. 1998. *Astronomy: From the Earth to the Universe.* 5th ed. Philadelphia: Saunders College Publishing. A nonmathematical, well-illustrated survey. www.williams.edu/Astronomy/jay.

Pasachoff, Jay M., and Alex Filippenko. 2000. *Journey Through the Universe.* Philadelphia: Saunders College Publishing. A simpler, shorter text. www.williams.edu/Astronomy/jay

FIRST GUIDES

Pasachoff, Jay M. 1998. *Peterson First Guide to Astronomy.* Boston: Houghton Mifflin Company. A short, simple introduction to astronomy, illustrated entirely in color. Includes simplified monthly sky maps and drawings of mythological constellation figures.

Pasachoff, Jay M. 1998. *Peterson First Guide to the Solar System.* Boston: Houghton Mifflin Company. A short, simple introduction to the planets, the sun, comets, and other objects in the solar system, entirely in color.

SKY ATLASES AND GENERAL REFERENCES

Arnold, H. J. P., P. Doherty, and P. Moore. 1997. *The Photographic Atlas of the Stars.* Bristol, UK: Institute of Physics.

Audouze, Jean, and Guy Israël. 1994. *Cambridge Atlas of Astronomy,* 3rd ed. New York: Cambridge University Press.

Cragin, Murray, James Lucy, and Barry Rappaport. 1993. *The Deep Sky Field Guide to Uranometria 2000.0.* Richmond, Va.: Willmann-Bell.

Hirshfeld, Alan, Roger W. Sinnott, and François Ochsenbein. 1991 (vol.1, 2nd ed.); Alan Hirshfeld and Roger W. Sinnott. 1985 (vol. 2). *Sky Catalogue.* Cambridge, Mass.: Sky Publishing Corp. Volume 1 is a list of stars; volume 2 provides lists of double stars, variable stars, galaxies, clusters, nebulae, and other objects. All positions are precessed to epoch 2000.0.

Meeus, Jean. 1998. *Astronomical Tables of the Sun, Moon, and Planets,* 2nd ed. Richmond, Va.: Willmann-Bell. Lists of astronomical phenomena, including planetary oppositions and conjunctions, eclipses, transits, etc.

Ridpath, Ian, ed. 1998. *Norton's Star Atlas and Reference Handbook (Epoch 2000.0).* 19th ed. New York: Longman. The old standard, updated.

Ridpath, Ian. 1998. *Eyewitness Handbooks: Stars and Planets.* New York: DK Publishing.

Sinnott, Roger W., ed. 1988. *NGC 2000.0: The Complete New General Catalogue and Index Catalogues of Nebulae and Star Clusters.* Cambridge, Mass.: Sky Publishing Corp. and New York: Cambridge University Press. A centennial reissue of Dreyer's work with updated data.

Sinnott, Roger W., and Michael A. C. Perryman. 1997. *The Millennium Star Atlas.* Cambridge, Mass.: Sky Publishing Corp. and European Space Agency.

Tirion, Wil, 1996. *Cambridge Star Atlas,* 2nd ed. New York: Cambridge University Press. A naked-eye star atlas in full color. A moon map, 24 monthly sky maps, 20 detailed star charts, and 6 all-sky maps.

Tirion, Wil, and Sinnott, Roger W. 1998. *Sky Atlas 2000.0,* 2nd ed. Cambridge, Mass.: Sky Publishing Corp. and New York: Cambridge University Press. Twenty-eight large-scale star charts, showing over 80,000 stars, down to magnitude 8.55 and about 2,700 deep-sky objects. The Atlas comes in three versions: white stars on black background for use outside, black stars on white background for use indoors, and colors on white background (deluxe). The first two versions, normally sold as loose charts, are also available as spiral-bound, laminated sets. The deluxe version is Wiro-bound.

Vehrenberg, Hans. 1984. *Atlas of Deep Sky Splendors,* 4th ed. Cambridge, Mass.: Sky Publishing Corp. and New York: Cambridge University Press. Color and black-and-white photographs, at a uniform scale, of the most interesting parts of the sky. Includes close-ups of Messier and other objects.

OBSERVING HANDBOOKS

Bishop, Roy L., ed. *Observer's Handbook*. Toronto: Royal Astronomical Society of Canada. Published annually, this popular guide to sky objects and events is available from the society at 136 Dupont Street, Toronto, Ontario, Canada, M5R 1V2.

Bone, Neil. 1998. *Meteors, Comets, Supernovae: Observing Transient Phenomena*. New York: Springer-Verlag.

Burnham, Robert, Jr. 1980. *Burnham's Celestial Handbook*. 3 vols. New York: Dover. Detailed constellation-by-constellation discussions, with photographs of a wide variety of objects. Out of print, but still a standard reference.

Chartrand, Mark R., III. 1990. *Skyguide*. New York: Golden Press. Observing hints and constellation maps, with beautiful illustrations by Helmut K. Wimmer.

Chartrand, Mark R., III, and Wil Tirion. 1995. *The Audubon Society Field Guide to the Night Sky,* revised. New York: Knopf.

Clark, Roger N. 1991. *Visual Astronomy of the Deep Sky*. New York: Cambridge University Press. How and what to observe.

Dickinson, Terence. 1998. *Nightwatch: A Practical Guide to Viewing the Universe,* 3rd ed. Willowdale, Ontario: Firefly Books.

Dickinson, Terence, and Jack Newton. 1997. *Splendors of the Universe: A Practical Guide to Photographing the Night Sky*. Willowdale, Ontario: Firefly Books.

Enright, Leo. 1999. *The Beginner's Observing Guide*. Toronto: Royal Astronomical Society of Canada.

Harrington, Philip S. 1990. *Touring the Universe Through Binoculars*. New York: John Wiley & Sons.

Harrington, Philip S. 1997. *Eclipse! The What, Where, When, Why, and How Guide to Watching Solar and Lunar Eclipses*. New York: John Wiley and Sons.

Harrington, Philip S. 1998. *The Deep Sky: An Introduction*. Cambridge, Mass.: Sky Publishing Corp.

Harrington, Philip S. 1998. *Starware: The Amateur Astronomer's Ultimate Guide to Choosing, Buying, and Using Telescopes and Accessories*. New York: John Wiley and Sons.

Jones, Kenneth Glyn. 1991. *Messier's Nebulae and Star Clusters,* 2nd ed. New York: Cambridge University Press. Messier's catalogue, discussed one by one.

Kepple, George R., and Glen W. Sanner. 1999. *Sky Observer's Guide*. 2 vols. Richmond, Va.: Willmann-Bell.

Kitchen, Chris, and Robert W. Forrest. 1998. *Seeing Stars: The Night Sky Through Small Telescopes*. New York: Springer-Verlag.

Levy, David H. 1998. *Observing Variable Stars: A Guide for the Beginner*. New York: Cambridge University Press.

Levy, David H. 1993. *The Sky: A User's Guide.* New York: Cambridge University Press. An introduction to observing.

Meeus, Jean. 1989. *Transits.* Richmond, Va.: Willmann-Bell.

Newton, Jack, and Philip Teece. 1995. *Guide to Amateur Astronomy,* 2nd ed. New York: Cambridge University Press.

North, Gerald. 1997. *Advanced Amateur Astronomy.* New York: Cambridge University Press.

O'Meara, Stephen James, and David H. Levy. 1998. *The Messier Objects.* New York: Cambridge University Press.

Pennington, Harvard. 1999. *The Year-Round Messier Marathon Field Guide.* Richmond, Va.: Willmann-Bell.

Rey, H. A., updated by Jay M. Pasachoff, 1989. *The Stars: A New Way to See Them.* Boston: Houghton Mifflin Company. Nontraditional constellation outlines, drawn to resemble actual objects more than the usual ways that stars are connected.

Rhoads, Samuel E. *The Sky Tonight: A Guided Tour of the Stars Over Hawaii.* Honolulu: Bishop Museum Press. Nice constellation-figure overlays. Suitable for equatorial latitudes.

Webb Society Deep-Sky Observer's Handbook, Hillside, N.J.: Enslow Publishers. Vol. 1: Double Stars, 2nd ed., 1986; vol. 2: Planetary and Gaseous Nebulae, 1979; vol. 3: Open and Globular Clusters, 1980; vol. 4: Galaxies, 1981; vol. 5: Clusters of Galaxies, 1982; vol. 6: Anonymous Galaxies, 1987; vol. 7: The Southern Sky, 1987; vol. 8: Variable Stars, 1990. All out of print.

Most of these publications are available from Sky Publishing Corp., 49 Bay State Rd., Cambridge, MA 02138, (800) 253-0245, www.sky-pub.com; Willmann-Bell, Inc., P.O. Box 35025, Richmond, Va. 23235, (800) 825-STAR, www.willbell.com; and through standard booksellers.

SKY BULLETINS

Astronomical Calendar (yearly). The changing sky and astronomical events such as eclipses. Available from Guy Ottewell, Astronomical Workshop, Furman University, Greenville, SC 29613.

Sky Calendar (monthly). Easy-to-use diagrams of the moon's phases and its daily changes in position against the starry background, plus diagrams of planetary conjunctions with bright stars, with the moon, and with other planets. Available by subscription from Abrams Planetarium, Michigan State University, East Lansing, MI 48824.

Skywatcher's Almanac (yearly). Computer-generated information on the visibility of the sun, moon, and other objects, tailored for the individual observer at a given latitude. Available from the Astronomical Data Service, 3922 Leisure Lane, Colorado Springs, CO 80917.

MAGAZINES

Astronomy (monthly). Monthly sky features and articles summarizing different fields of astronomy in lay terms. 21027 Crossroads Circle, P.O. Box 1612, Waukesha, WI 53187; (800) 446-5489; astronomy.com.

Sky & Telescope (monthly). The standard journal for amateur observers; includes popular articles on astronomical topics and sky events, in addition to regular monthly features. For subscription information, write to P.O. Box 9111, Belmont, MA 02478-9111; (800) 253-0245; www.skypub.com.

Mercury (bimonthly). Published by the Astronomical Society of the Pacific, 390 Ashton Ave., San Francisco, CA 94112; www.asp.org.

Odyssey (10 times a year). For children. Published by Cobblestone Publishing, Inc., 30 Grove St., Peterborough, NH 03458; www.odysseymagazine.com.

StarDate (bimonthly). Published by the McDonald Observatory, University of Texas, Austin, TX 78712; (800) STARDATE; stardate.utexas.edu.

Planetary Report. Published by The Planetary Society, 65 North Catalina Ave., Pasadena, CA 91106; tps@mars.planetary.org; www.planetary.org.

FOR CALCULATORS AND COMPUTERS

Duffett-Smith, Peter. 1996. *Easy PC Astronomy*. New York: Cambridge University Press.

Meeus, Jean. 1991. *Astronomical Algorithms*. Willmann-Bell, P.O. Box 35025, Richmond, VA 23235. Routines for equinoxes and solstices, conjunctions and oppositions, rising and setting of sun and moon, satellites of Jupiter and Saturn, Muslim and Hebrew calendars, etc.

Montenbruck, Oliver, and T. Pfleger (Storm Dunlop, translator). 1998. *Astronomy on the Personal Computer*, 3rd ed. New York: Springer-Verlag.

ELECTRONIC DETECTORS

Berry, Richard. 1994. *The CCD Camera Cookbook*. Richmond, Va.: Willmann-Bell. About CCDs.

Buil, C. 1991. *CCD Astronomy*. Richmond, Va.: Willmann-Bell.

Henden, Arne A., Ronald H. Kaitchuck, and Ryland J. Truax. 2000. *CCD Photometry*. Richmond, Va.: Willmann-Bell.

Ratledge, D. 1997. *The Art and Science of CCD Astronomy*. New York: Springer-Verlag.

American Association of Variable Star Observers (AAVSO), 25 Birch St., Cambridge, MA 02138; www.aavso.org. Finding charts on-line, in addition to a wide variety of other information.

American Meteor Society, Dept. of Physics and Astronomy, SUNY, Geneseo, NY 14454; www.serve.com/meteors.

The Association of Lunar and Planetary Observers, c/o Harry D. Jamieson, P.O. Box 171302, Memphis, TN 38187-1302; hjamieson@bellsouth.net; www.lpl.arizona.edu/alpo/.

The Astronomical League, the umbrella group of amateur societies. For their newsletter, *The Reflector,* write The Astronomical League, Executive Secretary, c/o Science Service Building, 1719 N St. N.W., Washington, D.C. 20030; www.astroleague.org.

The Astronomical Society of the Pacific, 390 Ashton Ave., San Francisco, CA 94112; www.aspsky.org.

British Astronomical Association, Burlington House, Piccadilly, London W1V 9AG, England; www.ast.cam.ac.uk/~baa.

International Dark-Sky Association, 3225 N. First Ave., Tucson, AZ 85719; ida@darksky.org; www.darksky.org.

The Planetary Society, 65 North Catalina Ave., Pasadena, CA 91106; tps@mars.planetary.org; www.planetary.org.

Royal Astronomical Society of Canada, 136 Dupont St., Toronto, Ontario, M5R 1V2, Canada; rasc@rasc.ca; www.rasc.ca.

PROFESSIONAL SOCIETY

American Astronomical Society, 2000 Florida Ave., N.W., Suite 400, Washington, D.C. 20009. For a booklet called "A Career in Astronomy," write to the AAS Education Officer at the above address; aas@aas.org; www.aas.org.

INDEX

The index provides references to objects, definitions, and general discussions of topics in astronomy. The most commonly observed individual objects are indexed, but not all objects on the Atlas Charts or their descriptions are included. The index should be used in conjunction with the tables and the glossary, the contents of which are not indexed. References to figures are in *italics*. References to tables showing which Sky Maps include given constellations are ***in bold italics***. References to Atlas Charts and the descriptions facing them are **boldfaced**. References to tables are followed by "t." References to Graphic Timetables are followed by "gt." Appendices are referred to by number following an "A." Data in appendices are referred to by Appendix number: A5, for example.

To find a constellation in the sky, use the underlined reference to get a list of Sky Maps on which the constellation appears. To look at the Atlas Chart showing a given constellation, use the boldfaced reference.

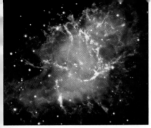

PETERSON FIRST GUIDES®

PETERSON FIELD GUIDE COLORING BOOKS

AUDIO AND VIDEO

EASTERN BIRDING BY EAR
cassettes 97523-9
CD 97524-7

WESTERN BIRDING BY EAR
cassettes 97526-3
CD 97525-5

EASTERN BIRD SONGS, Revised
cassettes 53150-0
CD 97522-0

WESTERN BIRD SONGS, Revised
cassettes 51746-X
CD 975190

BACKYARD BIRDSONG
cassettes 97527-1
CD 97528-X

EASTERN MORE BIRDING BY EAR
cassettes 97529-8
CD 97530-1

WATCHING BIRDS
Beta 34418-2
VHS 34417-4

PETERSON'S MULTIMEDIA GUIDES: NORTH AMERICAN BIRDS
(CD-ROM for Windows) 73056-2

PETERSON FLASHGUIDES™

ATLANTIC COASTAL BIRDS 79286-X
PACIFIC COASTAL BIRDS 79287-8
EASTERN TRAILSIDE BIRDS 79288-6
WESTERN TRAILSIDE BIRDS 79289-4
HAWKS 79291-6
BACKYARD BIRDS 79290-8
TREES 82998-4
MUSHROOMS 82999-2
ANIMAL TRACKS 82997-6
BUTTERFLIES 82996-8
ROADSIDE WILDFLOWERS 82995-X
BIRDS OF THE MIDWEST 86733-9
WATERFOWL 86734-7
FRESHWATER FISHES 86713-4

PETERSON FIELD GUIDES can be purchased at your local
bookstore or by calling our toll-free number, (800) 225-3362.

When referring to title by corresponding ISBN number,
preface with 0-395, unless title is listed with 0-618.